Manual of Echocardiography for Congenital Heart Diseases

Amal Paul

Manual of Echocardiography for Congenital Heart Diseases

Springer

Amal Paul
Consultant Cardiologist
MOSC Medical Mission Hospital
Thrissur, Kerala, India

ISBN 978-981-96-0242-1 ISBN 978-981-96-0243-8 (eBook)
https://doi.org/10.1007/978-981-96-0243-8

This Springer imprint is published by the registered company Springer Nature Singapore Pte Ltd.
The registered company address is: 152 Beach Road, #21-01/04 Gateway East, Singapore 189721, Singapore

Christian Medical College Vellore—my alma mater, which has been dedicated to the service of humanity with integrity, excellence and compassion for the past 125 years.

Department of Cardiology, CMC Vellore—my academic cradle, which has spearheaded cardiology training in India since 1965.

My dear parents, Shilpa and Abhay—my life and breath.

My dear teachers, friends and colleagues—my pillars and lamp posts in this academic journey.

My dear patients—my inspiration and drive.

Echocardiographers all round the world, who have wished atleast once in their life, for a ready reckoner in congenital heart diseases.

Foreword

A well-conducted echocardiographic examination remains the cornerstone of diagnosis and management in patients with congenital heart disease. As the art of clinical examination wanes in prominence for various reasons, the fusion of meticulous echocardiographic analysis with a thorough patient history becomes indispensable in assessing cardiac pathology. While advanced, non-invasive modalities such as CT and MRI offer better precision, their prohibitive costs place them beyond the reach of many. Moreover, with limited institutions offering specialized training in paediatric cardiology, echocardiographers often find themselves in a quandary when confronted with the complexities of congenital heart disease, thus impeding both diagnosis and recommendations for management.

In predominantly adult cardiology settings, there is often an insufficient emphasis on the nuanced instructions needed for the assessment of congenital heart disease, which further complicates the learning curve. Against this backdrop, this manual has clearly laid out its objectives, and the author has succeeded admirably in achieving them to perfection. With a strong focus on fundamental concepts and classifications, along with practical tips for basic skills such as probe positioning, this book stands as an indispensable manual for cardiology trainees and technicians involved in the echocardiographic imaging of congenital heart disease. Even seasoned cardiologists would find it valuable as a quick reference.

I extend my best wishes to the author, Dr. Amal Paul, and am confident that this book will ignite a deeper interest in the echocardiographic assessment of congenital heart diseases among countless emerging cardiology professionals across the globe.

Department of Cardiology
Christian Medical College Vellore,
Vellore, India

Paul V. George

Preface

Echocardiography of congenital heart diseases is often regarded as a specialty of its own, which most echocardiographers prefer to stay away from, whenever possible. The following real-life scenario sums it up all: The day is busy as usual in the echocardiography lab, with a long list of patients awaiting the scan. The work seems to be moving ahead smoothly at a reasonable pace. The next patient on the list walks into the suite. The echocardiographer hurriedly stares at the request slip—'Tricuspid atresia, pulmonary stenosis—S/P Fontan palliation'. An unwelcome panic follows.

It has been a few months since the last encounter with one such case. The specifics about the lesion seem to have faded out of memory. How was the Fontan circuit visualized? What were the parameters suggesting normalcy of the circuit? What are the complications to search for? What associated lesions could be expected?

It is not uncommon for the fast-paced proceedings of a busy echocardiography lab to become stalled when faced with a case of complex congenital heart disease. Congenital heart disease often turns out to be a stumbling block in an echocardiographer's routine for various reasons.

Firstly, congenital heart disease constitutes only a small proportion of the daily log of cases in an echocardiography service. Complex congenital heart diseases are even rarer. Hence, a veil of unfamiliarity prevails between the echocardiographer and a complex congenital cardiac defect. Secondly, in many training centres, congenital heart diseases are dealt with exclusively by specialists in paediatric cardiology, limiting the exposure of a trainee to the nuances of such cases.

However, faced with a case of complex congenital heart disease, no echocardiographer (be it a cardiology consultant, a resident on night call or a technician) can shy away from the situation without giving a report of reasonable competency. As a matter of fact, it is not an impossible task, either. One just needs to have awareness about the anatomic and physiologic attributes of the lesion(s) in question, the common associations to be searched for, and the relevant parameters and indices to be documented.

It is at this juncture that one hopes for a ready reckoner, a handy manual that presents all the relevant points in a concise format, which one could quickly glance through before or while performing the echocardiographic study. This burning felt need among the budding cardiologists and echocardiographers of Christian Medical College Vellore was the impetus behind this humble effort.

Although reference textbooks on congenital heart diseases are available, they do not come in handy while performing an echocardiographic study. The same holds true for echocardiography manuals as well, because the whole spectrum of congenital heart diseases is often relegated to a single chapter in such books.

The objective of this book is to empower an echocardiographer to confidently perform a complete two-dimensional echocardiographic examination on all types of congenital heart diseases (treatment naive as well as operated) and prepare a complete and meaningful report with all the relevant elements. I wish to acknowledge the contributions of my colleagues at Christian Medical College Vellore, whose whole-hearted support made this venture a reality.

Thrissur, Kerala, India Amal Paul

Acknowledgements

I am grateful to God almighty for the miraculous provisions made available from various quarters at each juncture, making this project a reality.

I fondly acknowledge the immense love and gratitude towards my beloved teachers—Dr. Paul V George, Dr. Oommen K George, Dr. Viji Samuel Thomson and Dr. Devi Manuel—for their affectionate guidance in the acquisition of echocardiographic skills during my training and the continual support thereafter towards the fulfilment of this undertaking.

I am thankful to my training partners—Dr. Gopal Chandra Ghosh, Dr. Killol Kumar Kaneria, Dr. Sharathbabu N M and Dr. Sudhakar P—and our contemporaries at Christian Medical College Vellore, for the whole-hearted encouragement and patronage during each phase of this journey together.

I am grateful to Dr. Jesu Krupa, Dr. Gayathri Kartha, Dr. Sruthi Irene Varghese and Dr. Ebenezer Daniel for providing images and video loops from the departmental archives.

I register my enormous gratitude towards my colleagues at MOSC Medical Mission Hospital Kunnamkulam—Mr. Anishkumar V G, Ms. Siya Xavier Derick and Sr Molly Shaji—who have been instrumental in compiling the echocardiographic data for this manual.

I thankfully acknowledge the selfless service and support rendered by Dr. Shilpa Elsa George throughout the preparation of this manual.

Last, but not least, I am grateful to all those who presented to us with congenital heart diseases and contributed to the completion of this manual in one way or another.

Contents

Contributors

Images

N. M. Sharath Babu, MD, DM

Cine Loops

Devi Manuel, MD, DM

Ebenezer Daniel, MD

Abbreviations

A

A2C	Apical two chamber
A3C	Apical three chamber
A4C	Apical four chamber
A5C	Apical five chamber
ALCAPA	Anomalous left coronary artery from pulmonary artery
ALVT	Aorto-left ventricular tunnel
AML	Anterior mitral leaflet
Ao	Aorta
AoV	Aortic valve
APW	Aortopulmonary window
AR	Aortic regurgitation
ARV	Atrialized right ventricle
ARVC/D	Arrhythmogenic right ventricular cardiomyopathy/dysplasia
ARVT	Aorto-right ventricular tunnel
Asc aorta	Ascending aorta
AS	Aortic stenosis
ASA	Atrial septal aneurysm
ASD	Atrial septal defect
ASH	Asymmetric septal hypertrophy
ASO	Arterial switch operation
ATL	Anterior tricuspid leaflet
AtrSO	Atrial switch operation
AV	Aortic valve
AV canal/concordance/connection/discordance	Atrioventricular canal/concordance/connection/discordance
AV valve	Atrioventricular valve
AVA	Aortic valve area
AVCD	Atrioventricular canal defect
AVSD	Atrioventricular septal defect
AVVI	Atrioventricular valve index
AW	Anterior wall

B

BAS	Balloon atrial septostomy
BcAV	Bicuspid aortic valve
BDG	Bidirectional Glenn
BT shunt	Blalock-Taussig shunt

C

CA	Coronary artery
CAA	Coronary artery anomaly
CAVV	Common AV valve
CCH	Criss-cross heart
ccTGA	Congenitally corrected transposition of great arteries
COA	Coarctation of aorta
CS	Coronary sinus
CSA	Cross-sectional area
CWD	Continuous wave Doppler

D

D-TGA	Transposition of great arteries with d-loop
DCRV	Double chambered right ventricle
DILV	Double inlet left ventricle
DIRV	Double inlet right ventricle
DORA	Double outlet right atrium
DORV	Double outlet right ventricle
DTA	Descending thoracic aorta

E

ECD	Endocardial cushion defect
EI	Eccentricity index

F

FAC	Fractional area change
fRV	Functional right ventricle

G

GOSE score	Great Ormond Street Echocardiography score

H

HCM	Hypertrophic cardiomyopathy
HLHS	Hypoplastic left heart syndrome
HOCM	Hypertrophic obstructive cardiomyopathy

I

IAS	Interatrial septum
IVC	Inferior vena cava
IVS	Interventricular septum
IW	Inferior wall

L

L-TGA	Transposition of great arteries with l-loop
LA	Left atrium
LAA	Left atrial appendage
LACV	Levoatrial cardinal vein
LAD	Left anterior descending artery
LAO	Left anterior oblique
LAVV	Left atrioventricular valve
LCA	Left coronary artery
LCCA	Left common carotid artery
LCx	Left circumflex artery
LIV	Left innominate vein
LLPV	Left lower pulmonary vein
LMCA	Left main coronary artery
LPA	Left pulmonary artery
LSCA	Left subclavian artery
LSVC	Left superior vena cava
LUPV	Left upper pulmonary vein
LV	Left ventricle
LVH	Left ventricular hypertrophy
LVNC	Left ventricular non-compaction
LVOT	Left ventricular outflow tract
LVOTO	Left ventricular outflow tract obstruction

M

MACS	Maximal aortic cusp separation
MAIF	Mitral aortic intervalvular fibrosa
MAPCA	Multiple aortopulmonary collateral arteries
MAPSE	Mitral annular plane systolic excursion
MPA	Main pulmonary artery

MR	Mitral regurgitation
MPI	Myocardial performance index
MS	Mitral stenosis
MV	Mitral valve

O

OS ASD	Ostium secundum atrial septal defect
OP ASD	Ostium primum atrial septal defect

P

PA	Pulmonary artery
PAH	Pulmonary artery hypertension
PAPVC	Partial anomalous pulmonary venous connection
PAPVD	Partial anomalous pulmonary venous drainage
PASP	Pulmonary artery systolic pressure
PA-VSD	Pulmonary valve atresia with ventricular septal defect
PBF	Pulmonary blood flow
PDA	Patent ductus arteriosus
PFO	Patent foramen ovale
PH	Pulmonary hypertension
PIG	Peak instantaneous gradient
PLAX	Parasternal long axis
PLSVC	Persistent left superior vena cava
PM	Papillary muscle
PML	Posterior mitral leaflet
PPHN	Persistent pulmonary hypertension of the newborn
PR	Pulmonary regurgitation
PRF	Pulse repetition frequency
PRIME	Pulmonary regurgitation index by M-mode echocardiography
PS	Pulmonary stenosis
PSAX	Parasternal short axis
PTL	Posterior tricuspid leaflet
PV	Pulmonary valve
PVC	Pulmonary venous confluence
PVR	Pulmonary vascular resistance
PW	Posterior wall
PWD	Pulse wave Doppler

R

RA	Right atrium
RAA	Right atrial appendage
RAIV	Retroaortic innominate vein
RAVV	Right atrioventricular valve
RCA	Right coronary artery
RLPV	Right lower pulmonary vein
RPA	Right pulmonary artery
RPDA	Right posterior descending artery
RSOV	Ruptured sinus of valsalva aneurysm
RSVC	Right superior vena cava
RUPV	Right upper pulmonary vein
RV	Right ventricle
RVOT	Right ventricular outflow tract
RVOTO	Right ventricular outflow tract obstruction
RVSP	Right ventricular systolic pressure
RWMA	Regional wall motion abnormality

S

SAM	Systolic anterior motion
SIV	Superior–inferior ventricle
SOV	Sinus of valsalva
SSN	Suprasternal notch
STL	Septal tricuspid leaflet
SV	Stroke volume/single ventricle
SV ASD	Sinus venosus atrial septal defect
SVC	Superior vena cava
SVAS	Supravalvular aortic stenosis

T

TAPSE	Tricuspid annular plane systolic excursion
TAPVC	Total anomalous pulmonary venous connection
TAPVD	Total anomalous pulmonary venous drainage
TCPC	Total cavo-pulmonary connection
TDI	Tissue Doppler imaging
TEE	Trans-esophageal echocardiography
TGA	Transposition of great arteries
TOF	Tetralogy of Fallot
TR	Tricuspid regurgitation
TTE	Transthoracic echocardiography
TV	Tricuspid valve

U

UH	Univentricular heart

V

VCR	Ventricular cavity ratio
VSD	Ventricular septal defect
VTI	Velocity time integral

1 Echocardiographic Principles

Echocardiography (or cardiac ultrasound) is the cardiac imaging modality that relies on the generation, propagation, reflection, and reception of sound waves. Sound waves with a frequency greater than 20,000 Hz (20 kHz or 0.02 MHz) are referred to as ultrasound waves. The velocity of sound in heart tissue is 1540 m/s [1].

1.1 Basic Principles

An ultrasound probe (transducer) functions both as a transmitter and receiver of ultrasound signals. The transducer consists of piezoelectric crystals, which generate the ultrasound waves when stimulated by an electric impulse. The ultrasound waves transmitted into the body are either absorbed or reflected by the tissues. The reflected ultrasound waves received by the transducer are called echoes [2].

Deformation of the piezoelectric crystal by the reflected ultrasound energy produces an electric impulse. Thus, the generation of an ultrasound image involves the transmission of ultrasound waves through the tissue, the reflection of the waves by the tissues, and the reception of the reflected waves by the transducer.

Transducer frequency: The frequency of diagnostic ultrasound waves is a property of the piezoelectric crystals used in the transducer and ranges from 2 to 12 MHz. The frequencies used in transthoracic echocardiography are:

- Adult transducers: 2–4 MHz
- Paediatric transducers: 7–12 MHz

Higher frequency is associated with shorter wavelength, less tissue penetration, and higher image resolution, whereas lower frequency is associated with greater wavelength, deeper tissue penetration, and lower image resolution [2, 3].

Harmonic imaging: The transducer receives frequencies at multiples of the emitted frequency (harmonic frequencies). This improves the signal-to-noise ratio and image quality [4].

Gain is the amplitude, or the degree of amplification, of the received signal. When gain is low, weaker echoes from the edge of the ultrasound beam may not be recorded [5].

Pulse repetition frequency (PRF): The echocardiographic transducer emits ultrasound waves in rapid pulses. The rate at which the ultrasound pulses are emitted is called pulse repetition frequency [3, 4]. PRF is commonly expressed as the number of pulses per second. PRF for different modes of imaging varies as follows:

- 3000–5000/s for 2D echo
- 1000–2000/s for M-mode

Frame rate refers to the number of still images displayed sequentially in a unit of time. A higher frame rate corresponds to better temporal resolution. Increasing the imaging depth and widening the imaging sector reduces the frame rate.

A. Paul, *Manual of Echocardiography for Congenital Heart Diseases*,
https://doi.org/10.1007/978-981-96-0243-8_1

Key point: Reducing the image depth and narrowing the imaging sector increases the frame rate, thereby improving the temporal resolution.

1.2 Echocardiographic Imaging

1.2.1 B-Mode (2D) Imaging

B-mode imaging refers to the imaging mode in which the brightness of each point in the image corresponds to the amplitude of the reflected wave received by the transducer. The vertical position of each dot depends on the time delay between the transmission and reception of the ultrasound pulse. The horizontal position of each dot on the image is determined by the location of the receiving transducer element [2, 3].

Imaging the heart in multiple two-dimensional planes aids in the three-dimensional reconstruction and visualization of the cardiac anatomy. By convention, structures closest to the transducer are displayed at the top of the screen in an echocardiographic image. However, in paediatric echocardiograms, recommendations suggest inversion of the image in such a way that the image apex is at the bottom of the screen during subxiphoid and apical imaging [6, 7].

The side of the ultrasound plane that corresponds to the notch on the transducer is displayed on the right side of the screen. By convention, the transducer is positioned in such a way that rightward structures are displayed on the left side of the image. The only exception to this convention is the parasternal long-axis view, in which the apex of the heart is displayed on the left side of the image in both levocardia and dextrocardia [7].

Image Resolution [8, 9]

- **Spatial resolution**: The minimum distance of separation between two points to be identified as two distinct points in the image. It has two components:
  - **Axial resolution**: The ability to differentiate two structures lying along the axis of the ultrasound beam (i.e. spatial resolution along the longitudinal axis parallel to the beam).
  - **Lateral resolution**: The ability to differentiate two structures lying side by side relative to the beam (i.e. spatial resolution along the lateral axis perpendicular to the beam).
- **Contrast resolution**: The ability to display different shades of grey within the image. This aids in distinguishing borders and textures of different tissues.
- **Temporal resolution**: The ability of an imaging system to determine the position of a structure at a particular instant in time. This aids in accurately tracking a moving target over time.

Imaging Artefacts [10]

Ultrasound imaging artefacts can generally result from defective ultrasound equipment, external interference from electronic equipment(s), or inappropriate ultrasound settings. However, most artefacts are due to physical interactions of the ultrasound beam with the tissue.

- **Attenuation artefact**: Results from the diminution in the intensity of the ultrasound beam with an increase in imaging depth [10].
  - 'Shadowing' behind a strong reflecting structure results in an echo-free space behind the reflector, erroneously creating the impression of a hypoechoic mass.
  - **Rib 'dropout' artefact**: An attenuation artefact caused by the bony rib cage.
- **Reverberation artefact**: Caused by internal reflections from echo-bright, fast-moving structures such as mechanical valves [10].
  - Seen as a duplication of the structure or as a trail of bright echo behind the structure (such as the mitral valve being seen in the echo-free space posterior to the heart) (Fig. 1.1).
  - **Comet-tail artefact**: Microreflections on the echo image created by an interface between tissues with marked differences in acoustic impedance (such as B-lines in the lung–fluid interface).
- **Side lobe artefact**: Echo-bright structures (such as wire or catheter) outside the path of the central ultrasound beam are erroneously

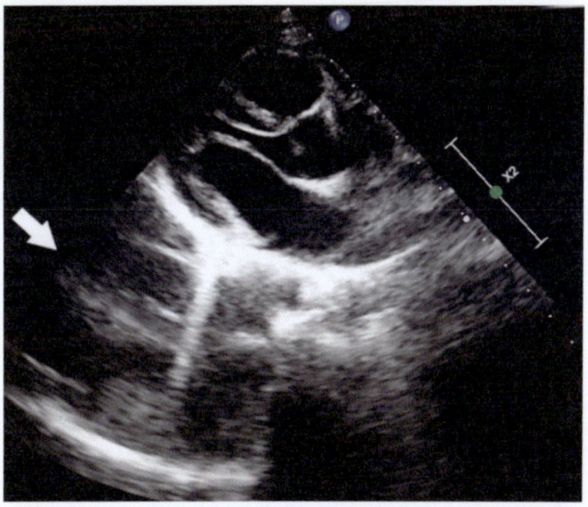

Fig. 1.1 Reverberation artefact: PLAX view with duplication of structures (denoted by the arrow) in the echo-free space posterior to the heart

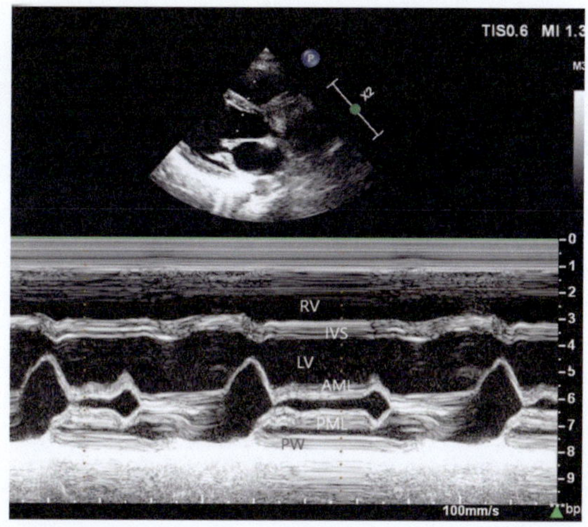

Fig. 1.2 M-mode image from the PLAX view along the mitral valve, depicting the motion and relative location of structures along the ultrasound beam. *RV* right ventricle, *IVS* interventricular septum, *LV* left ventricle, *AML* anterior mitral leaflet, *PML* posterior mitral leaflet, *PW* posterior wall

mapped onto the image and displayed in the echo-free spaces in or around the heart.

- **Speed displacement (propagation velocity) artefact**: Discontinuity and distortion of the displayed image resulting from an inaccurate depth measurement due to the interposition of tissues (such as bones or fat) with differential propagation velocities. An echogenic structure is seen to be displaced deeper than its actual position in the image.

1.2.2 M-Mode Imaging

M (motion)-mode echocardiography involves the reception of echoes from a moving tissue reflector over time [5]. An M-mode image depicts the structures along the path of a single line of the ultrasound beam (Fig. 1.2). This has been called the **'ice-pick view'** of the heart.

M-mode imaging has a high frame rate. With sampling rates of 1000–2000 pulses/s, as against a frame rate of 30–100 frames/s in B-mode imaging, the temporal resolution of an M-mode image is much higher. M-mode imaging is especially useful in:

- Identifying the relative location of mobile structures during the cardiac cycle and measuring their range of motion (such as systolic anterior motion of the anterior mitral leaflet in HOCM)
- Determining linear dimensions and estimating chamber sizes

1.3 Doppler Imaging

Doppler ultrasound is useful in assessing the velocity of moving objects, such as red blood cells and myocardium. While 2D echocardiography focuses primarily on the structural anatomy of the heart, Doppler imaging depicts the physiology and haemodynamics by targeting red blood cells [3]. Doppler imaging of the heart falls into two categories:

- Spectral Doppler (continuous wave Doppler/pulsed-wave Doppler)
- Colour flow Doppler

Doppler principle: The frequency of any waveform emitted from a moving object will be perceived as higher than or lower than the actual frequency, depending on whether the object is moving towards or away from the observer [3, 5].

Ultrasound waves reflected from moving red blood cells return to the transducer at a frequency different from that at which they were emitted. This difference in frequency, termed the **Doppler frequency shift**, depends on the speed of ultrasound through the medium and the velocity of blood flow, according to the equation:

$$\text{fd} = 2\text{ft} \times V \times \cos\theta / c$$

where fd is the Doppler shift frequency, ft is the transmitted ultrasound frequency, V is the velocity of blood flow, c is the speed of ultrasound in the tissue, and θ represents the angle of flow relative to the ultrasound beam (angle of insonation).

If the flow is towards the transducer, the frequency of the reflected waves is higher, whereas the waves return at a lower frequency if the flow is away from the transducer.

Key point: Doppler imaging is optimal when the ultrasound beam is aligned parallelly with the flow of blood, while 2D imaging is optimal when the beam and the target structure are at right angles.

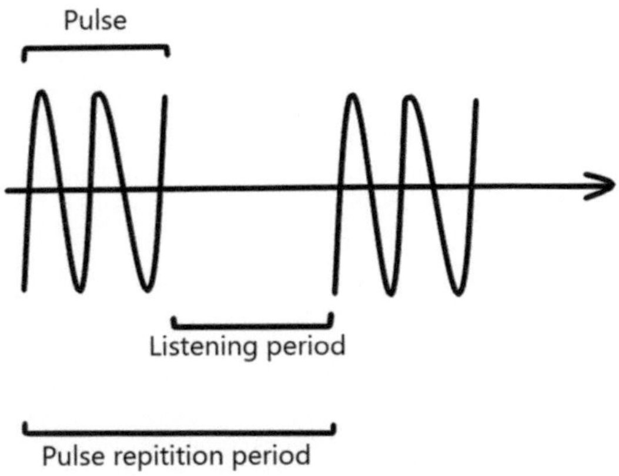

Fig. 1.3 Schematic representation of ultrasound waves emitted from the transducer as pulses

1.3.1 Pulsed-Wave Doppler (PWD)

Pulsed-wave Doppler utilizes the Doppler principle to estimate the velocity of blood at a particular location [11]. The transducer emits discrete pulses of ultrasound waves at a specific frequency (pulse repetition frequency, PRF) (Fig. 1.3). Each pulse is followed by a 'listening period' (pulse repetition period) to receive the reflected signals. This gating mechanism enables depth acuity, whereby the velocity of blood flow at a particular depth can be assessed. The sample volume is determined by the location of the cursor on the 2D image.

Pulse repetition frequency (PRF) is the sampling rate, which determines the depth at which the Doppler shift is evaluated. PRF is inversely proportional to the pulse repetition period. ***Deeper structures are imaged with a low PRF*** (and a longer listening period).

When the blood flow velocity is high, the Doppler shift frequency is also high. Hence, a higher sampling rate is needed to accurately sample that shift. However, a higher sampling rate (PRF) limits the depth of interrogation, as the next impulse is sent earlier. Thus, for a given sample volume and PRF, there is an upper limit to the velocity that can be interrogated accurately with PWD [11].

Nyquist limit refers to the maximum velocity that can be accurately quantified within a given sample volume. ***Nyquist limit is roughly one-half of PRF***, which in turn is inversely related to the distance from the sample volume to the transducer.

If the velocity of blood at the site of the sampling volume exceeds the Nyquist limit, the waveform appears to change direction and is displayed on the opposite side of the baseline, a phenomenon termed **aliasing** [12]. Hence, PWD is best suited for measuring lower velocities at a particular depth.

Key point: As the sample volume is placed farther away from the transducer, the pulse repetition frequency and the Nyquist limit decrease, and aliasing occurs at relatively lower velocities.

1.3.2 Continuous-Wave Doppler (CWD)

Continuous wave Doppler is characterized by continuous (rather than pulsed) transmission and reception of signals. The transducer has one set of piezoelectric crystals continuously emitting ultrasound waves, while another set of crystals continuously receives the returning signals [3].

CWD is not limited by factors such as PRF, aliasing, or Nyquist limit, as the ultrasound transmission is continuous. Hence, there is no limit to the peak velocities discernible with this technique. However, depth localization of the peak velocity is not possible, as the Doppler shift may occur anywhere along the ultrasound beam.

PEDOFF probe: **P**ulsed **E**cho **Do**ppler **F**low velocity metre (Pedoff) probe is a non-imaging CW Doppler probe used for estimating peak velocities, especially in cases of aortic stenosis [5].

Key point: PWD is useful in assessing relatively low-velocity flow (typically <1.5 metres/sec) present at a specific depth location, whereas CWD records the maximum velocity along the ultrasound beam without any depth specificity.

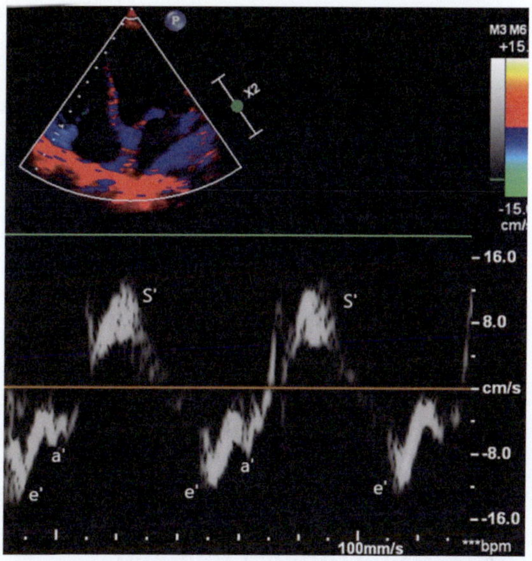

Fig. 1.4 Tissue Doppler imaging of the lateral tricuspid annulus, depicting the diastolic velocities (e' and a') and the systolic velocity (S') of the tricuspid annular myocardial tissue

1.3.3 High Pulse Repetition Frequency (HPRF) Doppler

- HPRF Doppler combines features of both PWD and CWD.
- HPRF can be used to record high velocities at particular depths, exceeding the Nyquist limit of PWD for the transducer.
- Multiple sample volumes are generated by increasing the PRF and capturing the returning signals (emitted by the previous ultrasound pulse) from greater depths. Thus, HPRF can quantify high-velocity flows, with 'partial depth ambiguity' [13].

1.3.4 Tissue Doppler Imaging (TDI)

- A variant of PWD, in which the Doppler technique is used to assess the velocity of excursion of myocardial tissue (Fig. 1.4) [14].
- The settings are different from conventional PWD to account for:
  - The lower velocity of the tissue (typically <20 cm/s) as compared to that of blood flow (measured in m/s)
  - The stronger reflection of the Doppler signal by the tissue (higher amplitude signals) compared with blood

1.3.5 Colour Flow Doppler

Colour flow Doppler is a PWD-based technique in which multiple sample volumes in a region of interest are interrogated simultaneously and displayed on a 2D image with a colour map [15]. Colour flow Doppler allows direct real-time visualization of the movement of blood in the heart.

By convention, ***flow that is moving away from the transducer is coded in blue, and flow that is moving towards the transducer is coded in red*** (Fig. 1.5).

- As the flow velocity increases, the colour code assumes progressively lighter shades until the Nyquist limit is reached, at which point aliasing occurs and the colour switches to the opposite one (i.e. red colour switches to blue, and blue to red).
- Flow acceleration and turbulence appear as multicoloured mosaic signals (usually green and yellow).

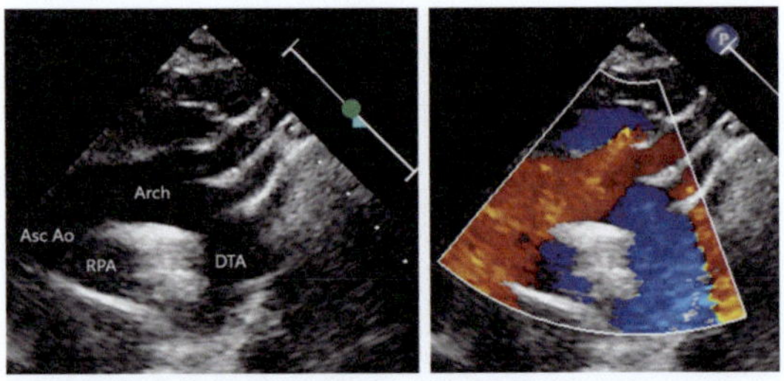

Fig. 1.5 Colour flow Doppler image of the aorta from the suprasternal window. It can be noted that the blood flow in the ascending aorta (Asc Ao), which appears red owing to its direction towards the probe, abruptly turns blue as the flow changes its direction away from the probe in the centre of the arch towards the descending thoracic aorta (DTA). *RPA* right pulmonary artery

1.4 Haemodynamic Assessment Using Doppler Imaging

1.4.1 Quantification of Flow Volumes

Basic principle: The volume of flow through an orifice in unit time (m^3/s) can be estimated by multiplying the area of the orifice (m^2) by the velocity of flow through the orifice (m/s). This principle may be applied to any orifice to measure blood flow.

1.4.1.1 Stroke Volume

Cardiac stroke volume can be estimated by multiplying the cross-sectional area (CSA) of the LVOT by the forward flow velocity of blood in the LVOT [16].

- Assuming the LVOT to be circular, LVOT CSA is obtained by measuring the LVOT diameter (D) from the PLAX view in mid-systole and applying it in the formula:

$$\text{LVOT area}\left(\text{cm}^2\right) = \Pi \times \left(D/2\right)^2$$

- Since the velocity of blood flow varies throughout the systolic ejection period, an average sum (or integral) of the velocities over the systolic ejection period (VTI or velocity-time integral) is used in the calculation [5].
- LVOT VTI (stroke distance) is obtained by tracing the PWD signal from the LVOT in the A5C or A3C views.

$$\text{Cardiac Stroke volume}\left(\text{cm}^3 \text{ or mL / beat}\right) = \text{LVOT area}\left(\text{cm}^2\right) \times \text{LVOT VTI}\left(\text{cm / beat}\right)$$

1.4.1.2 Shunt Quantification

Quantum of flow (Qp/Qs) across a shunt lesion can be determined by using the above principle [16].

- *Qs* (Systemic blood flow) is equal to the LV stroke volume, which is calculated as described above.
- *Qp* (blood flow across the pulmonary valve) is calculated by multiplying RVOT area by RVOT VTI (obtained from the PWD signal from the PSAX view at the level of the aortic valve). Alternatively, proximal PA diameter and VTI at the same level may be used.

$$Qp / Qs = \text{RVOT area} \times \text{RVOT VTI} / \text{LVOT area} \times \text{LVOT VTI}$$

- Since the areas of the RVOT and LVOT are estimated by squaring the diameter of these outflow tracts, minor errors in the measured diameter will lead to significant discrepancies in the final result. Inadequate windows may also render the measurement of RVOT diameter challenging.

1.4.2 Bernoulli Principle and Estimation of Pressure Gradients

Basic principle: The velocity of blood flow is determined by the pressure gradient between cardiac chambers, with higher gradients resulting in higher velocities. Hence, if the velocity of blood flow between two chambers is known, the pressure gradient between them can be inferred [16].

The Bernoulli principle is based on the '**Law of conservation of energy**'. Applying this principle to blood circulation, the total energy of the circulating blood must be conserved at all points. Hence, when blood flows across a valve or orifice, energy proximal to the valve must be equal to that distal to it [16].

Assuming that the loss of energy resulting from flow acceleration, viscous friction, and gravitational energy is negligible under most physiologic conditions:

- Energy proximal to the orifice (Pressure energy P1 + Kinetic energy K1) = Energy distal to the orifice (Pressure energy P2 + Kinetic energy K2).
- Therefore, the pressure gradient across the orifice, $P1 - P2(\Delta P) = K2 - K1$.
- The kinetic energy of blood is calculated by the formula $\mathbf{1/2\rho v^2}$, where ρ is the mass density of blood. The numerical value of 1/2ρ is approximately equal to 4.
- Thus, $P1\text{-}P2\ (\Delta P) = 4v2^2 - 4v1^2$.

The Bernoulli equation can be simplified to: $\mathbf{\Delta P = 4(v2^2 - v1^2)}$, where v1 and v2 are the velocities proximal and distal to the orifice, respectively.

- Since velocities proximal to an orifice or stenosis (V1) are generally negligibly low in comparison to the high velocities distal to the orifice (V2),V1 can usually be ignored.
- Thus, the final simplified equation would be: **$\Delta P = 4v^2$, where v is the maximal velocity of the jet across the orifice**.
- In situations where the proximal velocity (v1) is relatively higher than usual and the gradient is low, the use of the simplified equation would be inappropriate. The Bernoulli equation, in such cases, should be used as: $\mathbf{\Delta P = 4(v2^2 - v1^2)}$.
- Apart from determining the severity of valve stenosis, the Bernoulli equation can be used to estimate intracardiac pressures in patients with valvular regurgitation or ventricular septal defects (Table 1.1).
- For example, the blood flow velocity of tricuspid regurgitation (TR) reflects the pressure gradient between the RV and RA.

$$\text{RVSP} - \text{RA pressure} = 4 \times (\text{peak TR velocity})^2$$

Hence, **RVSP = 4 x (peak TR velocity)^2+ RA pressure**

In the absence of RV outflow tract obstruction (RVOTO), **RVSP = PASP**

Thus, PA systolic pressure can be inferred from TR velocity.

Key point: Although pressure gradients can be inferred from velocities based on the Bernoulli equation, absolute pressures within the chambers cannot be directly measured, as in cardiac catheterization.

Potential Sources of Error

- **Improper alignment**: When the incident angle between the ultrasound beam and the direction of blood flow is more than 20 degrees, a significant error is introduced into

Table 1.1 Clinical utility of the Bernoulli equation in echocardiographic haemodynamic assessment

Velocity measured using Doppler	Pressure estimated using Bernouli equation
Peak velocity of flow through stenotic aortic valve	LV diastolic pressure
TR jet velocity	RV systolic pressure
Peak velocity across a VSD	RV systolic pressure
Velocity through a PDA	Pulmonary artery systolic pressure
Peak velocity of PR jet	Pulmonary artery mean pressure
End-diastolic velocity of PR jet	Pulmonary artery diastolic pressure

the Doppler equation that results in an underestimation of true velocity [5, 16]. This error can be minimized by:
  - Using Colour Doppler imaging to facilitate better alignment by visualizing the jet
  - Using multiple imaging windows to choose the view providing the best alignment
- **Suboptimal image**: If the tip of the recorded Doppler envelope is not seen because of an incomplete signal, the recorded peak velocity will be inaccurate (Fig. 1.6). This can be rectified by:
  - Adjusting the CWD gain setting
  - Optimizing the beam alignment
  - Application of echo contrast agents to boost the signal

1.4.3 Continuity Equation and Estimation of Area

Basic principle: The continuity equation is based on the **'Law of conservation of mass'**. When applied to Doppler imaging, this principle implies that the flow in one region of the heart should be equivalent to flow in another region (assuming no intervening shunt). In simple terms, the volume of blood going in (inflow) must be equal to that coming out (outflow) (Fig. 1.7) [16].

- As discussed earlier, the flow rate across an orifice is the product of the Velocity Time Integral (VTI) at the orifice and the cross-sectional area (CSA) of the orifice.
- By estimating the flow proximal to a stenotic valve and the VTI through the valve, the area of the stenotic valve can be determined.

For example, Aortic valve area (AVA) can be estimated using this principle. According to the continuity principle, flow through the LVOT must equal the flow through the aortic valve (AV):

$$\text{LVOT VTI} \times \text{LVOTCSA} = \text{AV VTI} \times \text{AVA}$$

$$\text{AVA} = \left(\text{LVOT VTI} \times \text{LVOT area}\right) / \text{AV VTI}$$

Thus, AVA can be estimated by obtaining the following measurements:

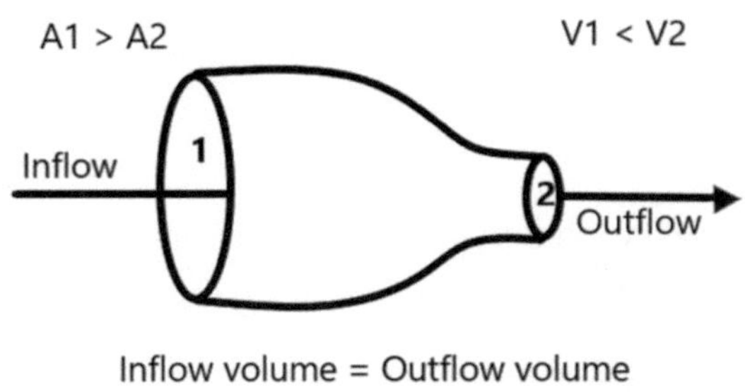

Fig. 1.7 Schematic representation of the continuity principle. The product of the cross-sectional area [A1×A2] and velocity time integral are the same for blood entering and leaving a cardiac chamber. *V1* Inflow velocity *V2* Outflow velocity

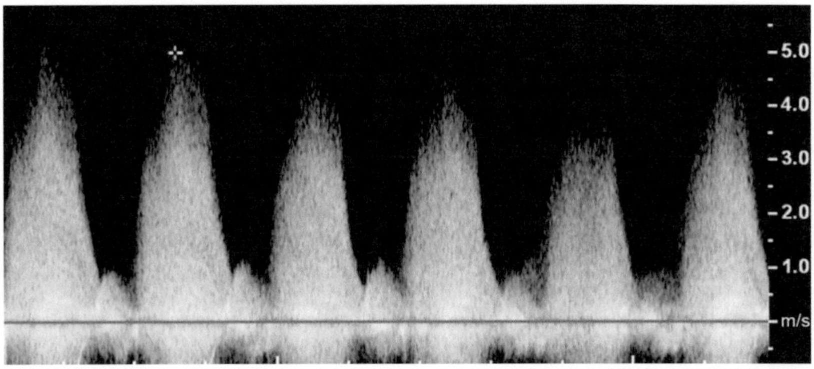

Fig. 1.6 CWD tracing from a VSD. The true peak velocity would be underestimated if the velocity were estimated from tracings where the tip of the Doppler envelope is not clearly seen

- LVOT VTI, measured using PWD just proximal to the stenotic valve
- LVOT diameter at the same point where the VTI was measured
- AV VTI, measured using CWD across the AV

Key point: Since the velocity of flow through a stenotic valve can be too high to be estimated with PWD, CWD can be used instead because the maximal velocity will correspond to the stenotic valve orifice, assuming that there is no other region of narrowing along the path of the ultrasound beam.

References

1. Goldman DE, Jueter TF. Tabular data of the velocity and absorption of high-frequency sound inmammalian tissues. J Acoust Soc Am. 1956;28:35–7.
2. Feigenbaum H, Zaky A. Use of diagnostic ultrasound in clinical cardiology. J Indiana State Med Assoc. 1966;59:140.
3. Burns PN. The physical principles of Doppler and spectral analysis. J Clin Ultrasound. 1987;15:567–90.
4. Kornbluth M, Liang DH, Paloma A, Schnittger I. Native tissue harmonic imaging improves endocardial border definition and visualization of cardiac structures. J Am Soc Echocardiogr. 1998;11:693–701.
5. Armstrong WF, Ryan T. Feigenbaum's echocardiography. 8th ed. Philadelphia: Wolters Kluwer; 2019.
6. Wasserman MA, Shea E, Cassidy C, Fleishman C, France R, Parthiban A, Landeck BF. Recommendations for the adult cardiac sonographer performing echocardiography to screen for critical congenital heart disease in the newborn: from the American Society of Echocardiography. J Am Soc Echocardiogr. 2021;34(3):207–22.
7. Lai WW, Geva T, Shirali GS, Frommelt PC, Humes RA, Brook MM, Pignatelli RH, Rychik J, Task Force of the Pediatric Council of the American Society of Echocardiography; Pediatric Council of the American Society of Echocardiography. Guidelines and standards for performance of a pediatric echocardiogram: a report from the Task Force of the Pediatric Council of the American Society of Echocardiography. J Am Soc Echocardiogr. 2006;19(12):1413–30.
8. Minich LL, Tani LY, Paul JJ. Technological advances to extend the echocardiographic imaging of pediatric patients. Pediatr Cardiol. 2002;23:257–65.
9. Hedrick WR, Hykes DL, Starchman DE. Ultrasound physics and instrumentation. St. Louis: Mosby; 1995. p. 1–90.
10. Le HT, Hangiandreou N, Timmerman R, Rice MJ, Smith WB, Deitte L, Janelle GM. Imaging artifacts in echocardiography. Anesth Analg. 2016;122(3):633–46.
11. Baker DW, Rubenstein SA, Lorch GS. Pulsed Doppler echocardiography: principles and applications. Am J Med. 1977;63:69–80.
12. Bom K, de Boo J, Rijsterborgh H. On the aliasing problem in pulsed Doppler cardiac studies. J Clin Ultrasound. 1984;12:559–67.
13. Snider AR, Stevenson JG, French JW, Rocchini AP, Dick M, Rosenthal A, Crowley DC, Beekman RH, Peters J. Comparison of high pulse repetition frequency and continuous wave Doppler echocardiography for velocity measurement and gradient prediction in children with valvular and congenital heart disease. J Am Coll Cardiol. 1986;7(4):873–9.
14. Waggoner AD, Bierig SM. Tissue Doppler imaging: a useful echocardiographic method for the cardiacsonographer to assess systolic and diastolic ventricular function. J Am Soc Echocardiogr. 2001;14:1143–52.
15. Aggarwal KK, Moos S, Philpot EF, Jain SP, Helmcke F, Nanda NC. Color velocity determination using pixel color intensity in Doppler color flow mapping. Echocardiography. 1989;6:473–83.
16. Madrazo JA, Arnold SV. Introduction to echocardiographic principles. In: Rasalingam R, Makan M, Perez JE, editors. The Washington manual of echocardiography. 1st ed. Philadelphia: Lippincott Williams & Wilkins; 2012. p. 1–11.

2 Echocardiographic Examination

2.1 General Considerations

2.1.1 Positioning the Patient

The majority of echocardiographic examinations for congenital heart disease are performed in children, including neonates and infants. Allaying the anxiety of the child prior to the study is an integral part of performing a complete echocardiographic scanning. This may be facilitated by:

- Providing the reassurance to the child that the echocardiographic probe is not painful by letting them to touch the probe and check for themselves
- Performing the examination in the presence of the parent(s)—sometimes by even positioning the child on their lap
- Distracting the attention of the child away from the examiner
- Confirming with the parents that the infant has been adequately fed prior to the study
- Warming the ultrasound gel before application minimizes the risk of awakening a neonate

Administering sedative medications may be necessary in some cases. The commonly used medications are Chloral hydrate and Triclofos.

The recommended dose of **Chloral hydrate** (oral/rectal) is:

- Age less than 3 months: 50 mg/kg
- Age 3–9 months: 75 mg/kg
- Age more than 9 months: 100 mg/kg (maximum dose of 1000 mg)

The recommended dose of oral **Triclofos** (1 g Triclofos = 600 mg Chloral hydrate):

- 25–75 mg/kg (100 mg/kg for deep sedation—a total maximum dose of 2 g)
- May be administered as a 50 mg/kg single dose; repeat after 30 minutes if required

2.1.2 Position of the Sonographer

The echocardiographer may be seated on either the left or right side of the patient. When seated on the patient's right side, scanning is performed with the probe in the right hand, and vice versa.

2.1.3 Transducer Manoeuvring [1]

- **Handling**: Stabilizing the transducer by fixing the examiner's little finger against the child's body prevents the transducer from slipping from its position.

Supplementary Information The online version contains supplementary material available at https://doi.org/10.1007/978-981-96-0243-8_2.

A. Paul, *Manual of Echocardiography for Congenital Heart Diseases*,
https://doi.org/10.1007/978-981-96-0243-8_2

- **Pressure**: While firm pressure is necessary for good contact, care must be taken to avoid pain, especially over bony surfaces.
- **Movements**: The movements of the transducer (translation, angulation, tilt, and rotation) must be slow and subtle.
- **Frequency**: Paediatric echocardiographic transducers typically have a frequency between 7 and 12 MHz.

2.1.4 Optimizing the 2D Image (Fig. 2.1a)

- **2D gain**: An optimum gain is indicated by a dark blood pool and distinct myocardial borders.
- **Depth**: An ideal imaging depth would be 1–2 cm beyond the cardiac border or the structure of interest. Greater depth compromises the temporal resolution [1].
- **Focus**: The focus of the ultrasound beam, indicated by an arrow outside the imaging sector, should be adjusted to the level of the area of interest.

2.1.5 Optimizing the Doppler Examination

- **Colour Doppler:**
  - **Gain**: Colour gain can be calibrated by placing the colour sector in extracardiac space, increasing the colour gain until noise appears, and adjusting the gain downward to the point where the noise just disappears.
  - **Colour sector**: Minimizing the area of the colour sector (sample volume) improves the temporal resolution.
  - **Nyquist limit**: 50–60 cm/s.
- **Spectral Doppler:**
  - **Spectral gain**: While recording spectral Doppler, optimal gain is indicated by a dark background and bright Doppler envelope. Measurements should be taken of the modal velocities (represented by the bright envelope) and not the spectral broadening (feathery appearance) [1] (Fig. 2.1b).
  - **Sweep speed**: Usually set at 50 mm/s; may be increased to improve temporal resolution.

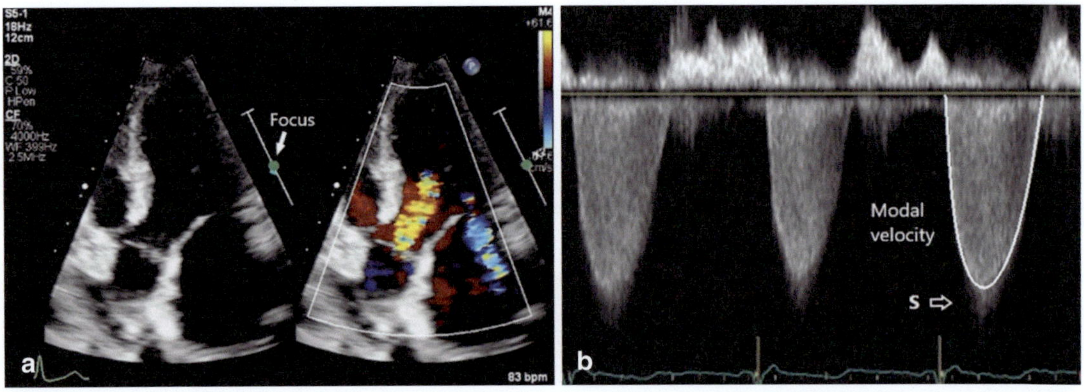

Fig. 2.1 (**a**) Normal apical five-chamber view with colour Doppler and appropriate gain settings. Note the position of the focus at the area of interest. (**b**) Continuous wave Doppler tracing depicting modal velocity (bright envelope) with spectral feathering (S)

2.2 Echocardiographic Windows

A thorough knowledge of the wide array of views and windows available in the echocardiographic armour, and the spectra of structures visualized in each view, is essential for performing a complete and meaningful examination. This becomes extremely important in congenital heart diseases, where anatomic variations and chest deformities may completely negate the utility of certain windows.

- A standard echocardiographic examination utilizes the parasternal, apical, subcostal, and suprasternal windows.
- Each window provides visual access to different imaging planes of the heart, namely sagittal, coronal, and transverse (axial).
- While performing an echocardiogram in a patient with congenital heart disease, it is a common practice to start with the subcostal window in order to follow the segmental approach (discussed in Chap. 3) in a systematic fashion.
- Sweeping the transducer through the nearly parallel planes within each acoustic window enables the echocardiographer to obtain parallel 'slices' within a given plane [1, 2].
- Thus, a wide array of **views** is obtained by **sweep**ing the transducer through different imaging **planes** from each **window.**

2.2.1 Parasternal Window

Patient position: Left lateral decubitus with the left shoulder extended

Sweeps: Parasternal long axis (PLAX), parasternal short axis (PSAX)

2.2.1.1 Parasternal Long-Axis (PLAX) Sweeps (Fig. 2.2)

Transducer position: Left parasternal area, at the third or fourth intercostal space, with the notch pointing towards the patient's right shoulder.

- Standard parasternal long-axis view
- RV inflow (Tricuspid valve or PLAX-RV) view: Obtained by tilting the transducer tail towards the left shoulder
- RV outflow (Pulmonary valve) view: Obtained by tilting the transducer tail towards the right hip
- High PLAX view: PLAX view taken one or two intercostal spaces above the standard PLAX view

2.2.1.2 Parasternal Short-Axis (PSAX) Sweeps (Fig. 2.3)

Transducer position: Rotate the transducer clockwise 90° from the PLAX view. Tilting the transducer tail towards the right shoulder opens up more apical views in the following order:

- Aortic valve level (superior basal) view
- Coronary artery level
- Mitral valve level
- Papillary muscle level (inferior) view
- Apical level

High PSAX or Ductal view: PSAX view with the transducer placed in the first or second intercostal space. This view is ideal for visualizing a patent ductus arteriosus and estimating its size (depicted in Chap. 4). It is also called the left subclavicular view.

2.2.2 Apical Window

Patient position: Left lateral decubitus with the left shoulder extended

Planes: Four chamber; two chamber, three chamber

2.2.2.1 Apical Sweeps (Figs. 2.4 and 2.5)

Transducer position: Near the apical impulse (normally left fifth intercostal space in the midclavicular line), with the notch pointing roughly towards the 2 o'clock position and the imaging plane oriented cranially. Tilting the transducer anteriorly and posteriorly opens up different imaging planes as follows:

- Apical four chamber (Standard-**A4C**) view
- Apical five chamber (Anterior-**A5C**) view
- Coronary sinus (Posterior) view

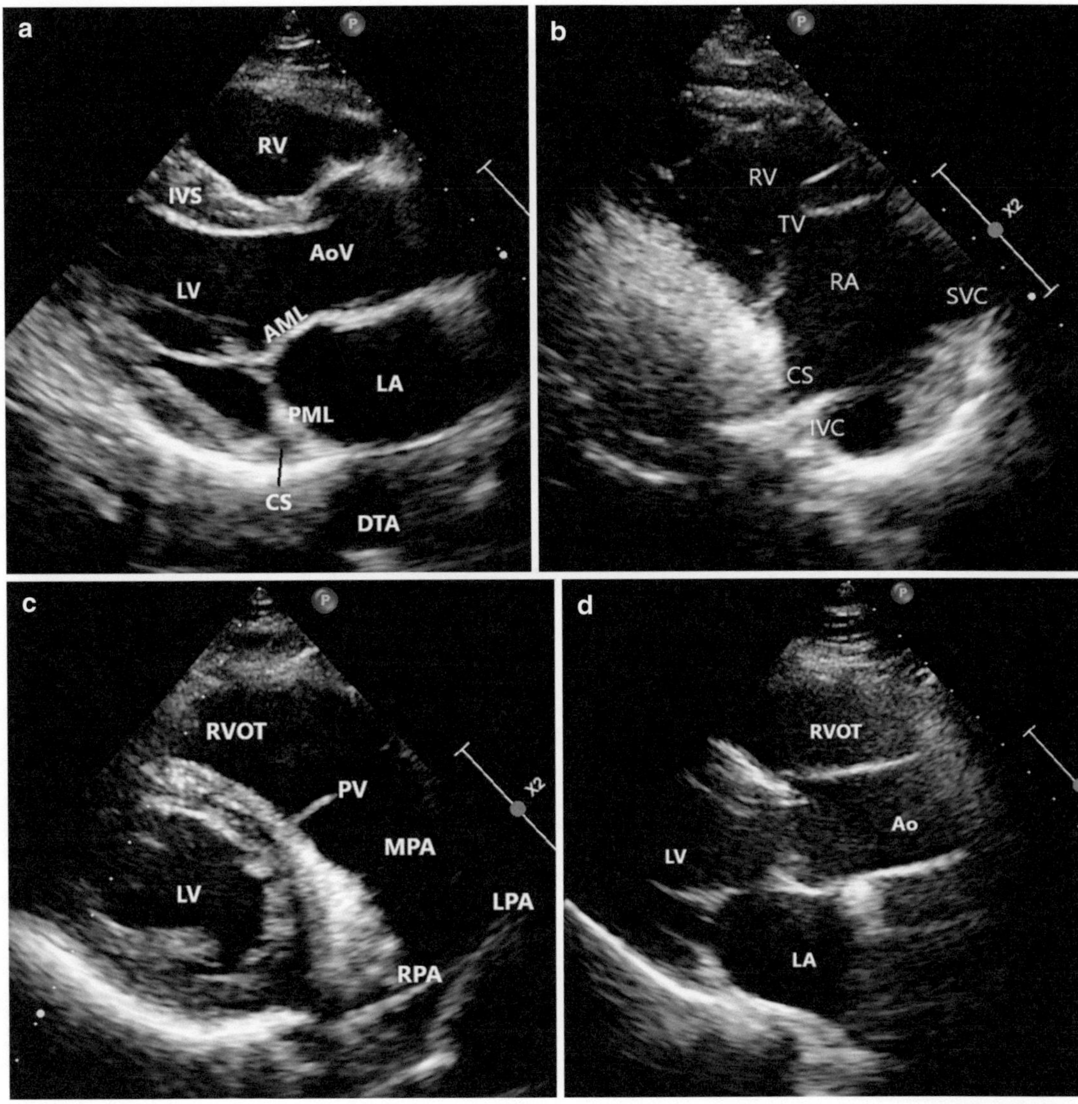

Fig. 2.2 Parasternal long-axis (PLAX) sweeps: (**a**) Standard PLAX view, (**b**) RV inflow view showing the orifices of the coronary sinus (CS), inferior vena cava (IVC), and superior vena cava (SVC). (**c**) RV outflow view showing the main pulmonary artery (MPA) and its branches. (**d**) High PLAX view with the probe positioned two intercostal spaces higher

Modified A4C view: From the standard A4C view, sliding the transducer medially (without any rotation) towards the sternal border allows an incident angle of 30–45° between the ultrasound beam and the atrial septum, facilitating the evaluation of atrial septal defects.

RV focused A4C view: From the standard A4C view, slide the transducer laterally such that the RV is placed in the centre of the image, while the LV apex remains central at the top of the image sector. Rotate the transducer such that the RV free wall is visualized in its entire length and maximum RV basal diameter is obtained.

- Apical two-chamber (**A2C**) view: Obtained by rotating the transducer about 30–60° counterclockwise from the A4C view
- Apical three-chamber (apical long axis-**A3C**) view: Obtained by rotating the transducer about 30–60° counterclockwise from the A2C view

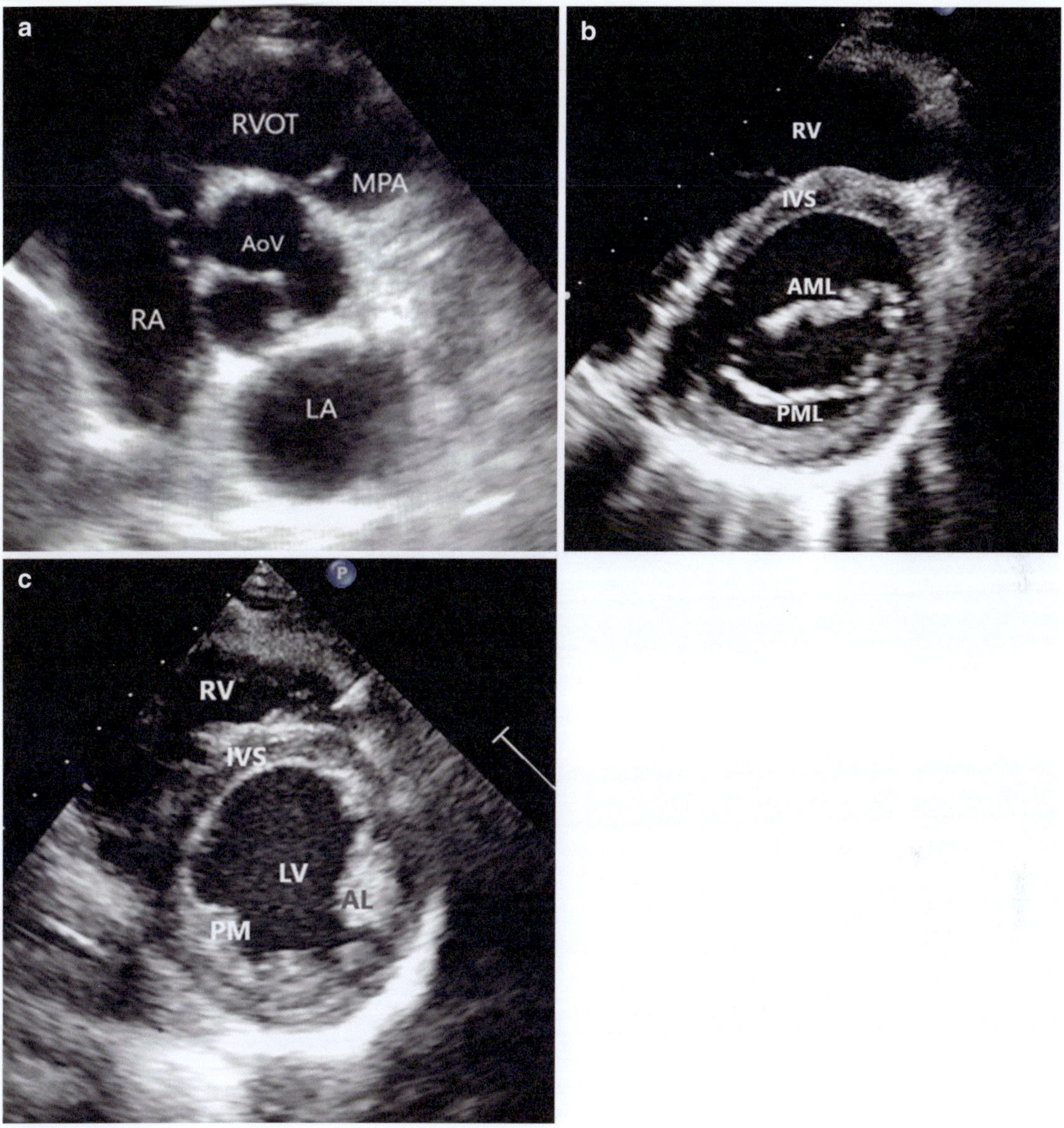

Fig. 2.3 Parasternal short-axis (PSAX) sweeps: (**a**) Aortic valve level. (**b**) Mitral valve level. (**c**) Papillary muscle level. *AoV* aortic valve, *MPA* main pulmonary artery, *AML* anterior mitral leaflet, *PML* posterior mitral leaflet, *AL* anterolateral papillary muscle, *PM* posteromedial papillary muscle

2.2.3 Subcostal Window

Patient position: Supine (with both the knees flexed in order to relax the abdominal muscles in older patients)

Planes: Long axis (coronal), Short axis (sagittal), Left anterior oblique

2.2.3.1 Subcostal Coronal (Long-Axis) Sweeps (Fig. 2.6)

Transducer position: Two finger-breadths below the xiphoid process, with the notch pointing towards the left side of the patient [3 o' clock position]. Tilting the transducer anteriorly (tail towards the right hip) with a slight clockwise rotation

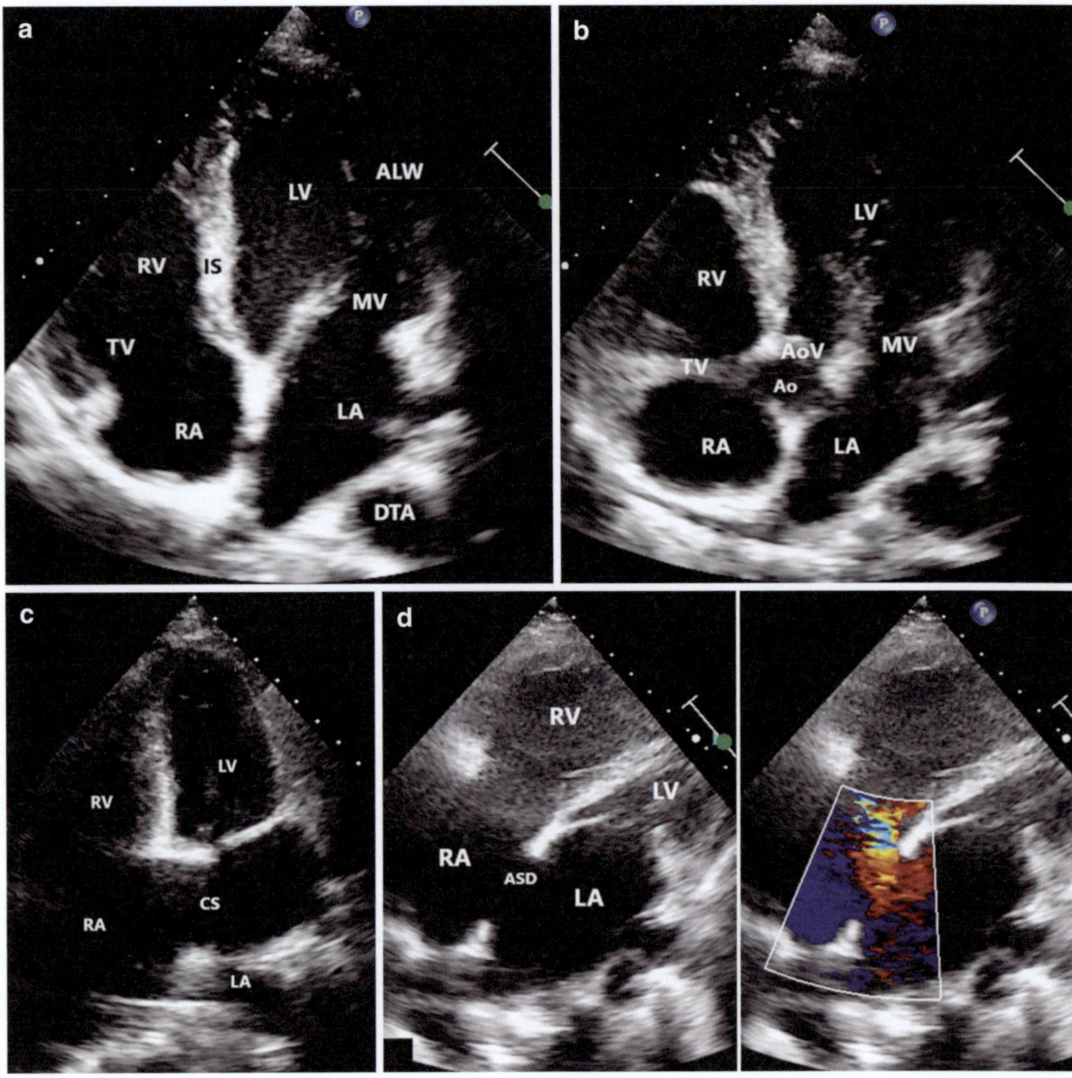

Fig. 2.4 Apical views: (**a**) Standard A4C view, (**b**) A5C view, (**c**) Coronary sinus view. (**d**) Modified A4C view to facilitate a less parallel incident angle for the ultrasound beam on the interatrial septum. True ostium secundum ASD is confirmed by colour Doppler. *RA* right atrium, *LA* left atrium, *RV* right ventricle, *LV* left ventricle, *Ao* aorta, *AoV* aortic valve, *MV* mitral valve, *TV* tricuspid valve, *DTA* descending thoracic aorta, *CS* coronary sinus, *ASD* Atrial septal defect

(directing the beam towards the left shoulder) opens up the following views sequentially:

- Abdominal short axis (transverse) view
- Coronary sinus (posterior) view
- Standard four chamber view
- Left ventricular outflow tract (anterior) view
- Right ventricular outflow tract (extreme anterior) view

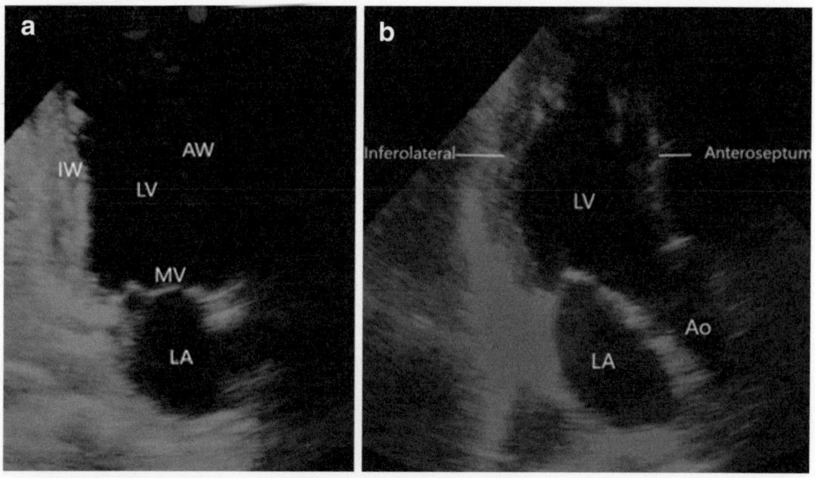

Fig. 2.5 (**a**) A2C view. (**b**) A3C view

2.2.3.2 Subcostal Sagittal (Short-Axis) Sweeps (Fig. 2.7)

Transducer position: Rotate the transducer clockwise from the coronal sweep until it aligns parallel to the cephalocaudal axis, with the notch pointing towards the 6 o' clock position. Tilting the transducer towards the right or left opens up the following views:

- Abdominal long-axis view
- Bicaval (rightward systemic venous return) view
- Left ventricular outflow tract (slightly leftward) view
- Right ventricular outflow tract (leftward) view
- Ventricular (extreme leftward) view

2.2.3.3 Subcostal Left Anterior Oblique View

Transducer position: Rotate the transducer approximately 45° clockwise from the subcostal coronal four-chamber view, such that the notch points towards the 4–5 o'clock position (left hip).

This view allows imaging of the atrial septum along its entire length, making it suitable for identifying various types of atrial septal defects, as well as for evaluating the proximal aorta.

2.2.3.4 Subcostal Right Anterior Oblique View

Transducer position: Rotate the transducer counterclockwise from the subcostal coronal four chamber view such that the notch points towards 2 o'clock position.

This view allows simultaneous imaging of RV inflow and outflow, along with en face view of the aortic valve, making it useful for evaluation of malalignment of conal septum in TOF. The fibrous continuity between tricuspid and aortic valves in perimembranous VSDs can also be demonstrated from this view.

2.2.4 Suprasternal Notch (SSN) Window

Patient position: Supine with the neck hyperextended (with a pillow under the shoulders for older patients). The chin may be slightly tilted to either side (without tensing the neck muscles) so that the movements of the probe are not hindered.

Planes: Long axis, Short axis

2.2.4.1 Suprasternal Long-Axis Sweeps (Fig. 2.8)

Transducer position: The transducer is placed above the suprasternal notch with a slight tilt of the transducer tail cranially and the transducer notch pointing towards the patient's head. Standard aortic arch view is obtained with the notch pointing towards the 1–2 o'clock position in the left aortic arch, and 11–12 o'clock position in the right aortic arch (discussed in Chap. 3). Rightward or leftward tilt of the transducer opens up the SVC and LPA views, respectively.

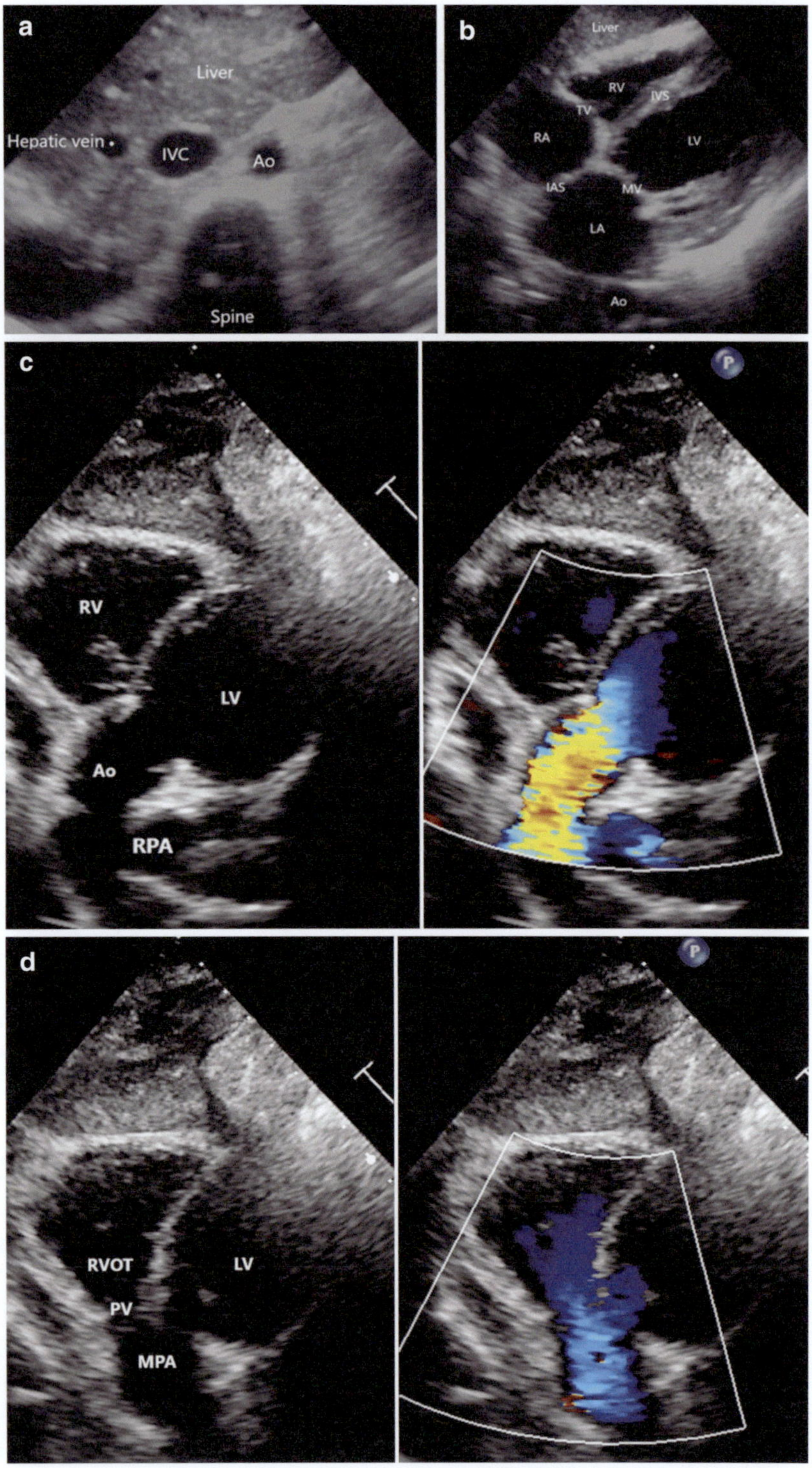

Fig. 2.6 Subcostal coronal sweeps: (**a**) Abdominal short axis view. (**b**) Standard four-chamber view. (**c**) Left ventricular outflow tract view. (**d**) Right ventricular outflow tract view. *RA* right atrium, *LA* left atrium, *RV* right ventricle, *LV* left ventricle, *Ao* aorta, *AoV* aortic valve, *MV* mitral valve, *TV* tricuspid valve, *DTA* descending thoracic aorta, *MPA* main pulmonary artery, *RPA* right pulmonary artery, *IVC* inferior vena cava, *RVOT* right ventricular outflow tract, *IVS* interventricular septum, *IAS* interatrial septum

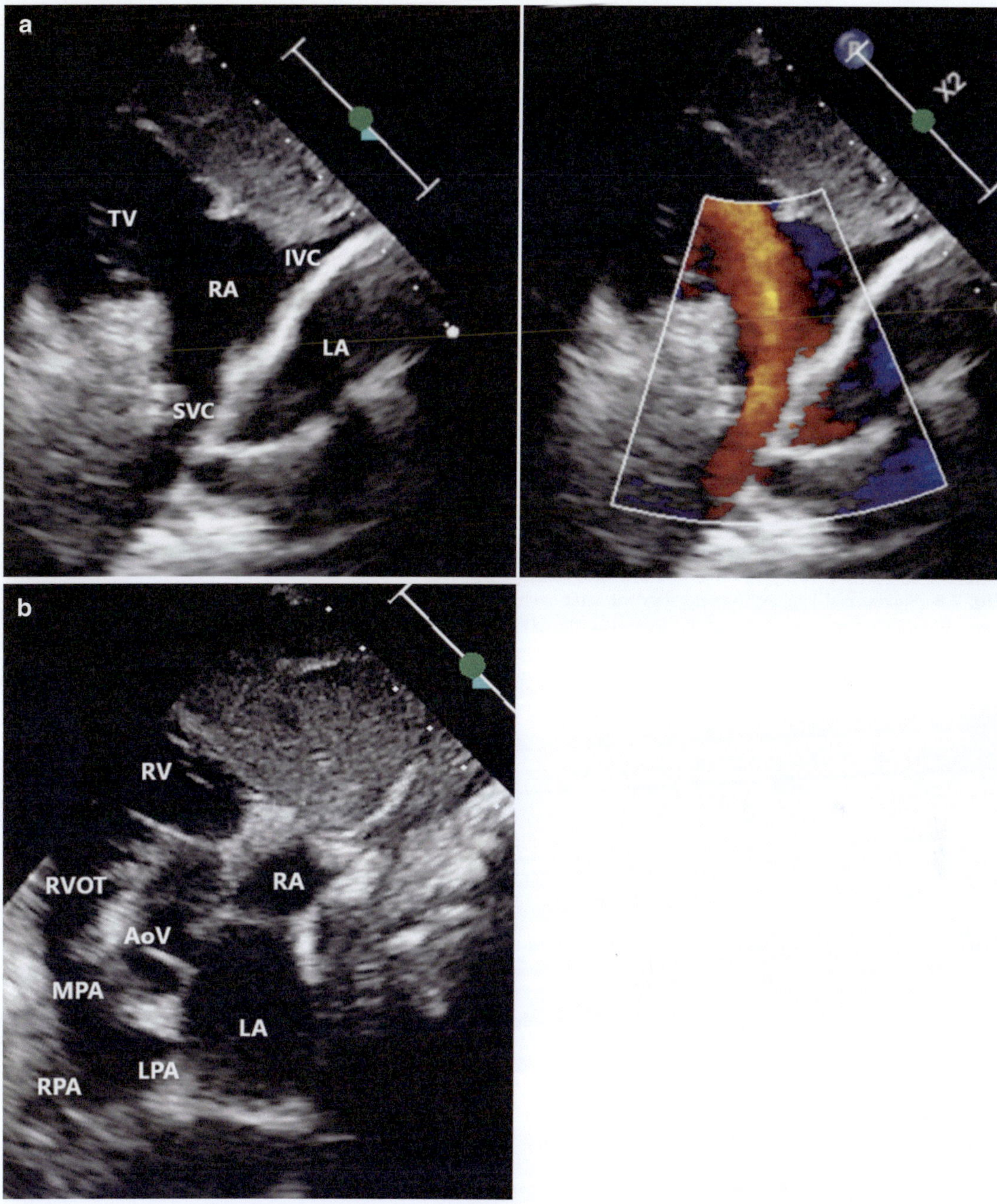

Fig. 2.7 Subcostal sagittal sweeps: (**a**) Bicaval view. (**b**) Right ventricular outflow tract view. *RA* right atrium, *LA* left atrium, *RV* right ventricle, *LV* left ventricle, *AoV* aortic valve, *TV* tricuspid valve, *IVC* inferior vena cava, *SVC* superior vena cava, *MPA* main pulmonary artery, *RPA* right pulmonary artery, *LPA* left pulmonary artery, *RVOT* right ventricular outflow tract

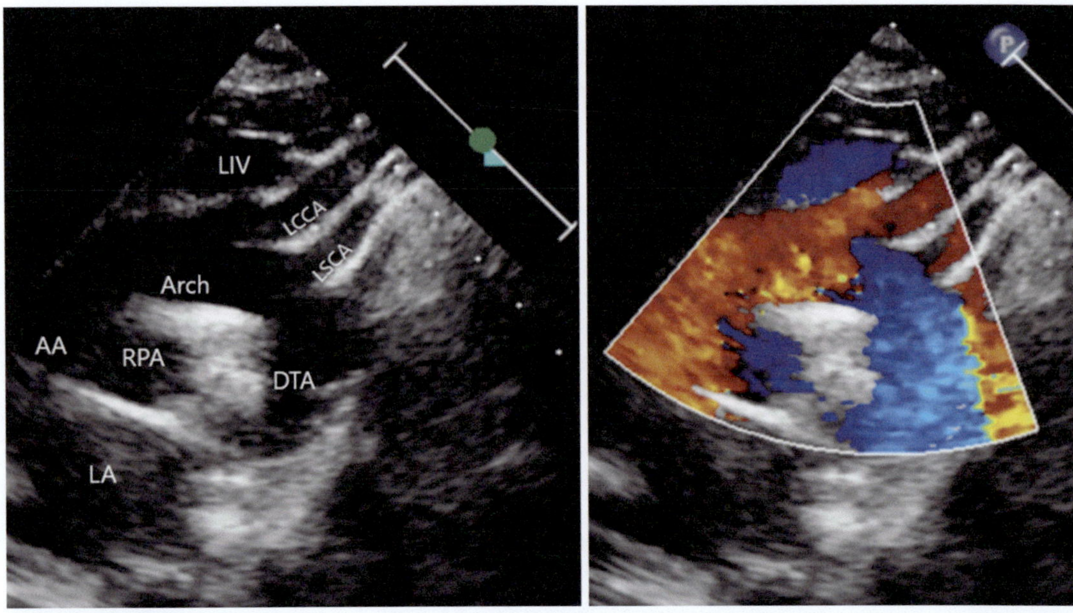

Fig. 2.8 Suprasternal long-axis sweep: standard aortic arch view. *AA* ascending aorta, *DTA* descending thoracic aorta, *LIV* left innominate vein, *LCCA* left common carotid artery, *LSCA* left subclavian artery, *RPA* right pulmonary artery, *LA* left atrium

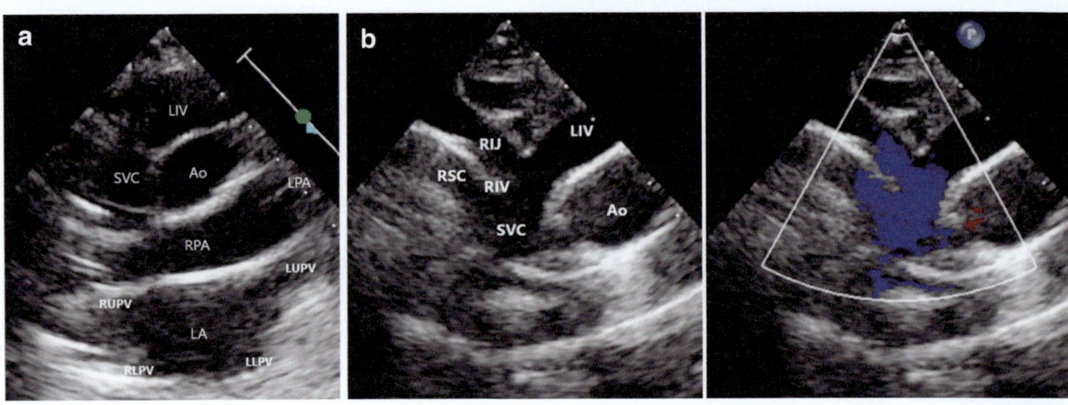

Fig. 2.9 Suprasternal short-axis sweeps: (**a**) standard (crab) view. (**b**) Vena cava and innominate vein view. *LIV* left innominate vein, *RIJ* right internal jugular vein, *RSC* right subclavian vein, *RIV* right innominate vein, *SVC* superior vena cava, *RPA* right pulmonary artery, *LPA* left pulmonary artery, *RUPV* right upper pulmonary vein, *RLPV* Right lower pulmonary vein, *LUPV* left upper pulmonary vein, *LLPV* left lower pulmonary vein, *Ao* aorta, *LA* left atrium

- Standard aortic arch view
- Superior vena caval (rightward) view
- Left pulmonary artery (leftward) view

2.2.4.2 Suprasternal Short-Axis Sweeps (Fig. 2.9)

Transducer position: Rotate the transducer clockwise 90° from the long axis sweep. Tilting the transducer from the most anterosuperior plane posteriorly opens up the following views:

- Vena cava and innominate vein (anterosuperior) view
- Right pulmonary artery and left atrial (standard or crab) view
- Descending aorta (posterior) view

2.2.5 Right Parasternal Window

Patient position: Right lateral decubitus position with the right arm extended beneath the head.

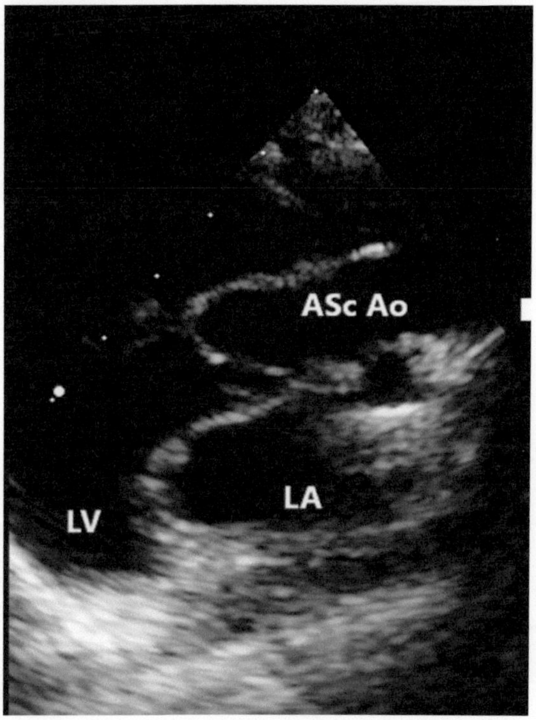

Fig. 2.10 Right parasternal long-axis view of ascending aorta (Asc Ao)

Transducer position: Right parasternal area, in the second or third intercostal space (usually one intercostal space above that of the left parasternal window).

Right parasternal sagittal view can be obtained with the notch pointing towards the 11-12 o'clock position (may vary between 10 o'clock in older patients and 1 o' clock in young patients). Angling the probe 45° towards the sternum visualizes the proximal ascending aorta in its long axis (Fig. 2.10). Angling the probe towards the patient's right side visualizes the atria with the interatrial septum, SVC, and IVC [bicaval view].

Right parasternal transverse view can be obtained by clockwise rotation of the probe from a slightly higher position in the chest, with the notch pointing towards the 3 o' clock position. This view aids in visualization of RUPV flow into the left atrium using colour Doppler.

Key point: Although the aforementioned standard views have enormous utility in the delineation of cardiac anatomy, imaging in orthogonal planes and deviation from the conventional views are necessary for a complete understanding of the cardiac anatomy and for tracing vascular channels and communications.

References

1. Huang PH. The comprehensive transthoracic echocardiographic examination. In: Rasalingam R, Makan M, Perez JE, editors. The Washington manual of echocardiography. 1st ed. Philadelphia: Lippincott Williams & Wilkins; 2012. p. 12–30.
2. Park MK. Noninvasive imaging tools. In: Park's pediatric cardiology for practitioners. 5th ed. Philadelphia: Saunders; 2014. p. 65–81.

3 Segmental Anatomy and Cardiac Malpositions

3.1 Segmental Anatomy

A systematic analysis of cardiac morphology aids in the precise understanding of complex congenital heart diseases. This process hinges on the accurate identification and categorization of cardiac segments—i.e. the three cardiac segments (namely atria, ventricles, and great vessels) and the two junctional segments (namely atrioventricular and ventriculo-arterial connections). This strategy, known as the segmental approach, simplifies the diagnosis of complex congenital heart diseases [1].

A complete echocardiographic report should contain a systematic description of the morphology, orientation, and spatial relationships of the following structural elements, followed by a succinct description of specific pathologies of structure and function:

- Thoraco-abdominal visceral situs
- Cardiac position
- Atrial morphology and situs
- Atrioventricular connection
- Ventricular morphology and looping
- Conus
- Ventriculo-arterial connection
- Great arterial relationship

3.2 Visceral Situs

Abdominal visceral situs is determined from the positions of (a) abdominal viscera and/or (b) aorta-inferior vena cava.

3.2.1 Abdominal Viscera

The position of abdominal viscera is best determined from the subcostal coronal abdominal short-axis (transverse) view by ascertaining the positions of the liver and the gastric air bubble.

- **Situs solitus**: The stomach is on the left and the liver is on the right.
- **Situs inversus**: The liver is on the left and the stomach is on the right.

3.2.2 Aorta and Inferior Vena Cava

The positions of the aorta and inferior vena cava are determined from the subcostal coronal abdominal short-axis view, with the probe pointing at a right angle to the spine.

The aorta (Ao) is identified as the non-compressible vessel with pulsatile flow, whereas the

A. Paul, *Manual of Echocardiography for Congenital Heart Diseases*,
https://doi.org/10.1007/978-981-96-0243-8_3

inferior vena cava (IVC) is the compressible vessel with continuous flow on pulse wave Doppler.

- **Situs solitus**: The aorta is to the left of the spine and the inferior vena cava to the right of the spine.
- **Situs inversus**: The aorta is to the right of the spine and the inferior vena cava to the left of the spine (Fig. 3.1).
- **Situs ambiguus**:
  - **Right isomerism**: Both the aorta and inferior vena cava are on the same side of the spine.
  - **Left isomerism**: Suspected when the intrahepatic inferior vena cava is interrupted (interrupted IVC), and azygous continuation is observed in the paravertebral gutter on either the left or the right side.

Key point: 'Abdominal visceral situs solitus' implies that the liver and IVC are on the right side, and the stomach, spleen, and aorta are on the left side of the spine, while the exact reverse of this is seen in 'abdominal visceral situs inversus'. Any other combination would fall under 'abdominal situs ambiguus'.

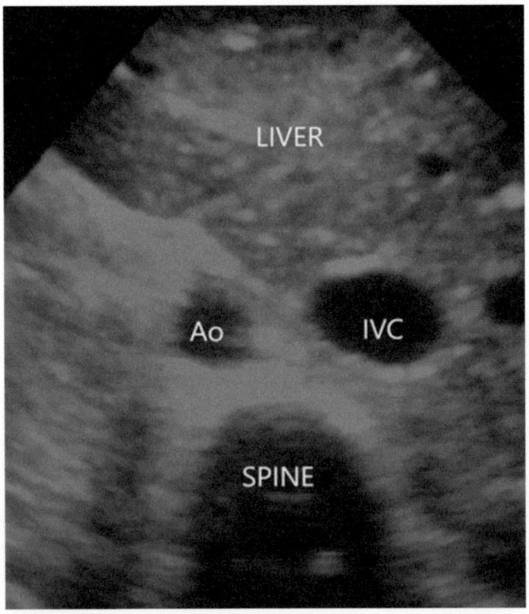

Fig. 3.1 Subcostal coronal abdominal short-axis (transverse) view: Situs inversus—aorta to the right, and IVC and liver to the left of spine

3.3 Cardiac Apex Position

The orientation of the cardiac apex (i.e. the base-apex axis) is best determined from the subcostal coronal four-chamber view (Fig. 3.2).

- **Levocardia**: The cardiac apex points to the left side. The heart is situated in the left hemithorax.
- **Dextrocardia**: The cardiac apex points to the right side. The heart is situated in the right hemithorax.
- **Mesocardia**: The cardiac longitudinal axis is in the mid-sagittal plane. The apex may not be distinct [2].

Dextroposition should be distinguished from dextrocardia

- **Dextroposition** refers to the displacement of the heart into the right hemithorax due to an extracardiac pathology.
- Causes of dextroposition include a mass on the left side, such as a diaphragmatic hernia, which may push the heart, or a deficiency on the right side, such as lung collapse, which may pull the heart to the right side.
- However, the apex remains pointed to the left, i.e. levocardia.

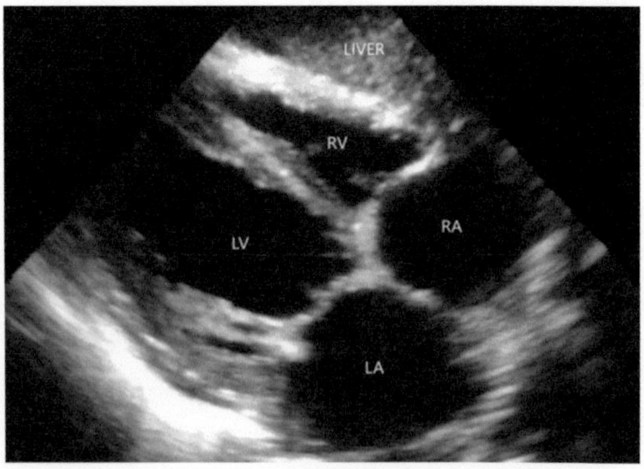

Fig. 3.2 Subcostal coronal four-chamber view: Dextrocardia with situs inversus (mirror image dextrocardia)—apex points towards right side; left atrium (LA) is to the right of right atrium (RA)

3.4 Atrial Situs

Atrial situs falls into one of the following diagnostic categories:

- Atrial situs **solitus (S)**: The morphologic right atrium is on the right side, and the morphologic left atrium is on the left side of the heart.
- Atrial situs **inversus (I)**: The morphologic right atrium is on the left side, and the morphologic left atrium is on the right side of the heart.
- Atrial situs **ambiguus (A)**:
  - Right atrial isomerism: Morphologic right atrium on both sides.
  - Left atrial isomerism: Morphologic left atrium on both sides.

The accurate identification of the morphologic right and left atria is the first step in determining atrial situs. Atrial morphology can be ascertained by evaluating:

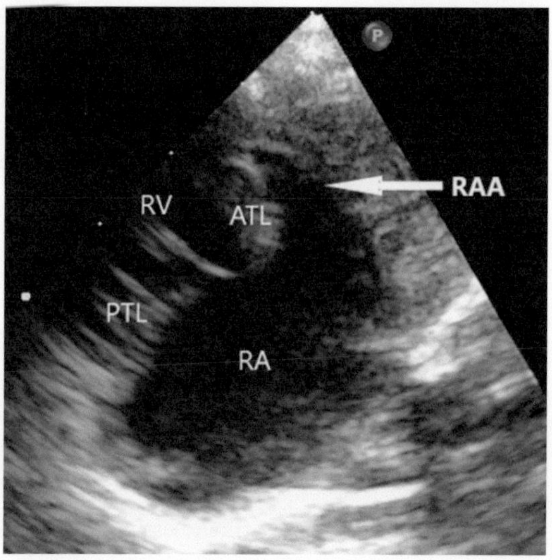

Fig. 3.3 Right atrial appendage (RAA) visualized from parasternal RV inflow view. *ATL* anterior tricuspid leaflet, *PTL* posterior tricuspid leaflet, *RA* right atrium, *RV* right ventricle

Atrial Appendages

- Right atrial appendage (RAA):
  - Short, fat, and broad-based
  - Best seen in the subcostal sagittal view and the PLAX (RV inflow) view (Fig. 3.3)
  - Pectinate muscles (raised parallel ridges resembling the teeth of a comb) of the RAA extend into the right atrium [3]
- Left atrial appendage (LAA):
  - Long, thin, and narrow-based.
  - Best seen in the **PSAX view** at the level of the aortic valve with slight clockwise rotation or downward tilt of the transducer; the **A5C view** with upward tilting of the transducer; and the **A2C view** with a slight lateral tilt or clockwise rotation of the transducer (Fig. 3.4).
  - Pectinate muscles of the LAA are confined to the appendage [3].

Septal Structures

- Right atrium: The embryonic valves—Eustachian and Thebesian—located on the right side of the atrial septum can be visualized in the subcostal coronal and sagittal views.
- Left atrium: The flap valve of the fossa ovalis, seen in the apical and subcostal coronal/sagittal views, is on the left atrial side of the interatrial septum.

Key point: Atrial situs follows abdominal situs in approximately 70% to 80% of cases. The atrium to which both the SVC and IVC are connected will be a morphologic RA because the right horn of the sinus venosus (i.e. the IVC and the SVC) cannot incorporate completely into the morphologic LA [4].

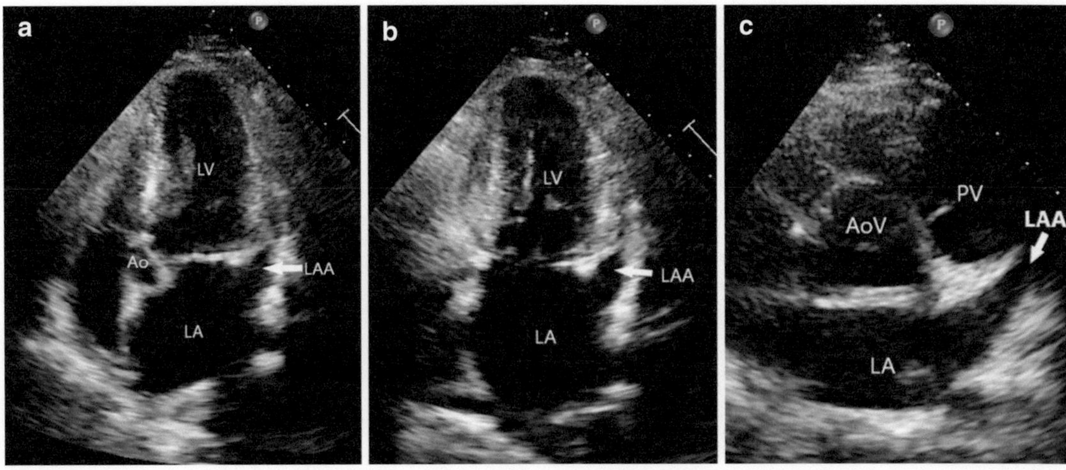

Fig. 3.4 Left atrial appendage (LAA) visualized from A5C view (**a**), A2C view (**b**), and PSAX view (**c**)

3.5 Atrioventricular Connection

Atrioventricular connection is determined by the relative positions of the annuli of the AV valves with the ventricles. AV valves always connect with their respective ventricles—i.e. a morphologic tricuspid valve connects to a morphologic RV, and a morphologic mitral valve connects to a morphologic LV.

- Identification of this arrangement by echocardiography allows determination of ventricular morphology.
- The atrioventricular relationship can be established, once the morphology and position of the ventricles are determined.

There are six possible AV connections:

- Biventricular: Concordant, Discordant
- Ambiguous
- Univentricular: Double Inlet, Single Inlet, Common Inlet

AV **concordance**: The morphologic RA drains into the morphologic RV, and the morphologic LA drains into the morphologic LV.

AV **discordance**: The morphologic RA drains into the morphologic LV, and the morphologic LA drains into the morphologic RV.

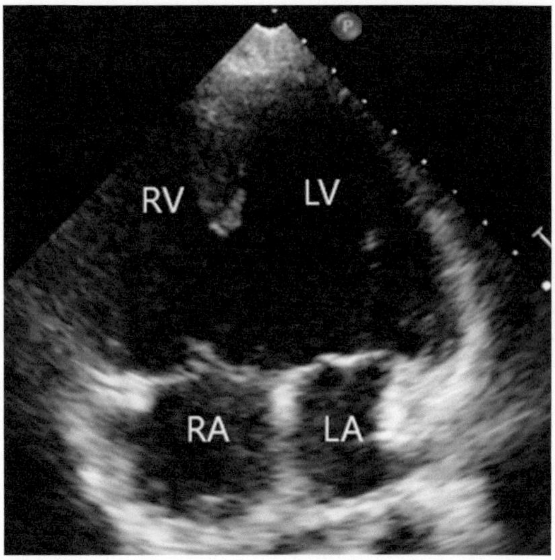

Fig. 3.5 A4C view of the heart depicting a large inlet VSD and malalignment of the atrial and ventricular septae. The right AV valve annulus is seen overriding the VSD, resulting in partial commitment of the valve to the left ventricle

- **Overriding** is defined as the biventricular emptying of an atrioventricular valve or the biventricular origin of a semilunar valve [5]. Partial commitment of an AV valve annulus to the contralateral ventricular chamber results from malalignment of the atrial and ventricular septa (Fig. 3.5).

- The percentage of **annular commitment** to a ventricular chamber with respect to the ventricular septum **determines the AV connection to a ventricular chamber**.
- An atrium is considered to be connected to the ventricle into which at least 50% of its valve empties. Thus, an annular commitment of 50% or more (of the total annular circumference) to a ventricular chamber establishes the valve connection to that ventricular chamber [6, 7].
- In contrast to annular overriding, **straddling** is a feature of the chordae tendineae and papillary muscles and results from their insertions into the contralateral ventricle through a ventricular septal defect (VSD) [7].

3.5.1 Tricuspid Valve

The tricuspid valve apparatus consists of the tricuspid valve annulus, leaflets, chordae tendineae, and papillary muscles [5].

- Three leaflets: anterior, posterior (inferior), and septal.
- Three papillary muscles, each attaching to two leaflets via chordae tendineae.
- Three commissures.
- The TV leaflets may have attachments to muscle bands, trabeculations, and multiple small papillary muscles in the RV [3].
- The chordae tendineae of the tricuspid valve have septal attachments (in contradistinction to the mitral valve).

Echocardiographic Assessment

While assessing the tricuspid valve by 2D echocardiography, it should be remembered that accurate identification of the visualized tricuspid leaflets needs a good understanding of the tricuspid valve anatomy and the surrounding structures. The following points help to identify the leaflets in each view [8].

PLAX (RV inflow) view:

- Anterior tricuspid leaflet (ATL) is consistently visualized in this view (on the side of the pointer), irrespective of the angulation, whereas the leaflet visualized on the opposite side is variable.
- In a standard RV inflow view (i.e. an RV two-chamber view, where neither the interventricular septum nor the coronary sinus is visualized), the posterior tricuspid leaflet (PTL) is visualized along with the ATL (Fig. 3.3).
- If either the coronary sinus or the interventricular septum is seen in the RV inflow view, the visualized leaflets are the septal tricuspid leaflet (STL) and the ATL.

PSAX view:

- In the PSAX view at the aortic valve level, a single leaflet (ATL) is consistently visualized.
- If the tricuspid valve annulus is dilated, two leaflets (i.e. ATL closer to the aortic valve, and PTL on the opposite side) are visualized.
- By angling the transducer beam apically towards the LV inflow, STL becomes visible (close to the septum) along with ATL on the opposite side.

Apical view:

- The septal tricuspid leaflet is consistently visualized in the A4C view, attached to the interventricular septum.
- In the standard A4C view, the other visualized leaflet is the ATL.
- Posterior angulation of the transducer (Apical coronary sinus view) results in visualization of the PTL, along with the STL.
- The size of the tricuspid annulus is assessed from the A4C view.

Subcostal coronal view: The abnormal anterior tricuspid leaflet in Ebstein's anomaly can be assessed from this view.

3.5.2 Mitral Valve

The mitral valve apparatus consists of the mitral annulus, leaflets, chordae tendineae, papillary muscles, and the underlying ventricular wall.

- The mitral annulus is saddle-shaped, with **two high points** (at the site of aortic–mitral continuity and at the posterior aspect of the valve) and **two low points** along the lines of the commissure between the mitral leaflets [5].
- The junction of the anterior mitral valve leaflet and aorta is called **annuloaortic fibrosa** or **mitral aortic intervalvular fibrosa** (MAIF).
- Two leaflets: the anterior or aortic, and the posterior or mural.
  - The anterior leaflet is in fibrous continuity with the non-coronary cusp of the aortic valve.
  - The posterior leaflet is anchored along the parietal part of the left atrioventricular junction.
- Each leaflet has three separate scallops: Anterior (**A1, A2, A3**) and posterior (**P1, P2, P3**).
  - The A1 and P1 scallops are anterolaterally located, nearest the left atrial appendage.
  - The A3 and P3 scallops are more posteromedial in location.
- Two major papillary muscles:
  - Anterolateral PM: Provides chordae to the anterolateral half of both mitral leaflets.
  - Posteromedial PM: Provides chordae to the posteromedial aspect of both leaflets.
  - The posteromedial papillary muscle is perfused by the right coronary artery, and the anterolateral papillary muscle has a dual blood supply.
- The chordae tendineae of the mitral valve have no septal attachment.

Echocardiographic Assessment

- The diameter of the mitral annulus is measured from the PLAX and A4C views (orthogonal planes).
- The papillary muscles are best visualized in the parasternal and subcostal short-axis views.
- Mitral stenosis is assessed in the PLAX and A4C views. MV area is determined by tracing the valve orifice in the PSAX view.
- Cleft mitral valve and double orifice mitral valve are best visualized in the PSAX view.

The **MV leaflet scallops** visualized in various transthoracic echocardiographic views are as follows [9]:

- PLAX:
  - Standard view: A2, P2
  - Tilting the transducer tail towards the right shoulder: A3, P3
  - Tilting the transducer tail towards the cardiac apex: A1, P1
- PSAX: A1, A2, A3 (anterolateral to posteromedial); P1, P2, P3 (anterolateral to posteromedial)
- Apical views:
  - Standard A4C: A3/A2; P1/P2
  - Anterior tilt (A5C): A1, P1, anterolateral commissure
  - Posterior tilt (coronary sinus view): A3, P3, posteromedial commissure
  - A2C: P1 (anterior), A2 (central), P3 (posterior), both commissures
  - A3C: A1, P2/P3

3.5.3 Common AV Valve

This AV connection may be represented by a single AV valve committed to both ventricles. The various anatomic variants of this entity are described in detail in the chapter on AV septal defects. This is best evaluated in the subcostal and parasternal short-axis views (Fig. 3.6).

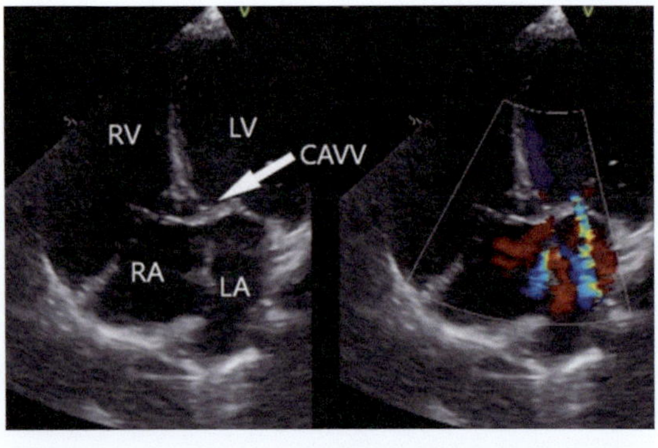

Fig. 3.6 Apical view of a case of transitional AVCD with common AV valve (CAVV) and two AV orifices. Left AV valve regurgitation and ostium primum ASD are seen

3.6 Ventricles

Normal ventricles consist of three portions: inlet, trabecular, and outlet.

- The inlet portion extends from the atrioventricular (AV) valve annulus to the level of papillary muscle insertion.
- The trabecular portion, which extends between the inlet and outlet portions, contains the ventricular apex.
- The outlet portion consists of the non-trabeculated region beneath the semilunar valve.

Accurate identification of morphologic right and left ventricles is crucial in the segmental assessment of cardiac morphology.

- Normal ventricular looping is designated as **d-loop (D)**, which implies that the morphologic right ventricle is rightward and anterior to the morphologic left ventricle.
- Any alteration in the relative position of the right ventricle is designated as **l-loop (L)**, where the morphologic right ventricle is located to the left of left ventricle.
- When the morphologic RV and LV are indistinguishable, the looping is designated as **indeterminate (X or I)**.

3.6.1 Right Ventricle

In a morphologic RV, the inlet and outlet portions are separated from one another by the crista supraventricularis [3]. The outlet (conus or infundibular) component of the RV consists of the infundibular septum (separating the aortic valve from the pulmonary valve), the parietal band (separating the tricuspid valve from the pulmonary valve), and the anterior and posterior limbs of the septal band [3, 5].

The morphologic RV has the following distinguishing features [3, 7]:

- Coarse septal endocardial surface with large apical trabeculations (trabeculated apex)
- Thin free wall
- Moderator band (trabecula septomarginalis), which runs transversely from the RV free wall to the septum in the inferior third of the RV cavity in the apical view
- Triangular shape of the cavity; crescentic in cross-section
- AV valve (tricuspid) features:
  - Trileaflet AV valve with a triangular orifice and three commissures
  - Septal attachment of the chordae tendineae
  - Lower (apical) septal attachment (hinge) of the tricuspid valve annulus (exaggerated in the presence of an Ebstein-like malformation and absent in the presence of an AV septal defect)
  - Multiple (usually three) small papillary muscles
- Muscular outflow tract (infundibulum) with no fibrous continuity between the inlet and outlet portions (tricuspid-pulmonary discontinuity)

Key point: The tricuspid valve is always 'attached' to the morphologic right ventricle.

3.6.2 Left Ventricle

The morphologic LV is an ellipsoidal-shaped structure with a fine trabecular pattern.

The morphologic LV has the following characteristics [3, 7]:

- Smooth septal endocardial surface with small apical trabeculations (smooth apex)
- Thick wall
- No moderator, septal, or parietal bands
- Ellipsoidal geometry; circular in cross-section
- AV valve (mitral) features:
  - Bileaflet AV valve with an elliptical orifice and two commissures
  - No septal attachment of the chordae tendineae
  - Higher (basal) septal attachment (hinge) of the annulus
  - Two large papillary muscles
- Fibrous continuity between the inlet and outlet portions (aorto-mitral continuity)

Key point: The mitral valve is always 'attached' to the morphologic left ventricle.

Echocardiographically, the LV can be divided into three segments along the longitudinal axis:

- **Basal** third extends from the atrioventricular groove to the tip of the papillary muscles.
- **Middle** third contains the papillary muscles.
- **Apical** third begins at the base of the papillary muscle and extends to the apex.
- **LVOT**: the area extending from the level of the free edge of the anterior mitral leaflet to the aortic valve annulus.

3.6.3 Functionally Single Ventricle

In a functionally single ventricle, both atria are predominantly connected to one ventricle (univentricular AV connection) by one or two AV valves (Fig. 3.7).

When a ventricular chamber is without a connecting AV valve, two factors that may aid in determining it as left or right ventricle are [6]:

- The relative position of the chamber:
  - A smaller ventricle lying posterior to a larger one is usually a morphologic LV.
  - A smaller ventricle lying anterior to a larger one is usually a morphologic RV.
- Ventricular looping:
  - L loop is more common in a double inlet left ventricle (i.e. the morphologic LV is situated to the right of the hypoplastic morphologic RV).

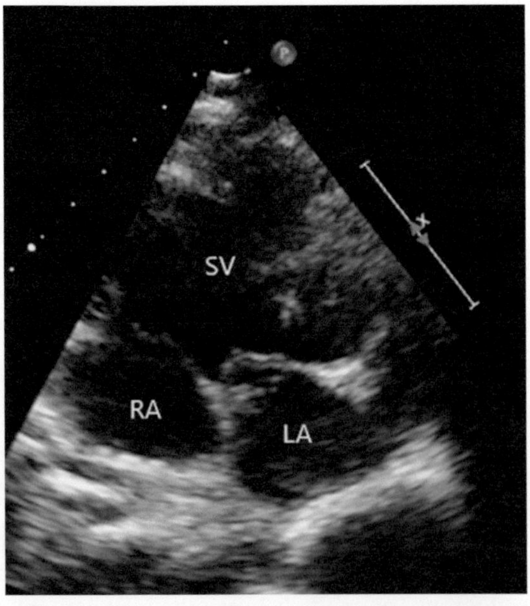

Fig. 3.7 Univentricular AV connection—both the AV valves are seen committed to a single ventricle (SV)

3.7 Conal Morphology

- The conus is a ring of muscle located under a great artery [3].
- In the normal heart, the subpulmonary conus (muscular separation between the tricuspid and pulmonary valves) is present, and there is no subaortic conus (mitral-to-aortic fibrous continuity).
- The presence of subaortic conus with an absent subpulmonary conus is seen in transposition of the great arteries (TGA).
- Both subaortic and subpulmonary conus are present in a double outlet right ventricle (DORV).
- The conus may be absent bilaterally, resulting in inlet-outlet fibrous continuity on both sides.
- The conus can be appreciated from the PLAX view.

3.8 Ventriculo-Arterial Connection

The most crucial step in determining the type of ventriculo-arterial connection is the accurate identification of the great arteries.

- The pulmonary artery is distinguished by its **early branching pattern** into RPA and LPA (Fig. 3.8).
- The aorta is distinguished by its 'candy cane' shape and the take-off of its three branches from the arch (innominate, carotid, and subclavian arteries).

Sometimes, there is a solitary trunk from either ventricle (truncus arteriosus), which may give rise to the head and neck vessels, as well as the pulmonary and the coronary arteries.

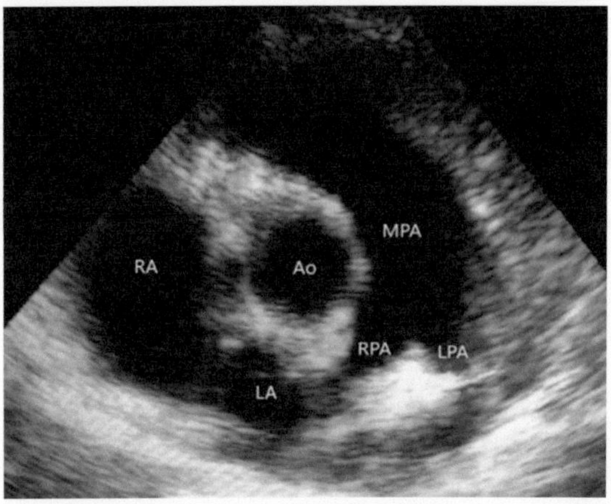

Fig. 3.8 PSAX view showing pulmonary artery bifurcation

Once the position of the great arteries is determined, the ventriculo-arterial relationship can be determined (Tables 3.1 and 3.2):

- Ventriculo-arterial **concordance**: The morphologic RV ejects into the pulmonary artery, and the morphologic LV ejects into the aorta.
- Ventriculo-arterial **discordance**: The morphologic RV ejects into the aorta, and the morphologic LV ejects into the pulmonary artery.

In the case of a semilunar valve overriding a VSD (Fig. 3.9), the ventriculo-arterial relationship is defined by the ventricle from which most

Table 3.1 Variants of visceroatrial situs inversus with different combinations of atrioventricular, ventriculo-arterial, and great arterial relationships

Atrioventricular	Ventriculo-arterial	Great arterial relation	Diagnosis
Concordant	Concordant	Aorta posterior and to the left of PA	Inverted-normal
Concordant	Discordant	Aorta anterior and to the left of PA	Inverted-TGA
Discordant	Discordant	Aorta anterior and to the right of PA	Inverted-cc TGA
Discordant	Concordant	Aorta posterior and to the left of PA	Inverted with the exception of normally related ventricles (TGA physiology)

Table 3.2 Variants of visceroatrial situs solitus with different combinations of atrioventricular, ventriculo-arterial, and great arterial relationships

Atrioventricular	Ventriculo-arterial	Great arterial relation	Diagnosis
Concordant	Concordant	Aorta posterior and to the right of PA	Normal
Concordant	Discordant	Aorta anterior and to the right of PA	TGA
Discordant	Discordant	Aorta anterior and to the left of PA	cc TGA
Discordant	Concordant	Aorta posterior and to the right of PA	Isolated ventricular inversion

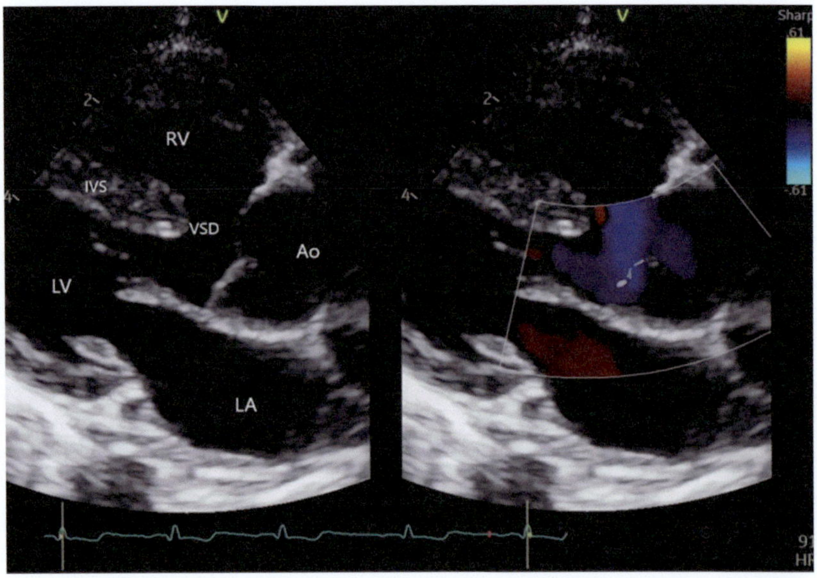

Fig. 3.9 Aortic valve overriding VSD. More than 50% of the aortic valve appears to be committed to the right ventricle. *LA* left atrium, *LV* left ventricle, *IVS* interventricular septum, *RV* right ventricle, *Ao* aorta

(>50%) of the overriding semilunar valve originates [5, 7].

- **Double-outlet right ventricle**: More than 50% of both the great arteries arise from the right ventricle.
- **Double-outlet left ventricle**: More than 50% of both the great arteries arise from the left ventricle.

The ventriculo-arterial relationship can be determined from subcostal coronal sweeps, PLAX view, and apical sweeps.

3.9 Great Arterial Relation

The spatial relationship between the great arteries must be determined. The origins of the great arteries are in orthogonal planes, creating a **'sausage and circle' appearance** in the PSAX view. This arrangement results from the developmental rotation of the RVOT and PA (the 'sausage') around the ascending aorta (the 'circle').

- ***Normal great arterial arrangement*** consists of:
  - Anterior and leftward pulmonary artery
  - Posterior and rightward aorta
  - Left-sided aortic arch and descending aorta
- Based on the aortic and pulmonary valve positions, ***eight types of great arterial relationships*** are possible (Fig. 3.10) [7]:
  - Right posterior aorta (normal or solitus (**S**))
  - Right lateral aorta (side by side)
  - Right anterior aorta (d-malposed (**D**))
  - Directly anterior aorta
  - Left anterior aorta (l-malposed (**L**))
  - Left lateral aorta (left side by side)
  - Left posterior aorta (inversus (**I**))
  - Directly posterior aorta

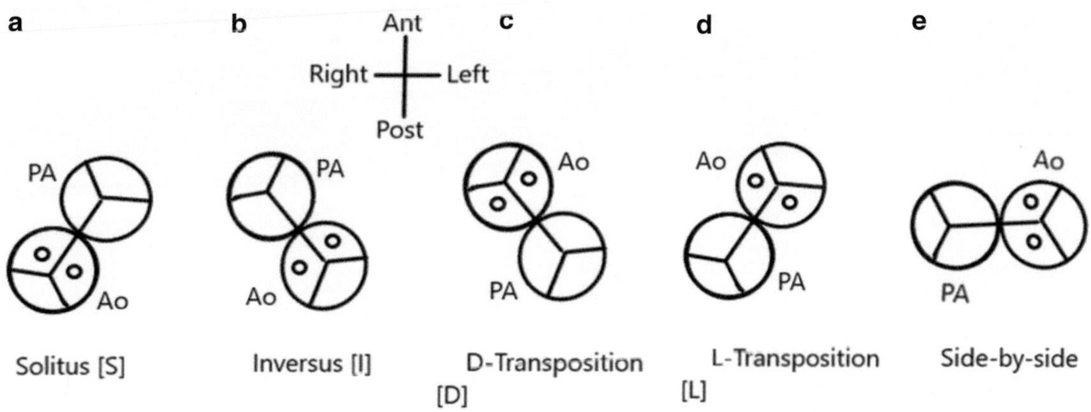

Fig. 3.10 Schematic representation of various types of great arterial relationship. *Ao* aorta, *PA* pulmonary artery

3.10 Aortic Arch Sidedness

- Normally, the aortic arch traverses over the left bronchus (left aortic arch).
- A right aortic arch traverses over the right bronchus.
- By echocardiography, **the first branch of the aorta defines the sidedness of the aortic arch** [3, 10]:
  - In a left aortic arch, the first branch of the aorta bifurcates to the right.
  - If the first branch of the aorta bifurcates to the left, a right aortic arch is present.
  - In a right aortic arch with mirror image branching, the first branch is a left-sided brachiocephalic artery, which divides into the left subclavian and left carotid arteries.
- **Visualizing the left-sided aortic arch:**
  - A left aortic arch is visualized from the suprasternal sagittal (long-axis) view with the ultrasound beam directed in a plane extending from the right sternal border (or right nipple) to the left scapula (i.e. with the transducer notch pointing towards the 1–2 o' clock position).
  - Branching of the right brachiocephalic trunk is confirmed from the suprasternal coronal (short-axis) view. After visualizing the aortic arch in its cross-section, gently sweep the transducer superiorly towards the right side of the neck. The cross-section of the right brachiocephalic trunk can be seen arising from the aortic arch. On sweeping further in the same direction, this vessel can be seen bifurcating.
  - From the suprasternal sagittal window, demonstrating the relative position of the transverse aortic arch to be on the left side of the tracheal rings (by a rightward sweep of the transducer) confirms the left-sidedness of the aortic arch.
- **Visualizing the right aortic arch:**
  - Inability to visualize the aortic arch in full profile from the standard transducer position is the first clue towards the possibility of abnormal aortic arch sidedness, as the standard suprasternal sagittal view will demonstrate only a part of the right aortic arch.
  - The transducer must be rotated counter-clockwise from the standard suprasternal position, such that the plane of the ultrasound beam extends from the left border of the sternum (left nipple) to the right scapula (i.e. with the transducer notch pointing towards the 10–11 o'clock position) in order to visualize a right aortic arch in full profile.
  - Branching of the left brachiocephalic trunk may be confirmed by sliding the ultrasound beam superiorly towards the left side of the neck from the suprasternal short-axis view.
  - From the suprasternal sagittal view, sweeping the transducer from right to left will demonstrate that the transverse aortic arch

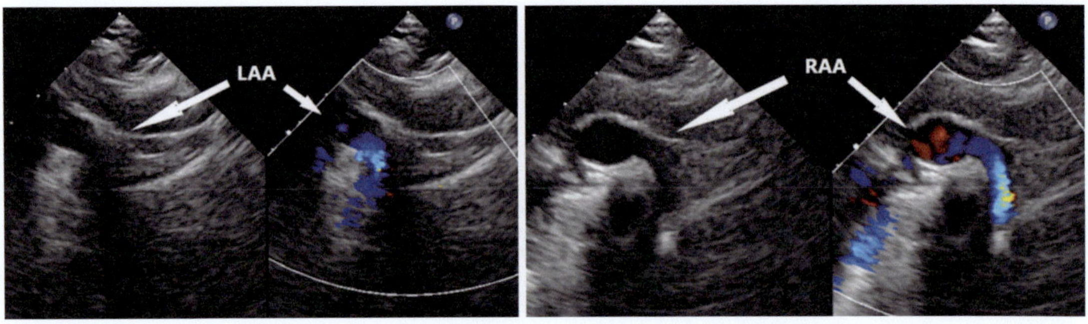

Fig. 3.11 Double aortic arch. Normal left aortic arch (LAA) is visualized from the standard suprasternal view. A counterclockwise rotation of the probe opened up the right aortic arch (RAA)

is positioned on the right side of the trachea in a right aortic arch.

- **Double aortic arch** (Fig. 3.11) results from failure of regression of the embryonic right aortic arch and persistence of the left aortic arch [10].
  - Both the aortic arches (right and left) can be visualized separately as described above.
  - A right-to-left sweep of the transducer from the suprasternal sagittal view will demonstrate transverse aortic arches on both sides of the trachea.

3.11 Coronary Arteries

The relatively superficial location of the coronary arteries and the resolution of the echocardiographic probe make visualization of the coronary arterial tree challenging. The origin of both the coronary arteries can be visualized from the PSAX view. Clockwise rotation of the transducer in the PSAX view allows for imaging of the LMCA bifurcation, while counterclockwise rotation facilitates imaging of the RCA.

3.11.1 LMCA and LAD

LMCA origin can be visualized from [11]:

- Modified PSAX view: Slight clockwise rotation and anterior tilt of the transducer from the PSAX view (at the aortic valve level) visualizes the LMCA (Fig. 3.12a).
- Modified A5C view: Visualization of the ascending aorta by anterior tilting of the transducer from the A5C view, followed by slight rotation, visualizes the origins of the LMCA (at around the 2–3 o'clock position) and RCA (at around the 9–10 o'clock position) (Fig. 3.12b).

LMCA bifurcation and Proximal LAD can also be visualized from:

- PSAX view at the coronary artery level
- Subcostal coronal view with anterior angulation
- Leftward sweep from the standard PLAX view
- A3C view

Mid LAD:

- Parasternal window: From the PSAX view at the papillary muscle level, sliding the transducer laterally by 2 cm allows visualization of the anterior interventricular groove. Sliding the transducer caudally by one intercostal space, followed by a tilt towards the base, facilitates visualization of coronary flow in the anterior interventricular groove using colour Doppler. Rotating the transducer by 70–90° to align the beam

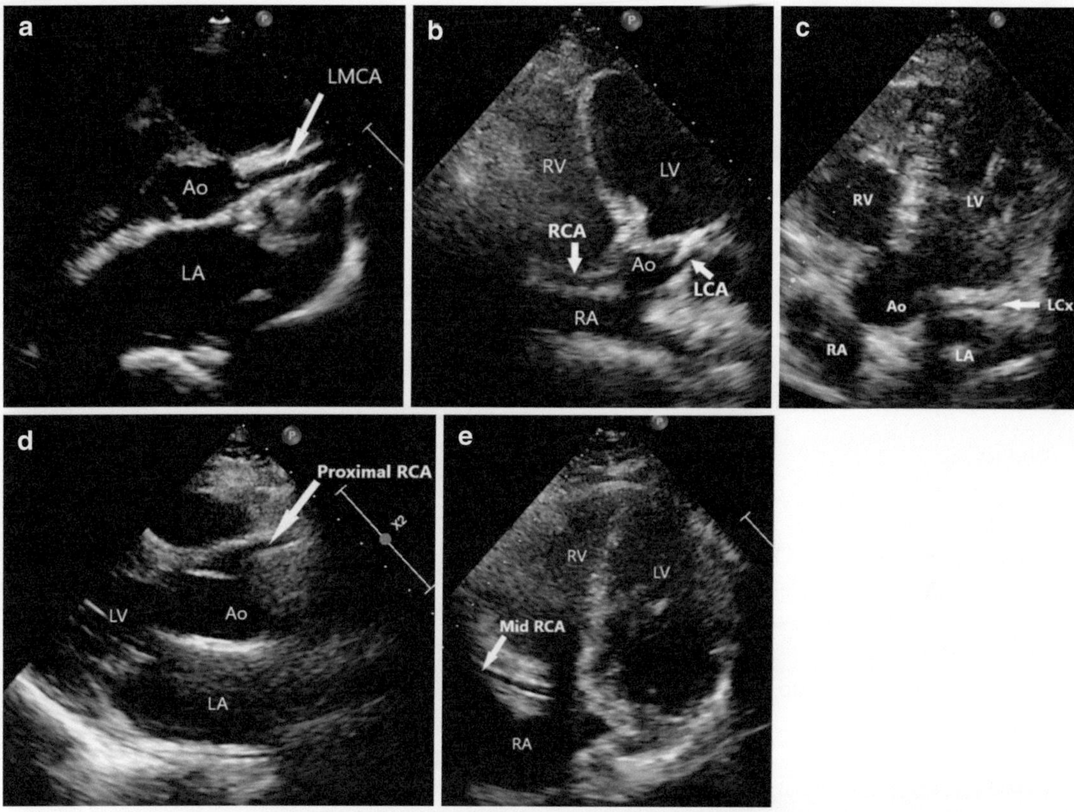

Fig. 3.12 Visualization of coronary arteries by transthoracic echocardiography. (**a**) PSAX view showing left main coronary artery (LMCA) origin. (**b**) Modified A5C view showing the origins of the left coronary artery (LCA) and right coronary artery (RCA). (**c**) Modified A5C view showing the proximal part of LCx. (**d**) PLAX view showing the RCA origin from the anterior aortic sinus. (**e**) Proximal-mid RCA visualized from the modified A4C view

parallel to the interventricular groove (modified PLAX view) visualizes the LAD along its long axis [11].

- Modified A5C view (traversing the anterior wall of the heart): LAD with some diagonal branches may be seen.

Distal LAD:

- Modified A3C view: From the standard A3C view, distal LAD flow can be detected in the interventricular groove close to the LV apex using colour Doppler. Sliding the transducer cranially by 1–2 intercostal spaces followed by an anterior tilt (foreshortened A3C view) visualizes the distal-most segments of the LAD [11].

3.11.2 Left Circumflex Artery (LCx)

Origin and proximal segment:

- PSAX view at the coronary artery level, just below the level of LA appendage
- Modified PLAX view: LCx is seen within the lateral wall of the LV
- Modified A5C view (with the imaging plane traversing the anterior wall of the heart): LCx is seen in the area lateral to the sinus of Valsalva [11]

Distal LCx in AV groove:

- PSAX view with the imaging plane traversing the mitral annulus
- A4C view with anterior angulation

3.11.3 Right Coronary Artery (RCA)

Origin and proximal segment:

- PSAX view at the coronary artery level
- Modified PLAX view: Proximal RCA is seen along the anterior wall of the aorta (Fig. 3.12)
- A4C view with anterior angulation: Proximal RCA is seen in the anterior AV groove
- Subcostal coronal view
- Modified A5C view: Origins of both RCA and LMCA are visualized (Fig. 3.12)

Mid RCA:

- Subcostal sagittal view: Mid RCA is seen along the anterior wall of RV at the AV groove
- Modified PLAX view angled towards the TV and lateral AV groove shows mid RCA in cross section
- Modified A4C view: Mid RCA is seen in the anterior AV groove (Fig. 3.12)

Distal RCA:

- A4C view with perpendicular angulation of the transducer to the chest: Distal RCA and right posterior descending artery (RPDA) can be seen in the posterior interventricular groove.
- Subcostal coronal view with posterior angulation.
- Modified A2C: From a standard A2C view, slight anticlockwise rotation followed by an anterior tilt of the transducer facilitates visualization of RPDA flow in the posterior interventricular groove using colour Doppler [11].

Coronary Artery Anomalies

- Common in complex congenital heart diseases such as TGA and TOF.
- The entire extent of the coronary artery must be interrogated from various imaging planes with simultaneous use of colour Doppler for detection of fistulous communications.
- MPA and branch PAs should be carefully imaged for an anomalously arising coronary artery.
- **Coronary cameral sinusoids**, seen within the RV in pulmonary atresia with an intact ventricular septum, are best visualized from the A4C view with simultaneous use of colour Doppler.
- The entire extent of the coronary artery should be interrogated for aneurysms and stenoses in Kawasaki disease.

Key point: Isolated dilation of a proximal coronary artery in the PSAX view should alert the possibility of a coronary artery fistula or anomalous coronary artery origin.

3.12 Venous Drainage

3.12.1 Pulmonary Venous Drainage

- The pulmonary veins are best visualized from the suprasternal short-axis plane (crab view), PSAX, and A4C views.
- The right pulmonary veins are best viewed in the A4C view.
- Left pulmonary veins, although seen in the A4C, are best visualized in the PSAX view.
- Colour Doppler with a low aliasing velocity can aid in visualizing the individual pulmonary veins.

Echocardiography
Drainage of each of the four pulmonary veins should be ascertained using multiple echocardiographic views [12].

Suprasternal short-axis view and crab view (Fig. 3.13)**:**

- The LA is visible below the long-axis view of the RPA.
- Slight cranial tilting and lateral angulation of the transducer reveal the pulmonary veins entering the LA.

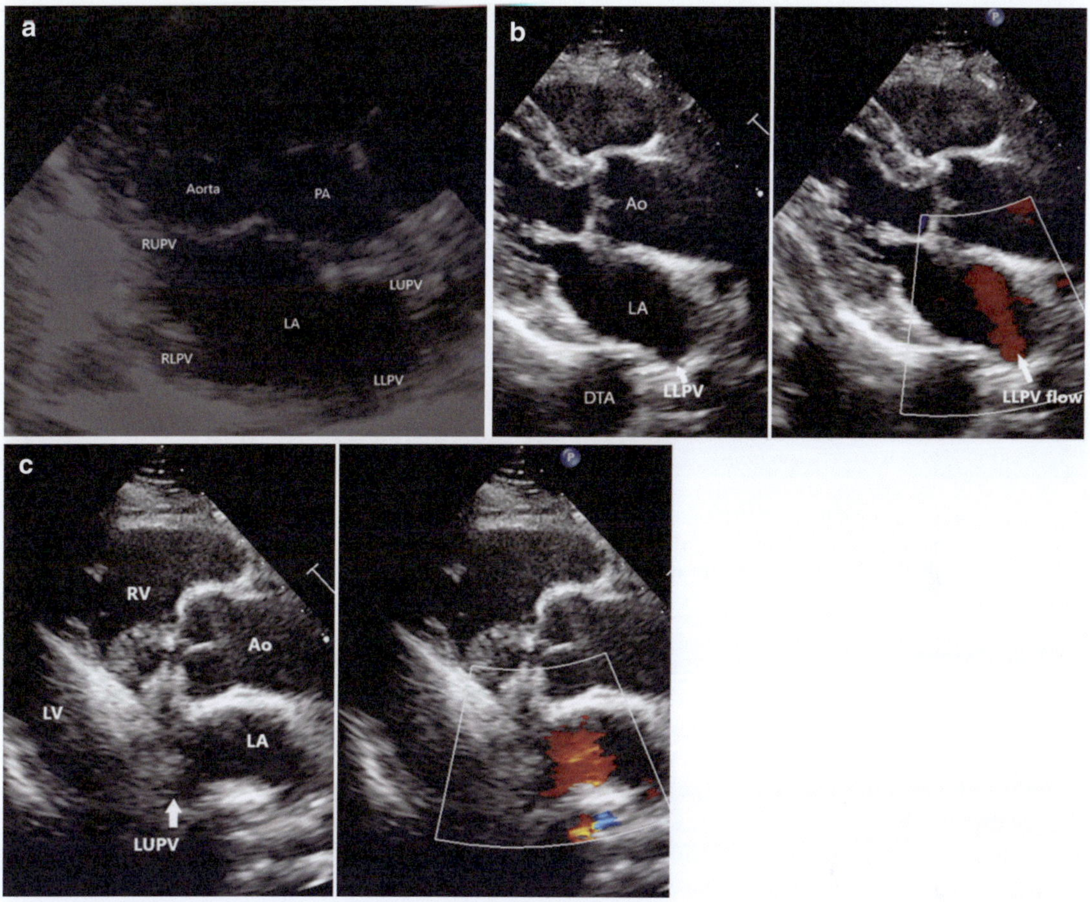

Fig. 3.13 (**a**) Suprasternal short-axis (crab) view showing the entry of pulmonary veins to the left atrium. (**b**) PLAX view showing left lower pulmonary vein (LLPV) flow into the LA. (**c**) PLAX view showing left upper pulmonary vein (LUPV) flow into the LA

- The **left pulmonary veins** enter the LA just below the level of the proximal RPA.
- The **right pulmonary veins** enter the LA at a level adjacent to the distal branching point of the RPA.
- The **upper pulmonary veins** are seen at a higher vertical level with the flow directed away from the transducer.
- The **lower pulmonary veins** are seen at a lower level with the flow directed towards the transducer.

Parasternal short-axis view:

- From the PSAX view at the aortic valve level, the LA appendage can be visualized by slight medial angulation or by moving the transducer slightly downward.
  - LUPV is seen close to the LA appendage with flow away from the transducer.
  - LLPV is seen close to the descending aorta with flow towards the transducer.

Parasternal long-axis view (Fig. 3.13):

- LUPV obliquely enters the LA from a left posterior to right anterior direction, with its flow away from the mitral valve.
- LLPV, seen adjacent to the descending aorta, enters the LA close to the LUPV.

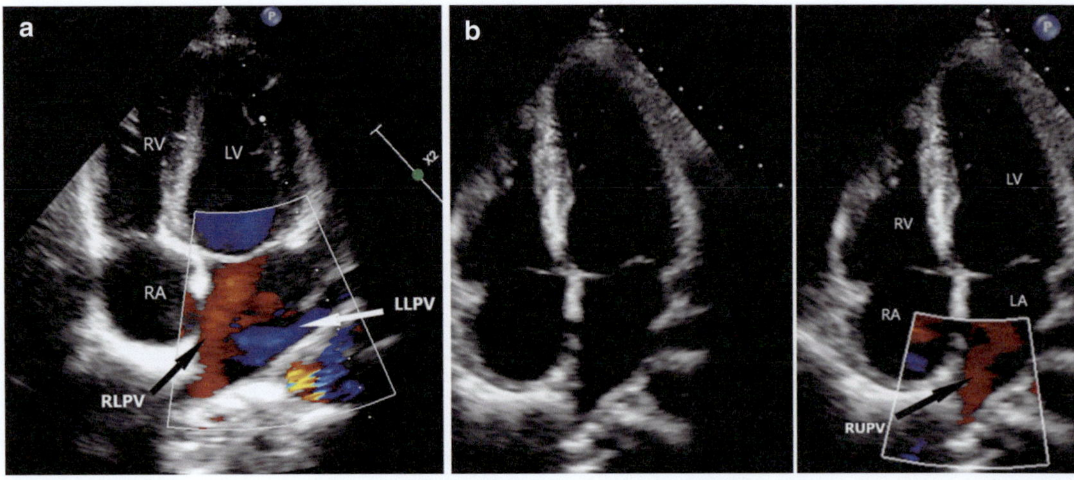

Fig. 3.14 Apical views demonstrating the normal pulmonary venous drainage into the left atrium. (**a**) RLPV flow seen parallel to the atrial septum; LLPV is also seen. (**b**) RUPV flow, visualized by slight anterior angulation of the probe, is seen directed towards the lateral mitral annulus

Apical views:

- RLPV is often best seen in the A4C, located in the roof of the LA close to the interatrial septum, with its flow parallel to the interatrial septum.
- A slight anterior angulation from the A4C (near 5-chamber or 4.5 view) shows the RUPV entering the right posteromedial aspect of the LA at an oblique angle, with its flow directed towards the lateral mitral annulus (Fig. 3.14) [12].
- The A2C view reveals the LUPV opening into the LA just below the LA appendage, adjacent to the anterior wall.

Subcostal views:

- RUPV and RLPV can be seen next to the interatrial septum in a subcostal coronal view, similar to A4C.
- Parasagittal imaging from the bicaval view can show the right pulmonary veins (with a rightward sweep) and the left pulmonary veins (with a leftward sweep) in cross-section as they course into the LA.
- An extreme rightward sweep in the subcostal sagittal plane reveals the right upper pulmonary vein.

Anomalous pulmonary venous drainage should be thoroughly interrogated from the suprasternal short-axis view (supracardiac drainage), parasternal and apical views (cardiac drainage), and subcostal coronal and sagittal views (infracardiac drainage).

Key point: Anomalous pulmonary venous pathways will have colour Doppler flow coursing away from the heart, in contrast to anomalous systemic venous channels, in which the flow is towards the heart.

3.12.2 Systemic Venous Drainage

3.12.2.1 Superior Vena Cava (SVC)

- Innominate veins are identified in the suprasternal short-axis view and traced downstream to the origin of the right superior vena cava (RSVC).
- The left innominate vein is traced upstream to look for the left superior vena cava (LSVC).

- A leftward sweep from the aortic arch in the suprasternal long-axis view demonstrates the LSVC in its long axis (discussed in Chap. 13).

The detection of LSVC is particularly important in:

- Any patient undergoing surgical repair using cardiopulmonary bypass, as both SVCs need to be cannulated [4].
- Single-ventricle physiology, where bilateral bidirectional cavo-pulmonary anastomoses will be required in the presence of bilateral SVC without an innominate vein.

The presence of an LSVC draining directly into the left atrium (LA) should be investigated using agitated saline contrast in patients with unexplained cyanosis, systemic embolization, or an absent innominate vein without a dilated coronary sinus [4].

3.12.2.2 Inferior vena cava (IVC)

- The IVC is identified in the subcostal coronal and sagittal views, **anterior to the aorta**.
- Interruption of the IVC should be suspected in the transverse abdominal view when a large venous channel is seen immediately **posterior to the aorta** (the azygous or hemiazygous vein). Absence of the IVC anterior to the aorta also flags this possibility (discussed in Chap. 13).

3.12.2.3 Coronary Sinus (CS)

The entrance of the CS into the RA can be seen in the subcostal coronal and the apical four-chamber views (discussed in Chaps. 2 and 13).

The size and possible unroofing of the CS can be assessed in a posterior sweep from the standard A4C and PLAX views.

Box 3.1 Essential Points to Be Included in a Complete Echocardiographic Report Describing the Segmental Anatomy of the Heart

Points to Report

- **Abdominal/Visceral Situs**: Solitus(S)/Inversus (I)/Ambiguus (A)
- **Systemic venous drainage: SVC**—Normal/LSVC, **IVC**—Normal/Interrupted, **CS**
- **Pulmonary venous drainage:** Normal/Anomalous—Total/Partial
- **Atrial Situs**: Solitus (S)/Inversus (I)/Ambiguus (A)
- **Direction of cardiac apex**: Levocardia/Dextrocardia/Mesocardia
- **Atrioventricular connection**: Concordant/Discordant/Common AV valve/Atresia/Criss-Cross/Double inlet
- **Ventricles and bulboventricular looping**: D/L/Single ventricle/X or I (indeterminate)
- **Conus**: Subaortic/Subpulmonic/Bilateral/Absent
- **Ventriculo-Arterial connection**: Concordant/Discordant/Double outlet/Single outlet
- **Great arterial relationship**: Solitus (S)/Inversus (I)/Transposed (D, L)/side-by-side
- **Coronary arteries:** Normal/Anomalous origin
- **Aortic arch:** Sidedness, Interruption/Coarctation
- **LV and RV size and function**
- **RV or LV Inflow/Outflow anomalies**

3.13 Cardiac Malpositions

Cardiac malposition refers to any abnormal position of the heart or its chambers. This includes abnormalities of:

- Cardiac base-apex axis (levo/dextro/mesocardia)
- Sidedness of the atria (situs solitus/inversus/ambiguus)
- Looping of the ventricles (d/l loops)

Heterotaxy refers to the abnormalities of lateralization, resulting in the loss of normal right-left asymmetry of the thoraco-abdominal viscera [13]. Any variation from normal situs solitus results in heterotaxy, which is either complete reversal of positions (situs inversus) or indeterminate random positioning not consistent with any standard category (situs ambiguus).

In many cases of visceroatrial situs ambiguus, a morphologic symmetry of normally asymmetric structures (isomerism) is observed. This may be of two types:

- Right isomerism or bilateral right-sidedness, characterized by asplenia
- Left isomerism or bilateral left-sidedness, characterized by polysplenia

The basic types of cardiac malpositions can be categorized as:

- Situs inversus with (a) dextrocardia (b) levocardia
- Situs solitus with dextrocardia
- Situs ambiguus
- Mesocardia

3.13.1 Situs Inversus with Dextrocardia

Situs inversus with dextrocardia is otherwise known as complete situs inversus (situs inversus totalis) or mirror-image dextrocardia, as the spatial orientation is a mirror image of the normal arrangements.

Echocardiography

- Subcostal coronal view (in supine position with normal transducer orientation, i.e. the notch pointing to the right) demonstrates the aorta to the right of the spine and the inferior vena cava to the left of the spine. The left-sided inferior vena cava connects to the left-sided morphologic right atrium.
- Patient positioning for imaging of dextrocardia from parasternal and apical views: Right lateral decubitus position with the right shoulder extended.
- Transducer position for various views:
  - **PLAX**: Right parasternal area, at the third or fourth intercostal space, with the notch pointing towards the patient's left shoulder.
  - **PSAX**: Rotate the transducer clockwise 90° from the PLAX view so that the notch points towards the left hip. This transducer position maintains the right-left spatial relationship (). The image will appear as an exact mirror image of a normal PSAX view [14]. Tilting the transducer tail towards the left shoulder opens up more apical views.
  - **Apical** window: The transducer is positioned near the apical impulse (usually right fifth intercostal space in the midclavicular line) with the notch pointing towards the left side and the imaging plane oriented cranially. Maintaining this position preserves the right-left spatial relationship of the image [14].
    The transducer orientation for A4C view would be as if the transducer was slid from the left side (from its usual position in levocardia) to the right side without any rotation. The image thus obtained will be a true mirror image of the one obtained from the left side in a case of levocardia-situs solitus.

The echocardiographer often gets confused with the image orientation of dextrocardia-situs inversus, as the image obtained with the recommended probe position is a mirror image of what is encountered in a normal heart. Hence, it is a common practice to invert the image in the right-left axis to screen for other abnormalities in a familiar orientation. Similarly, PSAX and apical images are often obtained by pointing the transducer notch towards the right side in order to obtain images oriented in a familiar fashion. However, it should always be remembered that the spatial relationship is distorted in such images.

Key point: Maintaining the transducer notch towards the left side maintains the left-right spatial relationship of the images in dextrocardia.

Variants of Situs Inversus with Dextrocardia (Table 3.1)

- **Situs Inversus with dextrocardia, AV and VA concordance, and left posterior aorta**
- One of the more common forms of dextrocardia.
- Atria, ventricles, and great arteries are all inverted in their spatial relationships.
- Usually associated with no significant functional abnormality of the heart.
- Rarely associated with VSD, OS-ASD, AVSD, Tetralogy of Fallot, and Pulmonary atresia [15].
- **Situs Inversus with dextrocardia, AV concordance, VA discordance, and left anterior aorta**
  - One of the common forms of dextrocardia with situs inversus in some series
  - Represents an 'inverted' form of complete TGA [7]
  - Physiologically similar to complete TGA with levocardia
- **Situs Inversus with dextrocardia, AV and VA discordance, and right anterior aorta**
  - An inverted form of congenitally corrected TGA
  - Associated anomalies such as RVOTO result in significant haemodynamic impairment
- **Situs inversus with dextrocardia, AV discordance, VA concordance, and left posterior aorta**
  - Rare form of situs inversus with dextrocardia
  - Structurally similar to isolated ventricular inversion [7]
  - Physiologically similar to complete TGA
  - Associated anomalies like unbalanced AVSD have been reported [15]

3.13.2 Situs Inversus with Levocardia

- Isolated levocardia refers to the presence of either atrial situs inversus or atrial situs ambiguus, in an otherwise normally positioned heart [13].
- Situs inversus with levocardia is usually associated with complex congenital heart disease.
- AV discordance is common.
- VA discordance or DORV with an anterior right aorta have been described.

3.13.3 Situs Solitus with Dextrocardia

- Situs solitus with dextrocardia is also known as 'Isolated dextrocardia'.
- It is uncommon to find a structurally normal heart in isolated dextrocardia [7, 13].

Common Variants (Table 3.2)

- **Isolated dextrocardia with AV and VA concordance, and right posterior aorta**
- Isolated dextrocardia with normal spatial orientation of atria, ventricles and great arteries is also known as 'Pivotal dextrocardia' and 'Dextroversion' [7].
- Commonly associated anomalies include ASD, VSD, AVSD, LSVC, aortic coarctation, and anomalous pulmonary venous drainage [16].
- **Isolated dextrocardia with AV and VA discordance, and left anterior aorta (cc-TGA)**
  - ccTGA is the most common complex cardiac malformation associated with isolated dextrocardia
  - Other associated anomalies include [15]:
    VSD, AVSD
    Pulmonary stenosis/atresia
    DORV (instead of TGA)

Double-inlet LV with hypoplastic subaortic RV
Mitral atresia (right AV valve)
Tricuspid (left AV valve) regurgitation with Ebsteinoid malformation
Anomalous pulmonary venous connections; LSVC

3.13.4 Situs Ambiguus

Situs ambiguus can present a pattern of abnormal symmetry in normally asymmetric structures (isomerism). Isomerism can be right-sided or left-sided. These are commonly referred to as heterotaxy syndromes, although heterotaxy includes situs inversus as well.

3.13.4.1 Right Isomerism (Bilateral Right-Sidedness)

- Characterized by bilateral morphologic right atria (with broad-based appendages) and asplenia
- Inferior vena cava and aorta are on the same side of the spine

The various anatomic associations reported are [7, 13, 15, 17]

- Dextrocardia/Levocardia/Mesocardia
- Systemic venous anomalies:
  - Bilateral superior venae cavae
  - Bilateral hepatic venous connections to the ipsilateral atrium
  - Azygous or hemiazygous continuation of the IVC to the SVC
  - Absent coronary sinus
- Pulmonary venous anomalies:
  - Total anomalous pulmonary venous connection (commonly supracardiac)
  - Bilateral direct pulmonary venous connections to the ipsilateral atrium
- Complete AV septal defect (often unbalanced); common atrium
- Univentricular AV connection
  - Right ventricular type with hypoplastic LV (commonly)
  - LV morphology with RV hypoplasia/tricuspid atresia
  - Uncertain or indeterminate morphology
- Ventricular relation:
  - Normally related (d-loop with right anterior aorta)
  - Ventricular inversion (l-loop with left anterior aorta)
- VA discordance or DORV
- Pulmonary stenosis or atresia

3.13.4.2 Left Isomerism (Bilateral Left-Sidedness)

- Characterized by bilateral morphologic left atria (with narrow-based appendages).
- The aorta and IVC lie anterior to or on the same side of the spinal column.
- Interruption of the inferior vena cava with azygous continuation is a distinctive feature.
- Polysplenia is common but not necessary.

The various anatomic associations reported are [7, 13, 15, 17]:

- Transverse liver
- Levocardia/Dextrocardia
- Systemic venous anomalies:
  - Bilateral superior venae cavae
  - Interrupted IVC
- Pulmonary venous anomalies:
  - Partial anomalous pulmonary venous connection
  - Bilateral direct pulmonary venous connections to the ipsilateral atrium
- Common atrium
- AV septal defect—complete or partial
- Common-inlet ventricle of RV morphology
- VA concordance or DORV
- Left ventricular outflow tract obstruction
- Pulmonary valve is occasionally stenotic but rarely atretic

Key point: Cardiovascular malformations found in the polysplenia syndrome are similar to, but less severe than, those found in the asplenia syndrome.

3.13.5 Mesocardia

- In mesocardia, the longitudinal axis of the heart is oriented cranio-caudally in the mid-sagittal plane (midline heart).
- There is no distinct cardiac apex [2].
- Most commonly associated with situs solitus and no other major abnormality.

The various anatomic associations described in mesocardia include [2, 17]:

- Mesocardia with situs solitus and d-bulboventricular loop is the most common variant.
- Mesocardia with situs inversus and l-bulboventricular loop is usually associated with significant cardiac anomalies.

VA discordance with or without AV discordance (i.e. d-TGA or l-TGA) has been described.

References

1. Van Praagh R, Vlad P. Dextrocardia, mesocardia and levocardia: the segmental approach to diagnosis in congenital heart disease. In: Keith J, RoweR VP, editors. Heart disease in infancy and childhood. 3rd ed. New York: Macmillan; 1978. p. 638.
2. Lev M, Liberthson R, Golden J, et al. The pathologic anatomy of mesocardia. Am J Cardiol. 1971;28:428–35.
3. Edwards WD, Maleszewski JJ. Cardiac anatomy and examination of cardiac specimens. In: Allen HD, Driscoll DJ, Shaddy RE, Feltes TF, editors. Moss and Adams' heart disease in infants, children, and adolescents. 8th ed. Philadelphia: Lippincott Williams & Wilkins; 2013. p. 1–31.
4. Geva T. Abnormal systemic venous connections. In: Allen HD, Driscoll DJ, Shaddy RE, Feltes TF, editors. Moss and Adams' heart disease in infants, children, and adolescents. 8th ed. Philadelphia: Lippincott Williams & Wilkins; 2013. p. 840–63.
5. Edwards WD, Maleszewski JJ. Classification and terminology of cardiovascular anomalies. In: Allen HD, Driscoll DJ, Shaddy RE, Feltes TF, editors. Moss and Adams' heart disease in infants, children, and adolescents. 8th ed. Philadelphia: Lippincott Williams & Wilkins; 2013. p. 32–51.
6. Earing MG, Hagler DJ, Edwards WD. Univentricular atrioventricular connection. In: Allen HD, Driscoll DJ, Shaddy RE, Feltes TF, editors. Moss and Adams' heart disease in infants, children, and adolescents. 8th ed. Philadelphia: Lippincott Williams & Wilkins; 2013. p. 1175–94.
7. O'Leary PW, Hagler DJ. Cardiac malpositions and abnormalities of atrial and visceral situs. In: Allen HD, Driscoll DJ, Shaddy RE, Feltes TF, editors. Moss and Adams' heart disease in infants, children, and adolescents. 8th ed. Philadelphia: Lippincott Williams & Wilkins; 2013. p. 1195–216.
8. Hahn RT. State-of-the-art review of echocardiographic imaging in the evaluation and treatment of functional tricuspid regurgitation. Circ Cardiovasc Imaging. 2016;9(12):e005332.
9. Robinson S, Ring L, Augustine DX, Rekhraj S, Oxborough D, Harkness A, Lancellotti P, Rana B. The assessment of mitral valve disease: a guideline from the British Society of Echocardiography. Echo Res Pract. 2021;8(1):G87–G136.
10. Weinberg PM, Natarajan S, Rogers LS. Aortic arch and vascular anomalies. In: Allen HD, Driscoll DJ, Shaddy RE, Feltes TF, editors. Moss and Adams' heart disease in infants, children, and adolescents. 8th ed. Philadelphia: Lippincott Williams & Wilkins; 2013. p. 758–98.
11. Krzanowski M, Bodzoń W, Dimitrow PP. Imaging of all three coronary arteries by transthoracic echocardiography. An illustrated guide. Cardiovasc Ultrasound. 2003;1:16.
12. Huang X, Huang Y, Huang T, Huang W, Huang Z. Individual pulmonary vein imaging by transthoracic echocardiography: an inadequate traditional interpretation. Eur J Echocardiogr. 2008;9(5):655–60.
13. Perloff JK, Marelli AJ. Cardiac malpositions. In: Clinical recognition of congenital heart disease. 6th ed. Philadephia: Saunders; 2012. p. 15–40.
14. Huhta JC, Hagler DJ, Seward JB, Tajik AJ, Julsrud PR, Ritter DG. Two-dimensiona lechocardiographic assessment of dextrocardia: a segmental approach. Am J Cardiol. 1982;50(6):1351–60.
15. Van Praagh R, Weinberg P, Smith S, et al. Malpositions of the heart. In: Adams F, Emmanouilides G, Riemenschneider T, editors. Heart disease in infants, children, and adolescents. 4th ed. Baltimore: Williams &Wilkins; 1989. p. 530.
16. Van Praagh R, Van Praagh S, Vlad P, et al. Diagnosis of anatomic types of dextrocardia. Am J Cardiol. 1965;15:234–43.
17. Stanger P, Rudolph A, Edwards J, Cardiac malpositions. An overview based on study of sixty-five necropsy specimens. Circulation. 1977;56:159–72.

4 Shunt Lesions

Cardiac shunt lesions are anomalous circulatory communications, mostly between the systemic and the pulmonary circulations, which facilitate mixing of oxygenated arterial blood with deoxygenated venous blood. The shunts can exist at the level of the atria, the ventricles, or the great arteries. Depending on the position of the receiving chamber (or vascular structure) relative to the tricuspid valve, shunt lesions can be classified as pre-tricuspid and post-tricuspid (Table 4.1).

4.1 Echocardiography of Shunt Lesions

Evaluation of a shunt lesion involves assessment of the anatomic attributes of the lesion, the physiological significance of the shunt, the complications/sequelae such as pulmonary hypertension, and evaluation for the need and options for treatment. The following overview is a general template for the echocardiographic evaluation and reporting of all shunt lesions.

Table 4.1 Classification of shunt lesions based on the position of the receiving structure relative to the tricuspid valve

Pre-tricuspid shunts	Post-tricuspid shunts
Atrial septal defect (ASD)	Ventricular septal defect (VSD)
Anomalous pulmonary venous drainage	Aortopulmonary window (APW)
Systemic AV fistulae	Patent ductus arteriosus (PDA)
Ruptured sinus of Valsalva to RA	Pulmonary AV fistulae
Coronary AV fistulae to RA	Coronary AV fistulae to RV, LA, LV
Gerbode defect (LV to RA shunt)	RSOV to RV, PA, LA, LV
	Aorta to LV tunnel
	Aortopulmonary collaterals

Supplementary Information The online version contains supplementary material available at https://doi.org/10.1007/978-981-96-0243-8_4.

A. Paul, *Manual of Echocardiography for Congenital Heart Diseases*,
https://doi.org/10.1007/978-981-96-0243-8_4

4.1.1 Anatomic Attributes of the Defect

- Size and shape of the defect:
  - The largest dimension of the defect was measured from multiple views.
  - In the case of oval lesions, dimensions in long and short axes should be documented.
- Adequacy/size of the margins: Vital for determining the suitability for device closure.
- Relationship with adjacent structures:
  - The distance between the aortic valve and the margins of a perimembranous VSD is vital for determining the treatment strategy [1].

4.1.2 Haemodynamic Attributes of the Defect

Shunt Quantification

- The quantum of shunt flow is proportional to the anatomic size of the defect and the relative resistances of the systemic and pulmonary vascular beds.
- The left-to-right shunt can be quantified as the ratio of pulmonary (Qp) to systemic (Qs) blood flow using the continuity equation: ***Qp/Qs = RVOT VTI × RVOT area/LVOT VTI × LVOT area*** (discussed in detail in Chap. 1).

Key point: In the case of PDA, the systemic blood flow is calculated using RVOT area and VTI, and the pulmonary blood flow is calculated using LVOT area and VTI.

Direction of Flow

- The direction of blood flow across a shunt lesion is determined by the use of colour Doppler.
- The blood flow is always in a left-to-right direction across an isolated uncomplicated shunt lesion.
- With the development of pulmonary hypertension and/or impaired RV compliance, the shunt flow becomes bidirectional before eventual reversal of direction.

Key point: Right-to-left shunting of blood across any shunt lesion suggests the presence of additional pathology, which needs to be thoroughly searched for.

4.1.3 Secondary Effects of a Shunt Lesion

Chamber dilation:

- Any significant shunt lesion will lead to volume overload and enlargement of the recipient chamber:
- A pre-tricuspid shunt causes enlargement of the RA and RV.
- A post-tricuspid shunt causes enlargement of the LA and LV.
- Enlargement of a chamber may have further consequences:
- Dilation of the mitral annulus secondary to a post-tricuspid shunt may lead to functional mitral regurgitation and further LV dilation.

Key point: Abnormal and unexplained dilation of cardiac chambers may be the first clue towards the diagnosis of an occult shunt lesion.

Exaggerated outflow tract gradient:

- Pre-tricuspid shunts may exaggerate the gradients across the RVOT due to excessive blood flow.
- Post-tricuspid shunts, such as VSD or PDA, may exaggerate the gradients across the left-sided valves.

Pulmonary arterial hypertension:

- Large left-to-right shunts eventually lead to elevated pulmonary arterial resistance.
- RV systolic pressure (RVSP) is equal to PA systolic pressure (PASP) in the absence of pulmonary stenosis. RVSP can be estimated using Bernoulli's principle (discussed in Chap. 1) in the following ways:

  RVSP = Peak TR gradient + RA pressure

  RVSP = LV systolic pressure – Peak gradient across a VSD (LV systolic pressure is almost equal to the systolic arterial blood pressure in the absence of aortic stenosis).

- Mean and diastolic pulmonary pressures can be estimated from the peak and end-diastolic velocities (pressure gradients) of pulmonary regurgitation, respectively.
- Significant pulmonary hypertension eventually leads to reversal of shunt flow (Eisenmengerization, Chap. 15).

Key point: A TR jet, which is suboptimal for satisfactory assessment of the spectral Doppler waveform velocity, may be enhanced with 10 mL of agitated saline contrast administered via an upper limb vein.

4.1.4 Ventricular Function:

- Ventricular function is a significant determinant of the pathophysiology of shunt lesions.
- Significant pulmonary hypertension can eventually lead to RV dysfunction.
- Closure of large ASDs at a later age may unmask previously undetected LV diastolic dysfunction, resulting in consequences such as acute pulmonary oedema.

4.1.5 Associated Lesions

The presence of other lesions can influence the physiological consequences of a shunt lesion:

- Shunt flow may be modified by the presence of distal obstruction:
  - Pulmonary stenosis can result in a diminished gradient across a VSD, and in severe cases, reversal of shunt flow.
  - Aortic stenosis or coarctation of the aorta may exaggerate the shunt flow across a VSD.
- The associated lesion(s), which were being masked by the shunt lesion, ***may become unmasked after closure of the shunt lesion*** [1]:
  - ASD closure may unmask a hitherto innocuous mitral valve lesion.
  - Aortic stenosis or coarctation of the aorta may be unmasked by closure of VSD or PDA.

Points to Report
- **Anatomic attributes of the shunt lesion:**
  - Location, size, margins
  - Distance from vital structures
- **Haemodynamic attributes:**
  - Quantum of shunt flow (Qp/Qs)
  - Direction of shunt
  - Chamber dilation
  - Pulmonary hypertension
  - Ventricular function
- **Associated lesions**

4.2 Atrial Septal Defect (ASD)

On the basis of their location in the interatrial septum, atrial septal defects are classified as (Fig. 4.1):

1. **Ostium Secundum (fossa ovalis) Atrial Septal Defect (OS ASD):**
   (a) The most common form of ASD.
   (b) Ostium secundum ASDs are true deficiencies in the septum primum.

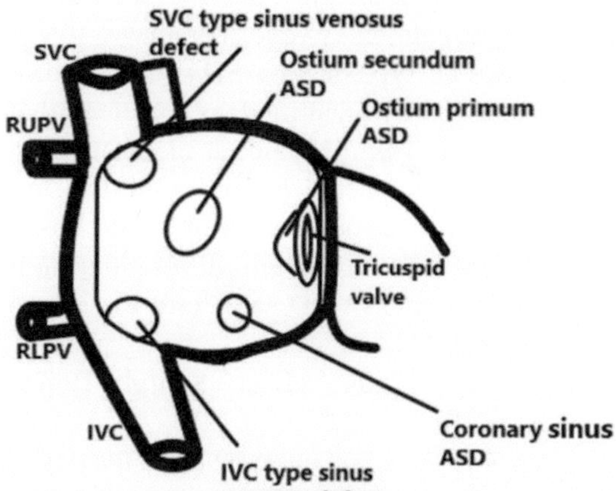

Fig. 4.1 An illustration showing the location of various atrial septal defects as viewed from the RA side. *ASD* atrial septal defect, *IVC* inferior vena cava, *SVC* superior vena cava, *RUPV* right upper pulmonary vein, *RLPV* right lower pulmonary vein

   (c) The longest dimension of these oval defects is usually in the cephalo-caudal plane [2].
2. **Ostium primum atrial septal defect (OP ASD):**
   (a) OP ASDs are deficiencies in the atrioventricular septum, resulting from the failure of fusion of the endocardial cushions (endocardial cushion defect) (discussed in Chap. 5).
   (b) OP ASD without inlet VSD is referred to as a partial AV septal defect.
3. **Sinus Venosus Atrial Septal Defect (SV-ASD):**
   (a) Sinus Venosus ASDs, otherwise called Sinus Venosus Defects (SVD), are deficiencies in the sinus venosus septum. They are of two types:
      SVC type: Septal defect between the right upper pulmonary vein (RUPV) and the SVC.
      IVC type: Septal defect between the right middle/lower pulmonary vein and the IVC.
   (b) The unroofing of the RUPV into the SVC, which occurs in the SVC type of SV-ASD, results in a left-to-right shunt from the LA to the RA across the RUPV-SVC communication.
   (c) SV-ASD is always associated with anomalous pulmonary venous drainage.
   (d) An IVC type of SV-ASD is echocardiographically identified as a defect that extends from the mouth of the IVC to the inferoposterior border of the left atrium, without any atrial septal tissue at the inferior margin, differentiating it from an OS-ASD with inferior extension [2].
4. **Coronary Sinus Atrial Septal Defect (CS-ASD):**
   (a) CS-ASD is not a true ASD, as the defect is actually an unroofed coronary sinus.
   (b) Unroofing of the coronary sinus results in a left-to-right shunt from the LA to the RA across the coronary sinus (discussed in Chap. 13).

4.2.1 Echocardiographic Diagnosis

The possibility of a haemodynamically significant pre-tricuspid shunt is suggested by ***features of RV volume overload***, such as:

- Dilated RA and RV (ideally appreciated in a perfectly oriented A4C view, where the apex occupies the centre and the IVS is vertical)
- Paradoxical septal motion (demonstrated in 2D and M-mode)
- Flattening of the IVS during diastole ('pancaking')

4.2.1.1 Visualization of the Defect

In the atrial septum by 2D imaging points towards the diagnosis of an ASD (Fig. 4.2). This may be facilitated by multiple imaging windows and planes.

- **Apical four-chamber (A4C) view** (Fig. 4.2a):
  - Ostium primum defects are best seen in the A4C view in relation to the AV valves.
  - A coronary sinus defect can be visualized by posterior angulation from the standard A4C view (coronary sinus view).
  - Measurement of the size of the defect is better avoided in this view.
  - Artefactual 'drop-outs' in the IAS may erroneously evoke the diagnosis of ASD in the A4C and PSAX views, where the ultrasound beam is parallel to the atrial septum.
  - Sliding the transducer medially from the apex towards the sternum (modified A4C view) improves the incident angle between the ultrasound beam and the IAS, thereby minimizing the artefactual drop-out (Fig. 4.2b).
  - T-artefact: Increased echogenicity with broadening of the edges of the defect in the A4C view may suggest the presence of a true defect.
- **Subcostal windows** are ideal for visualizing ASDs, as the ultrasound beam is perpendicular to the IAS:

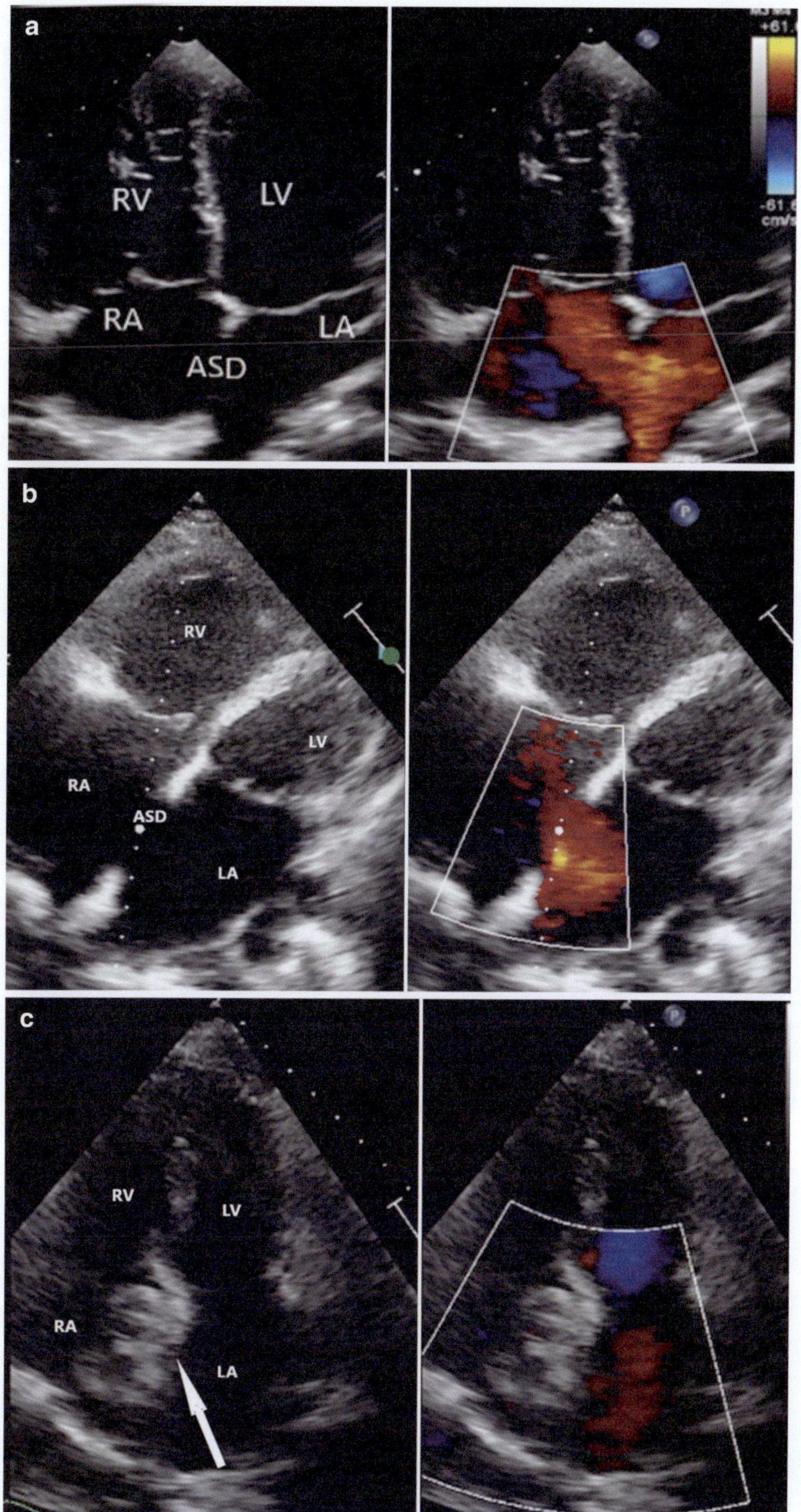

Fig. 4.2 (**a**) Apical four-chamber view of ostium secundum ASD. Colour Doppler demonstration of shunt flow through the drop-out confirms the presence of a true defect of the atrial septum (as opposed to an artefactual drop-out). (**b**) Modified A4C view obtained by sliding the transducer medially. An improved incident angle between the ultrasound beam and the atrial septum minimizes the possibility of artefactual drop-out. (**c**) A4C view after percutaneous device closure of ostium secundum ASD. The device (arrow) is seen in situ with no residual shunt across the atrial septum

- The subcostal coronal standard four-chamber view visualizes OS-ASDs and OP-ASDs (Fig. 4.3).
- The subcostal coronal coronary sinus view visualizes CS-ASD.
- The subcostal sagittal bicaval view visualizes SVC and IVC types of SVDs (Fig. 4.3).
- The subcostal left anterior oblique [LAO] view facilitates imaging of the atrial septum along its length, and the entrance of right-sided pulmonary veins into the LA [2].

Key point: Imaging with the breath held in deep inspiration may improve the subcostal window in adults.

- **Parasternal short-axis (PSAX) view:**
  - The aortic rim of the ASD can be identified and measured in this view.
  - Absence of the posterior rim ('bald' posterior wall) in the PSAX view is a hallmark of the IVC type of SV-ASD, which differentiates it from an inferiorly extending OS-ASD (which has a persistent posterior rim) [3].
- **High right parasternal view:**
  - Aids in the detection of sinus venosus defects
  - Useful in patients with inadequate subcostal views

4.2.1.2 Demonstration of Blood Flow across the Defect by Colour Doppler

Confirms the diagnosis of ASD (Fig. 4.2) and rules out the possibility of artefactual drop-outs.

- In an isolated large ASD, as the pulmonary vascular resistance increases by the third decade of life, the left-to-right shunt flow diminishes gradually and eventually reverses the direction.

4.2.1.3 Demonstration of a Classic Triphasic Left-to-Right Shunt by Pulse Wave Doppler

At the level of the defect:

- Characterized by flow waveforms in late systole and diastole, with pre-systolic accentuation (Fig. 4.4) [4].
- The distinguishing features of a normal venous flow pattern (as opposed to an ASD flow pattern) are: (a) earlier onset in systole; (b) distinct systolic, diastolic, and atrial reversal waves; and (c) inspiratory augmentation of flow.

Saline contrast echocardiography is useful in adults with poor echocardiographic windows:

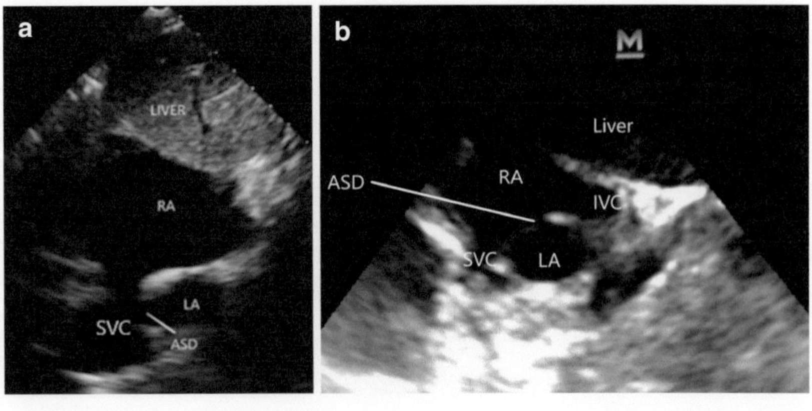

Fig. 4.3 Subcostal sagittal bicaval view demonstrating SVC-type of SV-ASD (**a**) and OS-ASD (**b**)

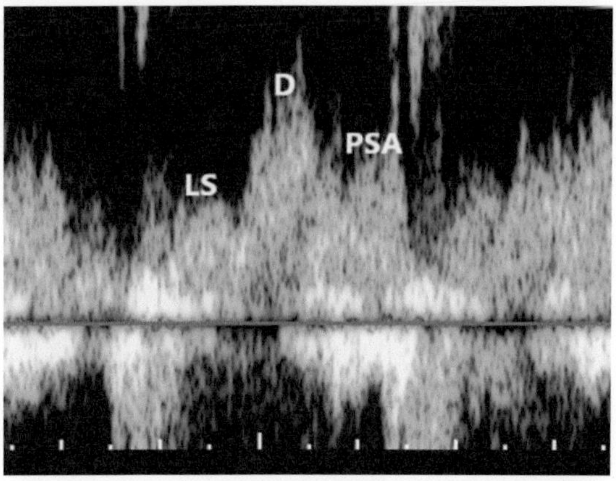

Fig. 4.4 Pulse wave Doppler of the shunt flow across an OS-ASD. The classic triphasic flow consists of late systolic flow (LS), diastolic flow (D), and pre-systolic accentuation

- Microbubbles are created by agitating a mixture of 8 mL of normal saline, 1 mL of the patient's blood, and 1 mL of air taken in a 10 mL syringe, using a stopcock and an empty 10 mL syringe. This agitated mixture is injected into an upper limb vein (usually an antecubital vein) while imaging the heart from an appropriate view.
- In an ASD with a left-to-right shunt, opacification of the RA with agitated saline will be accompanied by a negative contrast effect (i.e. clearing of bubbles) due to the shunt flow from the LA.
- In CS-ASD with left SVC (Raghib's syndrome), intravenous injection of agitated saline into a left upper limb vein results in the entry of bubbles into the left upper corner of the LA before their appearance in the RA [5].

4.2.2 Anatomy of the Defect

4.2.2.1 Size and Location of the Defect

- The size of the defect should be determined from multiple views.
- The greatest dimensions in the orthogonal planes should be documented.
- An ASD size >20 mm/m^2 body surface area is generally considered unsuitable for device closure.

4.2.2.2 Adequacy of Atrial Septal Rims

By convention, an OS-ASD has six anatomically named rims of surrounding tissue [2]:

- **Aortic rim** (antero-superior rim): Between the ASD and the AoV annulus/aortic root (***PSAX view***)
- **AV valve rim** (antero-inferior or mitral rim): Between the ASD and the AV valves (***A4C view***)
- **SVC rim** (postero-superior rim): Between the ASD and the SVC (***subcostal sagittal bicaval view***)
- **IVC rim** (postero-inferior rim): Between the ASD and the IVC (***subcostal sagittal bicaval view***)
- **Posterior rim**: Between the ASD and the posterior atrial wall (***PSAX view***)
- **Right upper pulmonary vein (RUPV) rim**: The posterior rim between the ASD and the RUPV (***subcostal coronal/LAO views***)

ASD rim length <5 mm is considered 'deficient' [2]:

- Having adequate SVC, RUPV, IVC, and AV valve rims is particularly important for successful transcatheter ASD closure.
- A deficient rim portends the risk of erosion.

4.2.2.3 Total Atrial Septal Length

Should be estimated from the subcostal coronal or left anterior oblique view:

- This determines the maximum possible diameter of the ASD closure device

Key point: Subcostal coronal/LAO views are useful in assessing atrial septal length as well as the adequacy of the RUPV rim.

4.2.3 Direction and Quantum of Shunt Flow

4.2.3.1 Direction of Shunt

- Assessed by colour Doppler: The settings should be optimized for the expected low velocity of shunt flow (i.e. 25–40 cm/s).

- Demonstration of PWD tracings on both sides of the baseline aids in the detection of bidirectional shunts.
- *A **Right-to-left shunt*** should immediately alert the echocardiographer towards the possibility of associated structural or functional abnormalities [1]:
  - Severe pulmonary arterial hypertension.
  - Severe pulmonary stenosis.
  - Obligatory right-to-left shunt, as seen in tricuspid atresia or TAPVC.
  - Hypoplasia of right-sided chambers, most often as a part of more complex anomalies.
- Bidirectional shunting occurs during the stage of progression of PAH and often with the onset of atrial fibrillation in the absence of PAH [6].

4.2.3.2 Shunt Quantification (Qp:Qs)

- Assessed by the continuity equation (as discussed earlier).
- A shunt with a colour flow width of 15 mm or more has been proposed to be associated with Qp:Qs > 2:1.
- The presence of a dilated RA and RV suggests a haemodynamically significant shunt.

4.2.4 Secondary Effects

- Dilated RA, RV, and MPA, with secondary TR and/or PR due to annular stretching.
- RVOT turbulence is common in ASD owing to the increased blood flow across the RVOT.
- Pulmonary arterial hypertension.

Key point: An RVOT gradient of up to 60 mm Hg has been reported in patients with ASD in the absence of pulmonary stenosis [7].

4.2.5 Associated Lesions

- Anomalous pulmonary venous drainage
- Inlet VSD, common AV valve, cleft mitral valve, and MR—in cases of OP-ASD
- Left SVC—especially in CS-ASD
- Other shunt lesions
- RV outflow tract obstruction

Key point: ASD closure is generally considered at any age if there is a left-to-right shunt with evidence of RV volume overload, although the ideal age is 2–4 years (4–5 years in SV-ASD).

Points to Report

- ASD: Location, size, direction of the shunt
- Secondary effects:
- Degree of RV dilation; presence of RV systolic dysfunction
- Pulmonary hypertension—RVSP or mean PA pressure
- Presence of functional TR
- Adequacy of rims, atrial septal length, and feasibility of percutaneous closure (for OS-ASD)
- Associated lesions
- Pulmonary venous drainage
- LV systolic and diastolic function
- Presence of Eustachian valve or Chiary network (important if device closure is being contemplated)
- **Post-repair:**
  - Patch/occluder integrity and presence of residual leak
  - RV size and function
  - Mitral and tricuspid valve regurgitation
  - RVSP or mean PA pressure
  - LV systolic and diastolic function

4.3 Patent Foramen Ovale (PFO)

Patent foramen ovale (PFO) is a separation between the septum primum and septum secundum, which results from the failure of closure of the valve of the fossa ovalis. Hence, a PFO is not a true deficiency of the atrial septum.

The following nomenclature has been suggested for anomalies involving the region of the fossa ovalis [2]:

- **PFO:** Right-to-left shunting of blood is demonstrable by colour Doppler or saline contrast, with no observable deficiency of the atrial septum. When the shunting of blood across a PFO is from left to right, it should be specified as '**PFO with left-to-right flow**'.
- **Stretched PFO:** The foramen ovale is stretched open by atrial haemodynamics, resulting in a defect with shunting of blood in either direction.
- **LA septal pouch:** Incomplete fusion of the septum primum and septum secundum results in a blind pouch-like anatomic region that often communicates with the LA cavity. This can serve as a nidus for thrombus formation [8].

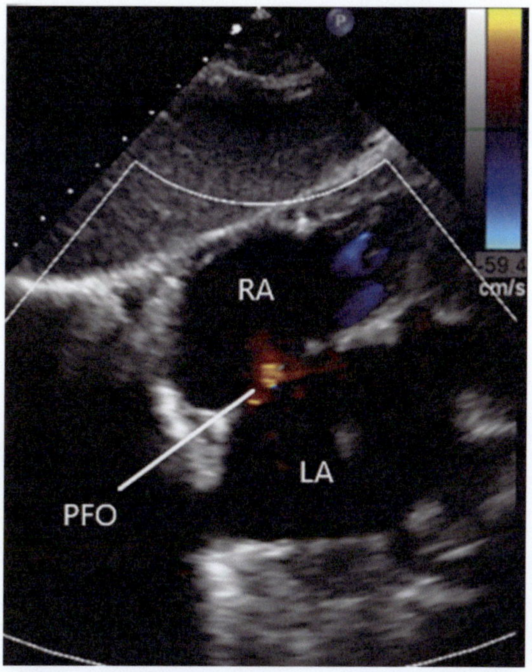

Fig. 4.5 Subcostal coronal view showing the turbulent flow across a patent foramen ovale

4.3.1 Echocardiographic Diagnosis (Fig. 4.5)

- Shunting across a stretched open PFO is often intermittent and might not be readily detectable using colour flow Doppler
- **Contrast echocardiography** with agitated saline aids in the detection of a PFO:
  - TEE is preferred over TTE for contrast echocardiographic detection of a PFO
  - Timing of appearance of microbubbles in the left atrium:
    - Early visualization (<3–6 cardiac cycles) suggests an intracardiac shunt (e.g. PFO)
    - Late visualization (after 3–6 beats) indicates intrapulmonary shunting
  - One of several grading systems based on the number of bubbles crossing over is as follows [9]:
    - Grade 1: ≤5 bubbles
    - Grade 2: 6 to 25 bubbles
    - Grade 3: >25 bubbles
    - Grade 4: Opacification of the chamber
- Provocative manoeuvres to transiently increase the RA pressure during contrast injection improve the detection of a PFO:
  - Valsalva manoeuvre
  - Pressure over the right upper quadrant of the abdomen
  - Cough
  - Intravenous hydration

Key point: Contrast saline injection through the lower limb improves the likelihood of detecting a PFO, as the eustachian valve naturally directs the blood flow towards the atrial septum [10].

4.3.2 Anatomic Assessment

Assessment of the anatomic attributes of a PFO should focus on the following features:

- Location of the PFO
- Length of the PFO tunnel
- Size of the PFO at the RA and LA ends (TEE-based classification) [9]:

- Small: <2 mm
- Medium: 2–3.9 mm
- Large: ≥4 mm
- Distance of the PFO from the vena cavae
- Total length of the atrial septum, as well as the thickness and extent of the septum secundum
- Presence of additional atrial septal defects
- The proposed high-risk PFO features for cryptogenic stroke include [9, 10]:
  - Presence of a concomitant atrial septal aneurysm
  - Tunnel-like appearance of the PFO
  - Size ≥3 mm
  - Spontaneous right-to-left shunting across the PFO (without provocative manoeuvres like Valsalva)

4.3.3 Atrial Septal Aneurysm (ASA)

The concomitant presence of a PFO with an ASA portends a greater risk of paradoxical embolism and stroke.

- ASA is defined as an excursion of the redundant valve of the fossa ovalis ≥10 mm from the plane of the atrial septum into either side (RA or LA), or a combined excursion to the right and left of ≥15 mm.
- **Olivares-Reyes classification** [11]:
  - Type 1R: ASA protrudes from the midline to the RA.
  - Type 2L: ASA protrudes from the midline to the LA.
  - Type 3RL: Maximal excursion is towards the RA, with lesser excursion towards the LA.
  - Type 4LR: Maximal excursion is towards the LA, with lesser excursion towards the RA.
  - Type 5: The excursion of the ASA is bidirectional and equidistant towards both sides.

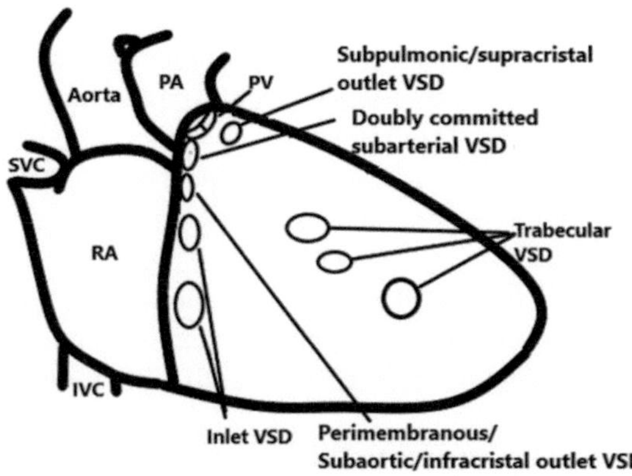

Fig. 4.6 Geographic locations of various types of ventricular septal defects as viewed from the RV

4.4 Ventricular Septal Defect (VSD)

Anatomically, the interventricular septum (IVS) consists of two parts: (a) small membranous septum and (b) large muscular septum (Fig. 4.6).

1. **Membranous septum**
   (a) The membranous septum is usually about 5 mm in diameter in the adult heart.
   (b) It consists of two portions, separated by the septal leaflet of the tricuspid valve [1]:
      Pars atrioventricularis (wedged between the TV and AV), where LV to RA shunts occur.
      Pars interventricularis, where interventricular communications are located.
   (c) Fibrous continuity between the aortic, mitral, and tricuspid valves is mediated by the membranous septum.
2. **Muscular septum** has three components:
   (a) Inlet septum: Inferior to the membranous septum, between the atrioventricular valves
   (b) Trabecular septum: Extends from the membranous septum to the apex

Table 4.2 Various classification systems and nomenclatures of VSDs

STS	ISNPCHD [12]	Anderson [13]	Van Praagh [14]	Other terminologies
Type 1	Outlet	Doubly committed subarterial	Conal septal	Infundibular Supracristal Subpulmonic
Type 2	Central	Perimembranous	Conoventricular	Paramembranous Infracristal Subaortic
Type 3	Inlet	[a]	AV canal type	AV septal defect
Type 4	Trabecular	Muscular	Muscular	

ISNPCHD International Society for Nomenclature of Paediatric and Congenital Heart Disease

[a] In Anderson's classification, there are only three categories of VSDs; the complete AV septal defect and the associated VSD are considered separately

(c) Outlet (conal or infundibular) septum: Immediately below the pulmonary valve (in a normal heart with subpulmonary conus)

Classification of Ventricular Septal Defects

Multiple classification systems have been proposed for the classification and nomenclature of VSDs (summarized in Table 4.2).

Border approach of classification (Anderson's classification) hinges on the anatomic structures immediately adjacent to the VSD [13]:

- Perimembranous defects are located adjacent to the membranous septum.
- Muscular VSDs are defects of the muscular septum with entirely muscular borders.
- Juxtaarterial or doubly committed VSDs are adjacent to the semilunar valves, such that the adjacent leaflets of the aortic and pulmonary valves are in fibrous continuity. The outlet septum and septal component of the subpulmonary infundibulum are deficient.

Geographical approach focuses primarily on the position of the defect within the ventricular septum and classifies VSDs as membranous, inlet, outlet, and muscular [14].

1. **Membranous VSD:**
   (a) Membranous VSDs are located just below the aortic valve and the septal leaflet of the tricuspid valve.
   (b) True defects of the membranous septum (with no extension into the adjacent muscular septum) are bordered completely by fibrous tissue. Membranous defects with extension into the inlet septum and/or outlet septum are called **perimembranous and paramembranous** VSDs [15].
   (c) Perimembranous VSDs involving the LVOT are called **subaortic, infracristal, and conoventricular** VSDs. These perimembranous VSDs have fibrous continuity between the aortic valve and the septal tricuspid leaflet. This is in contrast to the malalignment outlet VSDs with perimembranous extension, where the fibrous continuity is between the aortic valve and the anterior tricuspid leaflet [14, 15].
   (d) Membranous/perimembranous VSDs comprise nearly 80% of all VSDs. Spontaneous closure is common. They may also undergo anomalous closure by:
      Septal leaflet of the tricuspid valve (resulting in a ventricular septal aneurysm).
      Prolapse of an aortic cusp (resulting in LVOT obstruction and/or aortic regurgitation).
2. **Outlet VSD:**
   (a) The infundibular or outlet septum separates the right and left ventricular outflow tracts.
   (b) Outlet VSDs are located in the outlet septum, immediately beneath the aortic and pulmonary valves. These VSDs are also called **subpulmonic, supracristal, infundibular, conal septal, and doubly committed** subarterial VSDs.
   (c) Outlet VSDs comprise nearly 6% of all VSDs (reportedly up to 30% in Asian populations). Spontaneous closure is rare.

(d) A malalignment outlet VSD with perimembranous extension is characterized by a muscular conal septum, fibrous posteroinferior rim, and aortic valve-anterior tricuspid leaflet continuity [14, 15].
(e) An absent outlet septum results in doubly committed or juxta-arterial outlet VSD.
(f) Aortic cusp prolapse, commonly of the right coronary cusp, results in aortic regurgitation.

3. **Inlet (AV canal) VSD:**
   (a) Inlet VSDs are located in the inlet septum beneath the atrioventricular valves, usually with no muscular rim between the defect and the AV valve annulus.
   (b) Inlet VSDs comprise 8% of all VSDs.
   (c) VSDs in the inlet septum can be of different types:
      Inlet VSD associated with AV septal (endocardial cushion) defect.
      Inlet VSD with malalignment between IVS and IAS, and straddling the tricuspid valve.
      Inlet muscular VSD with completely muscular rims.
4. **Muscular (Trabecular) VSD:**
   (a) Muscular VSDs are located in the trabecular septum.
   (b) Based on the location in the trabecular septum, muscular VSDs are classified as:
      Anterior muscular VSD: Anterior to the septal band.
      Mid-muscular VSD: Posterior to the septal band.
      Apical muscular VSD: Inferior to the moderator band.
      Posterior muscular VSD: Beneath the septal tricuspid leaflet.
   (c) Inlet VSDs comprise 5–20% of all VSDs. Spontaneous closure is common.
   (d) Multiple, small, tortuous defects, described as 'Swiss-cheese' defects, are common in the muscular septum [15].

4.4.1 Echocardiographic Diagnosis

The diagnosis of ventricular septal defect is echocardiographically confirmed by:

- Detection of a potential defect in the ventricular septum using 2D imaging
- Demonstration of blood flow across the defect using colour Doppler

PWD at the level of the defect will reveal the flow pattern across a VSD:

- In patients with low PVR, the left-to-right shunt is predominantly systolic.
- With an increase in LVEDP, the left-to-right shunt extends into diastole as well.
- A typical 'M' shaped flow pattern is seen in non-restrictive VSD [6]:
  - Shunt flow occurs from left to right with the onset of systole.
  - Flow decreases during mid-systole (with equalization of RV and LV pressures).
  - Flow increases again in late systole as RV pressure drops earlier than that of LV.

Key point: The septum should be thoroughly interrogated along its entire length by sweeping from subcostal, apical, and parasternal views to look for multiple VSDs.

4.4.2 Location of the Defect

Multiple windows, planes, and sweeps should be used to assess the ventricular septum in its entirety.

Subcostal sagittal (short-axis) sweep delineates the VSDs at various locations:

- Perimembranous VSD is located behind the septal leaflet of the tricuspid valve.
- Muscular VSDs are seen within a completely muscular rim anywhere in the muscular septum.
- Doubly committed subarterial VSD is seen immediately below the semilunar valves with deficiency of the conal septum.
- Inlet (AV canal type) VSD is located in the vicinity of the AV valves.

PSAX view at the level of the aortic valve is useful in classifying the VSDs geographically (Fig. 4.7):

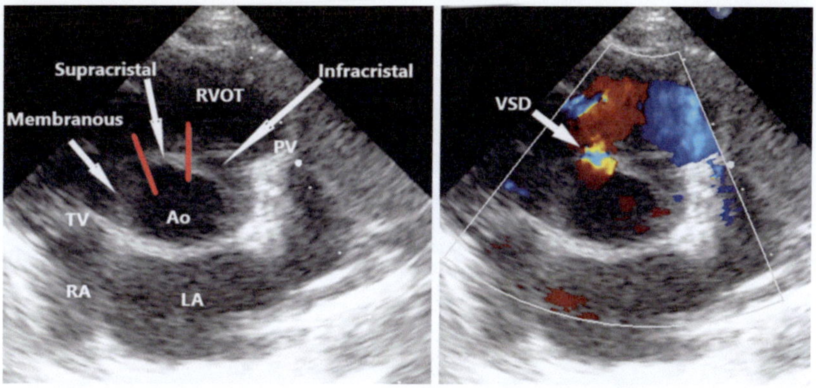

Fig. 4.7 PSAX view at the level of the aortic valve. The locations of various VSDs are depicted with demarcation of each type (red lines). Colour flow Doppler demonstrates turbulent flow between the 11 and 12 o'clock positions (suggestive of Perimembranous/subaortic/supracristal outlet VSD)

- Membranous VSD:
  - The location of the defect is between the 9 o'clock and 11 o'clock positions
  - Partial coverage of the defect by tricuspid valve tissue can result in underestimation of size
- Perimembranous VSD (infracristal or subaortic):
  - Located between the 11 o'clock and 12 o'clock positions
  - Results from the anterior extension of a membranous defect to the infracristal outlet (subaortic) septum
- Outlet VSDs (supracristal or subpulmonic):
  - Located between the 12 o'clock and 2 o'clock positions (pulmonary valve hinge point)

Apical window depicts various VSDs as shown in Fig. 4.8.

Focused Assessment of Various Types of VSDs
Perimembranous VSDs (Figs. 4.7 and 4.9):

- Visualized in the PLAX (standard and RV inflow), PSAX, A5C, A3C, and subcostal views as a defect in the region below the right border of the aortic annulus, adjacent to the septal tricuspid leaflet.
- The relationship of perimembranous VSD to the aortic valve can be assessed from the apical long axis and PLAX views.
- The distance between the margins of a perimembranous VSD and the aortic valve should be measured to assess the suitability for transcatheter closure [16].

Inlet VSDs:

- Best visualized in the standard A4C view (Fig. 4.10) and the PSAX view at the level of the AV valves (Fig. 4.12a).
- The position of the hinge points of the AV valve annuli distinguishes the AVSD-associated inlet VSD from the ones without AVSD. In AVSD, the hinge points are at the same level, whereas the tricuspid valve hinge point is more apical in inlet VSDs without AVSD.

Muscular VSDs (Fig. 4.11)**:**

- Anterior muscular VSDs can be visualized in the PLAX view by sweeping to the left.
- Mid-muscular VSDs can be seen in the standard PLAX, PSAX, and A4C views.
- Posterior muscular VSDs are best visualized in the PSAX view or in the A4C view by sweeping posteriorly.
- Apical muscular VSDs can be seen in the A4C view inferior to the moderator band.
- A basal-to-apical sweep from the PSAX view can profile the location of muscular VSDs (Fig. 4.12a, b):
  - Anterior muscular VSDs are seen between the 12 o'clock and 2 o'clock positions.

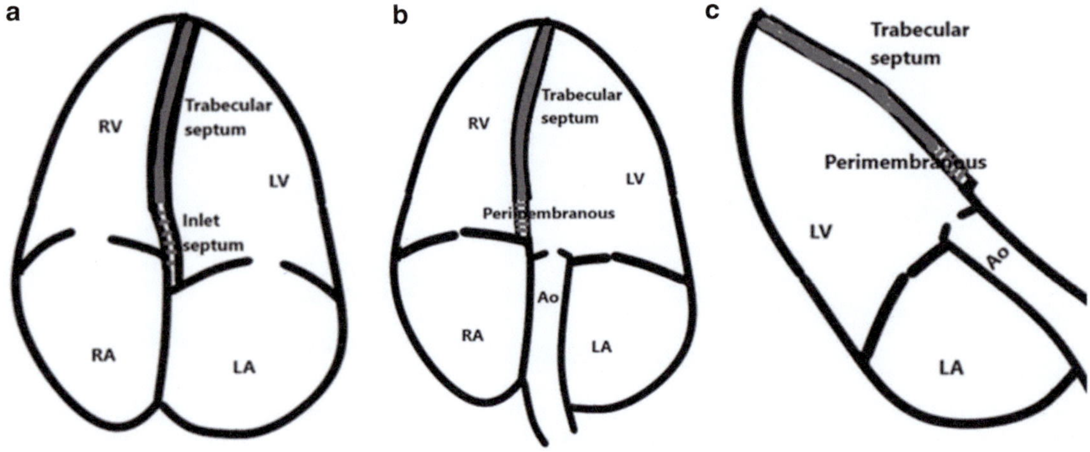

Fig. 4.8 Schematic representation of the location of various types of VSDs in the apical views: A4C (**a**), A5C (**b**), A3C (**c**)

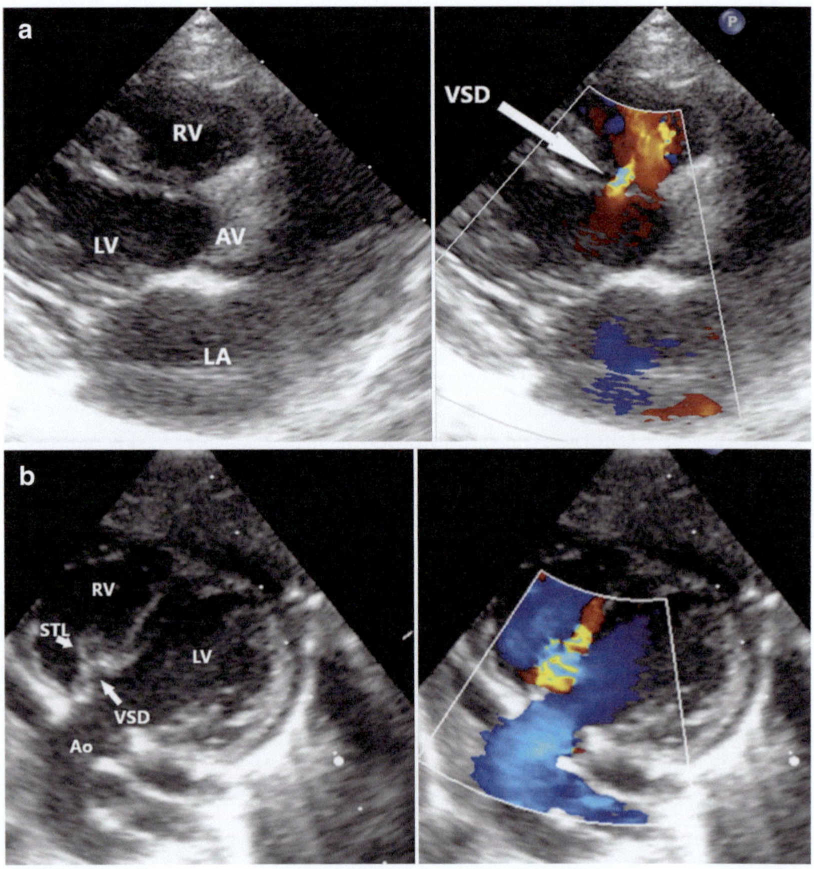

Fig. 4.9 Perimembranous VSD. (**a**) PLAX view depicting the shunt flow across a subaortic VSD. (**b**) Subcostal coronal view depicting a perimembranous VSD partially closed by the septal tricuspid leaflet (STL)

– Mid-muscular VSDs are seen between the 10 o'clock and 12 o'clock positions.
– Posterior muscular VSDs are seen between the 7 o'clock and 10 o'clock positions.

Outlet VSDs:

- Assessed in the PLAX view by sweeping left towards the pulmonary valve (RV outflow view), in the basal PSAX view (Fig. 4.7), and in the subcostal coronal and sagittal views.
- The supracristal (subpulmonic) defect is visualized above the crista supraventricularis in the RVOT, with the jet directed towards the 2 o'clock position in the PSAX view.

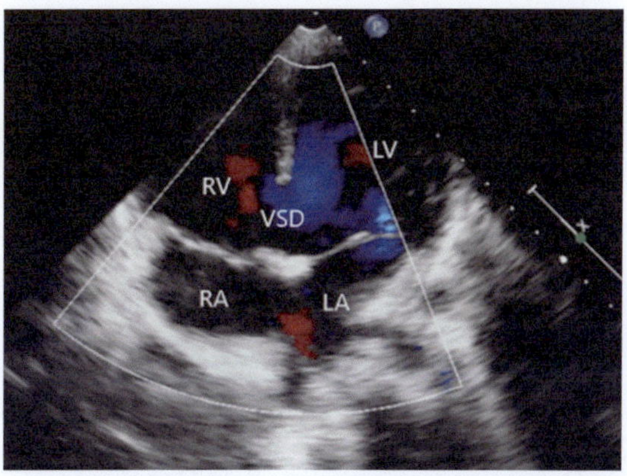

Fig. 4.10 A4C view of complete AV canal defect depicting inlet VSD and ostium primum ASD

- Subcostal and apical views allow assessment of malalignment VSDs (which occur as a result of deviation of the conal septum):
  – Anterior malalignment VSD: Associated with tetralogy of Fallot.
  – Posterior malalignment VSD: Associated with interrupted aortic arch and subaortic membrane.
- The severity of associated aortic regurgitation needs to be assessed.

4.4.3 Size/Severity of the Defect

4.4.3.1 Anatomic

The size of the VSD is reported in relation to that of the aortic valve annulus [15]:

- **Small VSD**: VSD diameter is less than 33% of the aortic valve annulus diameter.
- **Moderate VSD**: VSD diameter is more than one-third but less than 50% of the aortic annulus diameter.
- **Large VSD**: VSD diameter is >50% of the aortic annulus diameter.

Generally, VSDs of size <5 mm are considered small, and those ≥10 mm are considered large, although diameter indexed to body surface area would be more appropriate.

Fig. 4.11 (**a**) A4C view depicting a mid-muscular VSD. (**b**) Colour flow Doppler image of apical muscular VSD with two turbulent jets shunting left to right

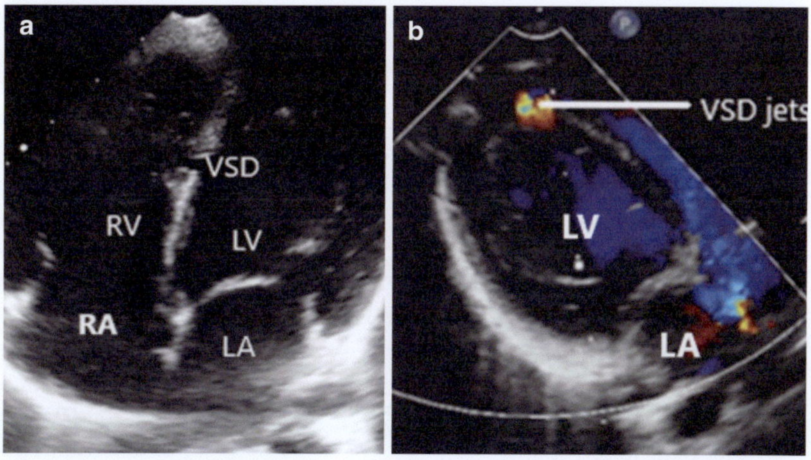

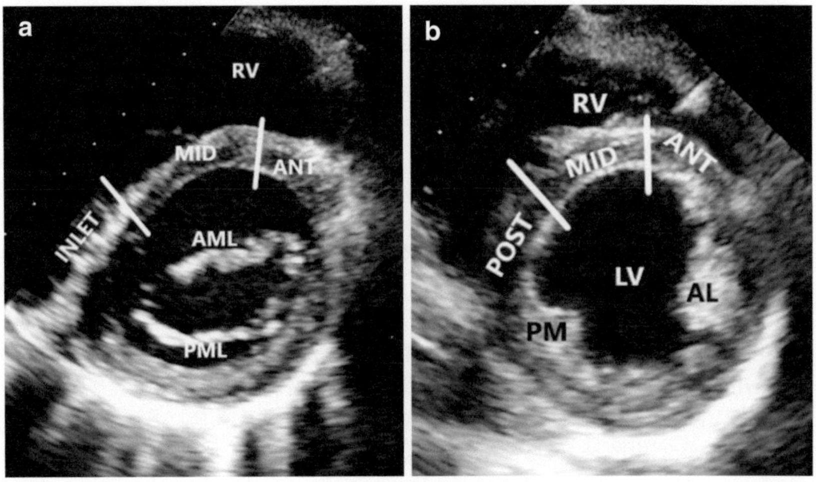

Fig. 4.12 PSAX view at (**a**) Mitral valve level and (**b**) Papillary muscle level, depicting the locations of inlet VSD and muscular VSDs. *AML* anterior mitral leaflet, *PML* posterior mitral leaflet, *AL* anterolateral papillary muscle, *PM* posteromedial papillary muscle, *ANT* anterior muscular VSD, *MID* mid-muscular VSD, *POST* posterior muscular VSD

4.4.3.2 Haemodynamic

Based on the peak pressure gradient across the VSD (i.e. the pressure difference between the LV and RV), calculated using Bernoulli's equation, VSDs are classified as:

- **Restrictive defect:**
  - Peak pressure gradient across the VSD >60 mmHg
  - Pulmonary-aortic systolic pressure ratio <0.33
  - Corresponds to small defects (Qp/Qs <1.5:1)
- **Moderately restrictive defect:**
  - Peak pressure gradient across the VSD 25–60 mmHg
  - Pulmonary-aortic systolic pressure ratio 0.33–0.66
  - Corresponds to moderate-sized defects (Qp/Qs of ≥1.5 to 2:1)
- **Non-restrictive defect:**
  - Peak pressure gradient across the VSD <25 mmHg
  - Pulmonary-aortic systolic pressure ratio >0.66
  - Corresponds to large defects (Qp/Qs >2.1)
- **Eisenmenger VSD:**
  - Systolic pressure ratio of 1
  - Qp/Qs of less than 1:1
  - Net right-to-left shunt

While estimating the pressure gradient across the VSD using CWD, the ultrasound beam should be well aligned with the VSD jet for accurate estimation of the interventricular gradient. In the case of a lengthy defect, the viscous frictional forces make the application of the modified Bernoulli's equation inappropriate [6].

4.4.4 Direction and Quantum of Shunt Flow

- Determined using colour flow Doppler
- PWD, CWD, and colour M-mode may also indicate the direction of flow
- With a restrictive VSD, left-to-right shunting occurs throughout systole
- Bidirectional or right-to-left shunting can occur with both restrictive and non-restrictive VSDs with an increase in pulmonary vascular resistance
- Shunt quantification (Qp:Qs) may be done by using the continuity equation (as discussed earlier)
- Haemodynamic significance:
  - The presence of LA and/or LV dilation suggests that the VSD is haemodynamically significant, with a Qp:Qs > 2:1
  - Increased RV wall thickness (>5 mm from the subcostal view) and systolic septal flat-

tening reflect elevated RV systolic pressure

4.4.5 Secondary Effects

- The RVOT gradient and mitral valve gradient may be exaggerated in the presence of VSD, even without pulmonary stenosis or mitral stenosis [6]
- Pulmonary arterial hypertension gradually results in reversal of shunt flow
- In perimembranous VSD, the findings suggestive of increased susceptibility to AR are [17]:
  - Aortic cusp override in A5C and subcostal coronal views
  - Abnormal aortic cusp movements
  - Flow across the ventricular septum in the standard PLAX view
  - Dilated aortic root

4.4.6 Associated Lesions

- Pulmonary stenosis; progressive RVOT obstruction (***Gasul's phenomenon***)
- Aortic regurgitation resulting from prolapse of the right or non-coronary cusp (see above)
- DCRV
- Gerbode defect
- Subaortic ridge
- Malalignment of the outlet septum
- Straddling AV valves

4.4.7 Suitability for Device Closure

- Percutaneous closure can be done for mid-muscular VSD and perimembranous VSD.
- The defect should be at least 5 mm away from the AV valves and semilunar valves.
- Associated defects requiring cardiopulmonary bypass should be ruled out.

Key point: Echocardiographic indications to consider VSD closure include haemodynamically significant shunts (Qp:Qs ≥1.5:1) with evidence of LV volume overload and/or pulmonary hypertension; perimembranous or supracristal VSD with aortic cusp prolapse and worsening AR; and progressive significant RVOT obstruction.

Points to Report

- VSD: Location, size, direction of the shunt, peak pressure gradient
- Distance between the defect and the AV/semilunar valves
- Shunt quantification
- Secondary effects:
  - LV size and function
  - LA size
  - Pulmonary hypertension—RVSP or mean PA pressure
  - Aortic valve prolapse/regurgitation
- Associated lesions (see above)
- **Post-repair:**
  - Patch/occluder integrity and presence of residual leak
  - LV size and function
  - Aortic valve regurgitation
  - RVSP or mean PA pressure

4.5 Patent Ductus Arteriosus (PDA)

Ductus arteriosus is the normal fetal conduit that connects the descending aorta and the main pulmonary artery. Functional closure of the ductus is usually complete by the second week of extra-uterine life. Persistent postnatal patency of this conduit beyond the first few weeks of life is referred to as PDA. Anatomical closure of the ductus is usually complete by 2–3 weeks after birth [18]. The closure of some small PDAs may be delayed up to 3 months in full-term infants and 12 months in premature infants, beyond which spontaneous closure is unlikely [19].

- The pulmonary orifice of the ductus is usually at the left of the MPA bifurcation, near the origin of the LPA.
- The aortic orifice is located immediately distal to the origin of the left subclavian artery, on the opposite side of the aortic wall.
- The aortic end (ampulla) is usually larger than the pulmonary end, giving the ductus the shape of a truncated cone or funnel.
- Spontaneous closure of the ductus usually begins at the PA end. A sealed pulmonary orifice with a patent aortic end results in a ductal aneurysm [18].
- Large PDAs are prone to develop pulmonary hypertension (Eisenmengerization) within one year of life.

The **Krichenko classification**, an anatomical classification based on the angiographic appearance of the ductus, categorizes PDA as Type A (conical: the most common type), Type B (window: short and wide), Type C (tubular: long with no constriction), Type D (complex with multiple constrictions), and Type E (elongated with remote constriction) [20]. Additionally, type F ductus (foetal type: large, long, and tortuous) has been described in premature children.

PDA, which occurs in the setting of complex congenital heart disease, has been categorized as [21]:

- **Group A**:
  - Conical ductus from proximal DTA to the dome of MPA
  - Short and straight course
  - Associated with pulmonary atresia with intact IVS, tricuspid atresia, critical PS
- **Group B**:
  - Vertical ductus from the underside of the aortic arch opposite the origin of the LSCA (or its equivalent in a right aortic arch)
  - Commonly inserts onto the proximal part of the LPA in a left aortic arch
  - Associated with pulmonary atresia-VSD, TGA, single ventricle, TOF-pulmonary atresia
- **Group C**:
  - Intermediate origin between A and B
  - Arises from the side of the aortic wall opposite the origin of the LSCA
  - Less vertical take-off than in group B
  - Usually long and tubular with a more tortuous course
- **Group D**:
  - Originates from the subclavian artery on the side of the aortic arch
  - The long tubular ductus joins the PA perpendicularly
  - There is usually a tight constriction at the site of pulmonary insertion
  - Appearance resembles that of a BT shunt

The three major factors that determine the magnitude of shunting are:

- Diameter and length of the ductus, which determine the resistance to the shunt flow
- Pressure gradient between the aorta and PA
- Systemic and pulmonary vascular resistances

4.5.1 Echocardiographic Diagnosis

- The ductus is best demonstrated from the PSAX (or high ductal) view (Fig. 4.13) and the suprasternal window.
- On 2D imaging, the MPA with ductal attachment may appear to 'trifurcate' (into RPA, ductus, and LPA).

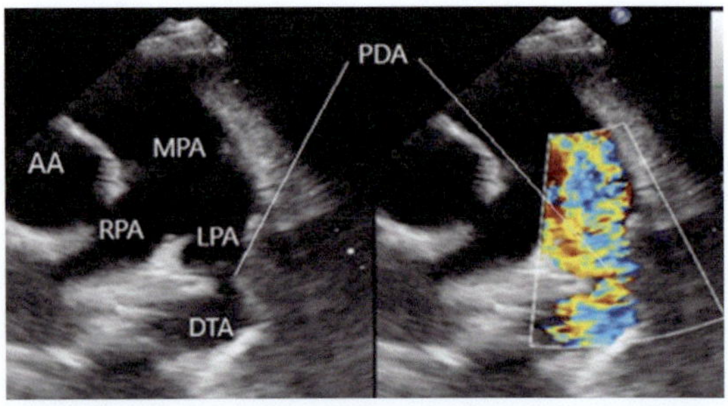

Fig. 4.13 PSAX view showing the turbulent flow through the PDA into the left pulmonary artery (LPA) and main pulmonary artery (MPA). *AA* ascending aorta, *DTA* descending thoracic aorta

- The PA may be dilated, aneurysmal, and calcified.
- Colour Doppler imaging demonstrates the continuous left-to-right shunt flow through the PDA.
- A narrow jet of turbulent flow in the proximal PA may be the first indication of a small PDA.
- The high PSAX view demonstrates the PDA flow as a retrograde mosaic jet entering the MPA (usually along its lateral wall) through the LPA.
- Diastolic flow reversal is observed in the descending aorta, similar to that in aortic regurgitation.
- CWD reveals a continuous high-velocity flow, which peaks in late systole (Fig. 4.14). In the absence of surgical systemic-to-PA shunts, this pattern of CWD tracing is diagnostic of PDA.
- Swirling of blood within the dilated MPA may result in low-velocity retrograde flow in late systole.

Parasternal Windows

- **PSAX view at the level of the aortic valve:**
  - The ductal flow is seen along the left lateral border of the MPA.
- **High left PSAX (ductal) view:**
  - After visualizing the PA bifurcation, slow anticlockwise rotation opens up the DTA so that the entire ductus may be visualized (LPA will disappear from the view) [22].
  - This view visualizes the origin of the left subclavian artery in neonates and infants.
  - The posterior shelf of aortic coarctation can also be visualized from this view.

Suprasternal Windows

- **Suprasternal long-axis view:**
  - This is the ideal view for visualizing the aortic orifice of a vertical ductus in pulmonary atresia [22].
  - Demonstration of the PA orifice may require further anterior tilt due to the tortuous nature of the ductus.
  - This view demonstrates the ductus in its entire length in cases of TGA.
  - The increased diastolic flow in the aortic arch is seen from this view.
- **Suprasternal short-axis view:**
  - This is ideal for visualizing group D PDAs, which arise from the base of the left subclavian artery and descend perpendicularly to insert into the LPA [21, 22].
  - Group D PDA can usually be visualized in its entire length from this view.
- **Modified ductal view** [6, 22]:
  - Anticlockwise rotation of the transducer from the suprasternal long-axis view, followed by a slight anterior tilt, visualizes the ductus from the ampulla to its insertion.
  - The entire length of the ductus can be visualized from this view.
  - The view is similar to a lateral angiogram.

Key point: The echocardiographic detection of PDA relies mainly on the colour Doppler demonstration of abnormal retrograde flow into the pulmonary artery.

Differential Diagnosis

When abnormal flow is detected in the MPA, the other diagnostic possibilities to be considered are:

- Aortopulmonary window
- Coronary AV fistula draining into the PA
- Ruptured sinus of Valsalva aneurysm into the PA
- Anomalous coronary artery origin from the PA (as in ALCAPA)
- Normal helical flow in the MPA:
  - An early systolic retrograde flow (red jet) is often seen along the medial border of the MPA
  - Differentiated from the ductal flow by its location and timing

Key point: The diagnostic features favouring PDA over the other differentials are: (1) origin of the shunt from the descending thoracic aorta and (2) entry of the jet at the site of the PA bifurcation.

PDA with Right-to-Left Shunt

- A large PDA with right-to-left shunt is difficult to identify by colour Doppler, as the direction of flow through the PDA is the same as that in the PA and descending aorta.
- When a right-to-left shunt is suspected across the ductus owing to differential cyanosis or significant pulmonary hypertension [23]:
  - The abdominal aorta should be imaged in the subcostal view after injecting intravenous agitated saline into an upper limb vein.
  - ***Appearance of microbubbles in the descending aorta without opacification of the ascending aorta*** suggests the presence of a right-to-left shunt across the ductus.
- It needs to be remembered that ***right-to-left shunting can occur even in a small PDA*** in the setting of:
  - Suprasystemic pulmonary artery pressure [as in obstructed TAPVC] [18].
  - Duct-dependent systemic circulation.

4.5.2 Dimensions

The modified ductal view is the ideal view for the measurement of dimensions. The various dimensions of the ductus to be measured and documented are:

- Narrowest diameter: Usually at the PA end
- Diameter of the ampulla: At the aortic end
- Length of the duct: Determines the feasibility of transcatheter closure by coil/closure device

The narrowest width of the colour flow across the duct may be used when 2D images are inadequate, although there is a possibility of overestimation of size.

4.5.3 Haemodynamic Significance

The haemodynamic significance of the shunt may be inferred from the echocardiographic features as follows [22]:

- **Small PDA with normal PA pressure**:
  - No enlargement of LA or LV
  - Normal LV function
  - Colour Doppler: Mosaic flow from the descending aorta to the pulmonary artery
  - CWD: Continuous flow towards the transducer with peak in late systole (Fig. 4.14)
- **Moderate to large PDA with normal or mildly elevated PA pressure**:
  - LA/LV enlargement
  - Varying degree of impairment in LV function
  - Colour Doppler: Predominantly laminar flow from the aorta to the PA with minimal aliasing
  - Evidence of significant run-off from the aorta:
    - The presence of holodiastolic flow reversal in the abdominal aorta
    - Increased diastolic flow in the branch PAs

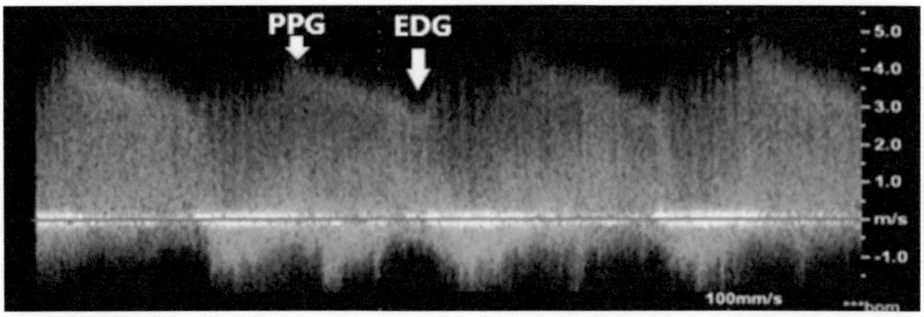

Fig. 4.14 CWD tracing of the flow through ductus arteriosus. Note the continuous high-velocity flow with peak pressure gradient (PPG) in late systole. *EDG* end diastolic gradient

- **Large ductus with significant pulmonary arterial hypertension:**
  - Bidirectional shunting with right-to-left flow in early systole
  - As pulmonary pressure increases:
    Left-to-right shunt is limited to late systole and early diastole
    Duration of right-to-left shunt flow in diastole increases
    Right-to-left shunt begins in late diastole and extends into early systole
    Peak of right-to-left shunt appears earlier in systole

The size of LA and LV may be quantitatively estimated to identify enlargement of these chambers.

- **LA enlargement:**
  - Indicated by an increased LA diameter-to-aortic root (Ao) diameter ratio
  - Diameters of LA and aortic root are obtained from the 2D image in the PLAX view or by M mode
  - Normal LA:Ao ratio in infants is between 0.8 and 1
  - A ratio >1.2 suggests LA enlargement [22]
- The size of the LV is indexed to the infant's body surface area

4.5.4 Pulmonary Arterial Pressure

- The peak pressure gradient across the ductus permits a quantitative estimate of PA systolic pressure (PASP).
- If the ductus is relatively long (>7 mm), the simplified Bernoulli equation may be inaccurate.
- Eisenmenger physiology is indicated by:
  - Features of pulmonary hypertension (flattening of the IVS in systole, RA/RV dilation, right ventricular hypertrophy, and systolic notching of PA Doppler tracing—Chap. 15).
  - Net right-to-left shunting across the ductus.

4.5.5 Associated Defects

- The patency of the ductus may be obligatory in certain cardiac anomalies, collectively known as ***'duct-dependent lesions'***. Duct dependent lesions can be categorized as:
  - **Duct-dependent pulmonary circulation:**
    Pulmonary Atresia with intact ventricular septum
    Pulmonary Atresia with VSD
    Pulmonary Atresia with single ventricle
    Critical pulmonary stenosis
    Severe Ebstein anomaly
    Tricuspid atresia with no or small VSD
  - **Duct-dependent systemic circulation:**
    Interrupted aortic arch
    Hypoplastic Left Heart syndrome
    Critical Aortic stenosis
    Critical coarctation of the aorta
  - **Duct-dependent admixture lesion**: Transposition of the great arteries with intact IVS
- Aortic arch anomalies:
  The sidedness of the arch, branching pattern, and size should be assessed from the suprasternal sagittal view.
- A left duct in a right aortic arch with an aberrant left subclavian artery forms a vascular ring.

4.5.6 Post-PDA Closure Follow-up

- Rule out residual shunts.
- Device closure:
  - The device is visualized between the PA bifurcation and the inferior margin of the DTA.
  - It should not be protruding into either lumen.
- Ductal ligation in premature infants: PDA size prior to ligation directly correlates with the likelihood of post-operative low cardiac output.

Points to Report

- PDA: Location/morphology, direction of the shunt
- Measurements: Narrowest diameter, diameter of the ampulla, length of the duct
- Secondary effects:
  - LV and LA size
  - Pulmonary hypertension—RVSP or mean PA pressure
- Associated lesions (see above)
- LV systolic and diastolic function
- **Post-repair:**
  - Presence of residual leak
  - LV size and function
  - RVSP or mean PA pressure

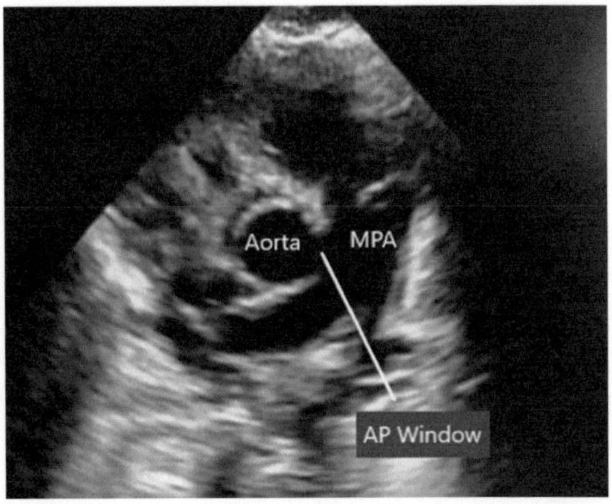

Fig. 4.15 Aortopulmonary window: PSAX view depicting the communication between ascending aorta and main pulmonary artery (MPA)

4.6 Aortopulmonary Window (Aortopulmonary Septal Defect)

Aortopulmonary window or aortopulmonary septal defect is a discrete deficiency in the aortopulmonary septum, resulting in a communication between the ascending aorta and the MPA.

Mori's classification [24]:

Type I: A small defect midway between the semilunar valves and the pulmonary bifurcation (most common type)

Type II: A more distal defect. The pulmonary bifurcation forms the distal border. Associated with an aortic origin of the RPA

Type III: A large, confluent defect involving the entire aortopulmonary septum (most rare type)

4.6.1 Echocardiographic Diagnosis

- PSAX view at the level of the great vessels, subcostal coronal view of the LVOT, and the suprasternal views aid in the visualization of this defect (Fig. 4.15).
- A 'T' artefact at the edges of the defect helps to distinguish it from an artefactual drop-out [25].
- ***Abnormal, continuous, forward flow in the pulmonary arteries*** indicates the presence of an aortopulmonary communication [18].
- Significant ***retrograde diastolic flow is found in proximal aortic arch and abdominal aorta*** due to run-off (In contrast to PDA, where the pre-ductal proximal arch has antegrade diastolic flow).
- Large defects have laminar, low-velocity, bidirectional flow, whereas smaller defects have a continuous high-velocity left-to-right jet [25].
- Left-to-right shunt results in dilated branch PAs, LA, and LV.
- ***Ascending aorta is small in diameter*** in cases of:
  - A very proximal defect.
  - Associated arch anomalies.

4.6.2 Associated Lesions

50% of the cases are associated with other cardiovascular anomalies such as [26]:

- Type A interruption of the aortic arch
- Tetralogy of Fallot
- Aortic origin of RPA
- Anomalous origin of the right or left coronary artery from the pulmonary artery
- Right aortic arch
- Ventricular septal defect
- Pulmonary or aortic atresia, tricuspid atresia
- D-TGA

Key point: Continuous forward flow in the pulmonary arteries and diastolic flow reversal in the proximal aortic arch are the key features suggesting the possibility of an aortopulmonary window.

4.7 Gerbode Defect

Gerbode defect refers to an abnormal communication between the left ventricle and the right atrium (**ventriculo-atrial defect**), and the consequent ventriculo-atrial shunt [27]. This can be congenital or acquired (due to iatrogenic causes, infection, trauma, or ischaemia). These LV-RA shunts can be anatomically categorized as [28, 29]:

- Type 1 (Infravalvular or Indirect) defects result from a combination of VSD with tricuspid regurgitation. These defects result in an indirect LV-RA shunt, as the blood flowing from the LV to the RV through the VSD regurgitates into the RA through the TR jet.
  - VSD in these cases can be one of [30]:
    Anterior defect within the membranous septum.
    Central defect involving both membranous and muscular septa
    Isolated inlet VSD.
  - May be associated with anomalies of the septal tricuspid leaflet, such as a cleft, widened commissural space, perforation, or abnormal chordae [30].
  - Closure of the VSD by adjacent tricuspid valve tissue may result in aneurysmal transformation.
- Type 2 (Supravalvular or Direct) defects are direct communications between the LV and RA through the superior part of the membranous septum (pars atrioventricularis), above the tricuspid valve (Fig. 4.16).
  - Results in direct shunting of blood from the LV to the RA.
  - This is the most common type of Gerbode defect.
- Type 3 (Intermediate) defects comprise a combination of supravalvular and infravalvular defects. This is the least common type of Gerbode defect.

4.7.1 Echocardiographic Diagnosis

The direct type of Gerbode defect is characterized by a high-velocity systolic flow (>4 m/s) through the superior part of the membranous septum towards the RA [31].

- The shunt can be visualized by colour Doppler imaging from parasternal, apical, and subcostal windows (Fig. 4.16).
- CWD demonstrates the high gradient between the high-pressure LV and the low-pressure RA.
- Volume overload from the LV-RA shunt results in dilation of the RA and RV (similar to ASD), which is unusual in a case of isolated VSD.

4.7.2 Differential Diagnosis

1. ***Ruptured sinus of Valsalva aneurysm (RSOV)***:
   (a) Gerbode defect produces a left-to-right shunt, which is predominantly systolic.
   (b) RSOV produces a continuous turbulent flow, spanning across systole and diastole.
2. ***Tricuspid regurgitation:***
   (a) The turbulent jet of a type 2 Gerbode defect may be mistaken for a TR jet. However, a precise identification of the origin of the jet resolves the confusion. The jet in a direct Gerbode defect is from the LV to the RA through the membra-

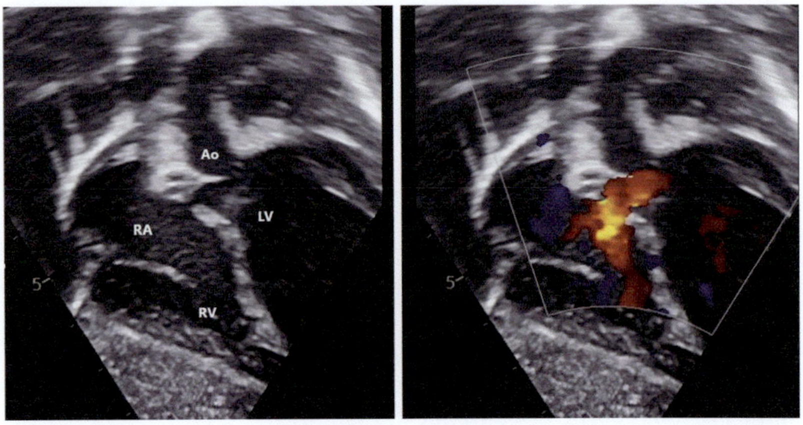

Fig. 4.16 Subcostal coronal view depicting shunting of blood from LV to RA across a defect in the superior part of membranous septum (direct Gerbode defect)

nous septum, whereas a TR jet is from the RV to the RA through the tricuspid valve.

(b) Misinterpreting the high-velocity systolic jet of a Gerbode defect as TR can result in an incorrect overestimation of PA systolic pressure. However, this error can be identified by the discrepancy with the other findings:

PA diastolic and mean pressures estimated from the PR jet will be disproportionately low as compared to the 'high PASP' estimated from the Gerbode jet.

Apart from volume overload of the RA and RV, there is a conspicuous absence of other findings of significant pulmonary hypertension.

4.8 Aorto-Left Ventricular Defect (Tunnel)

Aorto-LV tunnel (ALVT) is a rare vascular connection between the aorta (usually distal to the right sinus of Valsalva) and the left ventricle. Aorto-RV tunnel (ARVT) is rarer.

The aortic opening of the tunnel may be located anywhere above the right or left coronary sinus or the intervening commissure. The tunnel is located anteriorly, just posterior to the RVOT and PAs. It passes through the anterior upper part of the IVS and becomes contiguous with the LVOT just below the aortic annulus. It is unusual for the tunnel to pass through the myocardium to reach the ventricular cavity. It is usually short and can be aneurysmal [32].

Classification [33]:

- Type 1: Slit-like opening at the aortic end with no valve distortion
- Type 2: Large extracardiac aneurysm
- Type 3: Intracardiac aneurysm of the septal portion of the tunnel, with or without RVOT obstruction
- Type 4: Combination of types 2 and 3

4.8.1 Echocardiographic Diagnosis

Echocardiography with colour Doppler reveals the retrograde flow of blood from the aorta into the LVOT. A5C, A3C, and PLAX views aid in visualizing the defect. The location of the aortic orifice of the tunnel, the length of the tunnel, and the location of the ventricular opening of the tunnel need to be ascertained.

The flow of blood from the aorta to the LV is diastolic, with or without concomitant aortic valve regurgitation [34]. The severity of nonvalvular and valvular regurgitation should be estimated using CWD-derived pressure half-time. The aortic root and LV are usually dilated.

There may be posterior deviation of the membranous septum resulting in malalignment VSD and LVOT obstruction [35].

4.8.2 Associated Anomalies [32, 35]

- Aortic valve abnormalities: Bicuspid aortic valve with or without aortic stenosis; aortic valve dysplasia or atresia; aortic cusp perforation and valve regurgitation
- Coronary artery anomalies: Anomalous coronary artery origin from the tunnel; atresia of coronary artery ostia; coronary artery fistula to the tunnel
- RVOT obstruction: Valvular pulmonary stenosis; subvalvular obstruction due to the tunnel

4.8.3 Differential Diagnosis and Distinguishing Features

1. ***Aortic valve regurgitation:***
   (a) The retrograde flow in the Aorto-LV tunnel is not through the aortic valve, distinguishing it from valvular aortic regurgitation
2. ***Coronary artery-LV fistula:***
   (a) Normal undilated coronary arteries can be demonstrated, separate from the vascular tunnel
   (b) Unlike a coronary artery fistula, the tunnel usually does not traverse the myocardium to reach the LV cavity
3. ***Ruptured sinus of Valsalva aneurysm (RSOV):***
   (a) The origin of the tunnel is usually from the sinotubular junction (i.e. distal to the sinus of Valsalva)
   (b) The tunnel runs anterior to the aortic sinuses
   (c) The sinuses of Valsalva are not dilated in ALVT
4. ***Ventricular septal defect:***
   (a) There is no left-to-right shunt in aorta-LV tunnel
   (b) The flow in the aorto-LV tunnel is predominantly diastolic

4.9 Ruptured Aneurysm of Sinus Of Valsalva (RSOV)

Sinus of Valsalva aneurysm results from a localized weakness of the wall of the sinus of Valsalva. This may be congenital or acquired.

- The usual site of the aneurysm is just above the aortic annulus at the leaflet hinge-point.
- The right aortic sinus (65%) is commonly involved, followed by non-coronary sinus (25%), and the left aortic sinus [36].

Sinus of Valsalva aneurysms can rupture into any cardiac chamber, resulting in a fistulous communication between the sinus and the cardiac chamber [36, 37].

- Rupture into the right heart produces a left-to-right shunt
- The two most common sites of rupture are:
  - Right coronary sinus aneurysm into the right ventricle
  - Non-coronary sinus aneurysm into the right atrium
- Rupture into the pericardium is rare

4.9.1 Echocardiographic Diagnosis

RSOV is visualized from the parasternal long- and short-axis views, as a classic windsock deformity projecting from the sinus of Valsalva into the chamber of drainage, with one or more openings near the end of the windsock.

- Rupture into right heart chambers result in turbulent high-velocity continuous flow demonstrable by colour flow Doppler (Fig. 4.17).
- An unruptured Sinus of Valsalva aneurysm must be distinguished from a membranous septal aneurysm by delineating the location from PLAX and A5C views:

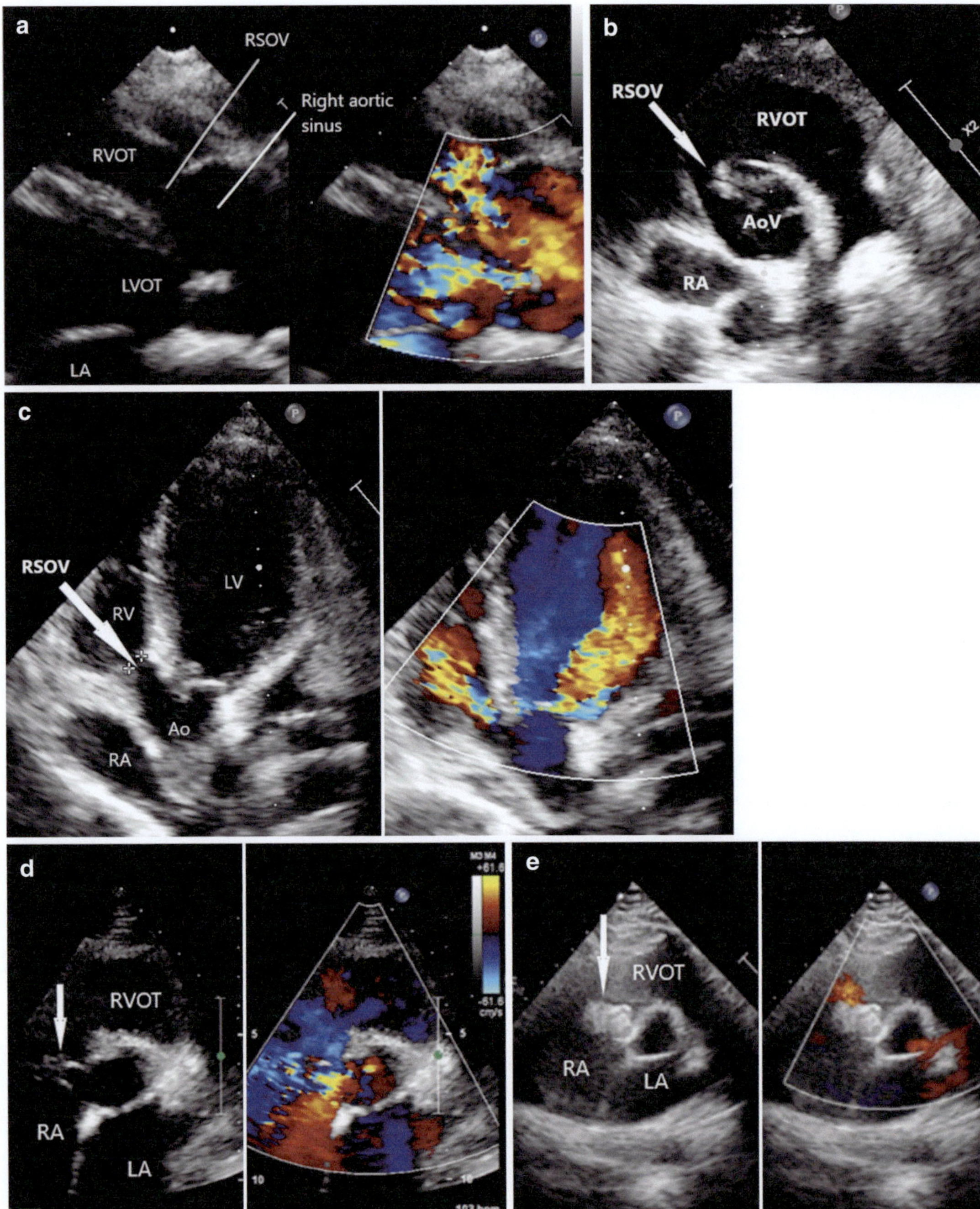

Fig. 4.17 Ruptured right sinus of Valsalva aneurysm into right ventricle. (**a**) PLAX view showing shunting from right aortic sinus to RVOT across ruptured sinus of Valsalva aneurysm. Note that the root of the aneurysm is located above (distal to) the aortic annulus. (**b**) PSAX view depicting the windsock deformity of right aortic sinus. (**c**) Modified A5C view showing turbulent flow through the RSOV aneurysm into the RV. A posteriorly directed eccentric jet AR is also seen. (**d**) RSOV from non-coronary sinus to right atrium just below the tricuspid valve. PSAX view depicting the windsock deformity (arrow). (**e**) PSAX view from the same patient after percutaneous device closure of the defect. The device is seen in situ (arrow) with no residual shunt

- The root of the SOV aneurysm is located above the aortic annulus.
- The membranous septal aneurysm is located below the aortic annulus.

4.9.2 Differential Diagnosis

1. ***Right Coronary AV fistula*** (discussed in Chap. 14):
   (a) Both coronary artery fistulae and RSOV are located above the aortic annulus.
   (b) There is no dilation of the aorta above the sinotubular junction in either condition.
   (c) RSOV is usually saccular, whereas the coronary AV fistula is usually tubular.
   (d) Right coronary AV fistula usually protrudes in a cephalad direction to the right (following the course of RCA), whereas RSOV usually protrudes caudally and to the left.
   (e) The coronary artery can be visualized separately from the shunt flow in RSOV, whereas the flow in the dilated coronary artery is continuous with the shunt flow in a coronary artery fistula.
   (f) Colour Doppler and spectral Doppler imaging demonstrate turbulent continuous flow in the RVOT in RSOV-to-RV, whereas RVOT flow is not turbulent in an isolated coronary AV fistula.
2. ***VSD with aortic regurgitation:***
   (a) VSD jet is located below the plane of the aortic annulus, whereas the RSOV jet is above the annulus.
   (b) Colour and spectral Doppler patterns in the RVOT and LVOT help to distinguish VSD-AR from RSOV:

      RSOV-to-RV demonstrates continuous high-velocity turbulent flow in the RVOT.

      VSD is associated with turbulent systolic flow in the RVOT, and AR is associated with diastolic flow in the LVOT.

4.9.3 Associated Anomalies [36]

- Ventricular septal defect:
  - Commonly associated with aneurysms of the right aortic sinus
  - Outlet VSDs are common
  - Subpulmonic VSD is associated with prolapse of the aortic valve cusp and progressive aortic regurgitation
- Coarctation of aorta, bicuspid aortic valve
- ASD, PDA
- Tetralogy of Fallot

4.9.4 Complications

- Mechanical effects: RVOT obstruction; compression of the left coronary artery
- Aortic regurgitation
- Infective endocarditis: Commoner in smaller fistulae

References

1. Sachdeva R. Atrial septal defects. In: Allen HD, Driscoll DJ, Shaddy RE, Feltes TF, editors. Moss and Adams' heart disease in infants, children, and adolescents. 8th ed. Philadelphia: Lippincott Williams & Wilkins; 2013. p. 672–90.
2. Silvestry FE, Cohen MS, Armsby LB, Burkule NJ, Fleishman CE, Hijazi ZM, et al. Guidelines for the echocardiographic assessment of atrial septal defect and patent foramen ovale: from the American Society of Echocardiography and Society for Cardiac Angiography and Interventions. J Am Soc Echocardiogr. 2015;28(8):910–58.
3. Snarr BS, Liu MY, Zuckerberg JC, Falkensammer CB, Nadaraj S, Burstein D, et al. The parasternal short-axis view improves diagnostic accuracy for inferior sinus venosus type of atrial septal defects by transthoracic echocardiography. J Am Soc Echocardiogr. 2017;30(3):209–15.
4. Perloff JK, Marelli AJ. Atrial septal defect: simple and complex. In: Clinical recognition of congenital heart disease. 6th ed. Philadephia: Saunders; 2012. p. 212–72.

5. Raghib G, Ruttenberg HD, Anderson RC, Amplatz K, Adams P, Edwards JE. Termination of left superior vena cava in left atrium, atrial septal defect, and absence of coronary sinus: a developmental complex. Circulation. 1965;31:906–18.
6. Awasthy N, Radhakrishnan S. Stepwise evaluation of left to right shunts by echocardiography. Indian Heart J. 2013;65(2):201–18.
7. Perloff JK, Marelli AJ. Patent ductus arteriosus aortopulmonary window. In: Clinical recognition of congenital heart disease. 6th ed. Philadephia: Saunders; 2012. p. 368–93.
8. Krishnan SC, Salazar M. Septal pouch in the left atrium: a new anatomical entity with potential for embolic complications. JACC Cardiovasc Interv. 2010;3:98–104.
9. Diener HC, Akagi T, Durongpisitkul K, Thomson VS, Prabhakar AT, Sharpe R, Albers B, Lewalter T, Oki K, Sharma VK. Closure of the patent foramen ovale in patients with embolic stroke of undetermined source: a clinical expert opinion and consensus statement for the Asian-Pacific region. Int J Stroke. 2020;15(9):937–44.
10. Homma S, Sacco RL. Patent foramen ovale and stroke. Circulation. 2005;112:1063–72.
11. Olivares-Reyes A, Chan S, Lazar EJ, Bandlamudi K, Narla V, Ong K. Atrial septal aneurysm: a new classification in two hundred five adults. J Am Soc Echocardiogr. 1997;10:644–56.
12. Lopez L, Houyel L, Colan SD, Anderson RH, Béland MJ, Aiello VD, et al. Classification of ventricular septal defects for the eleventh iteration of the international classification of diseases-striving for consensus: a report from the International Society for Nomenclature of Paediatric and Congenital Heart Disease. Ann Thorac Surg. 2018;106(5):1578–89.
13. Anderson RH. Ventricular septal defects. In: Wilcox BR, Anderson RH, editors. Surgical anatomy of the heart. New York: Raven Press; 1985. p. 6.5–6.10.
14. Van Praagh R, Geva T, Kreutzer J. Ventricular septal defects: how shall we describe, name and classify them? J Am Coll Cardiol. 1989;14(5):1298–9.
15. Rubio AE, Lewin MB. Ventricular septal defects. In: Allen HD, Driscoll DJ, Shaddy RE, Feltes TF, editors. Moss and Adams' heart disease in infants, children, and adolescents. 8th ed. Philadelphia: Lippincott Williams & Wilkins; 2013. p. 713–21.
16. Butera G, Carminati M, Chessa M, et al. Transcatheter closure of perimembranous ventricular septal defects: early and long-term results. J Am Coll Cardiol. 2007;50:1189–95.
17. Eapen RS, Lemler MS, Scott WA, Ramaciotti C. Echocardiographic characteristics of perimembranous ventricular septal defects associated with aortic regurgitation. J Am Soc Echocardiogr. 2003;16:209–13.
18. Moore P, Brook MM. Patent ductus arteriosus and aortopulmonary window. In: Allen HD, Driscoll DJ, Shaddy RE, Feltes TF, editors. Moss and Adams' heart disease in infants, children, and adolescents. 8th ed. Philadelphia: Lippincott Williams & Wilkins; 2013. p. 672–90.
19. Gittenberger-de Groot AC, Van Ertbruggen I, Moulaert AJ, et al. The ductus arteriosus in the preterm infant: histologic and clinical observations. J Pediatr. 1980;96:88–93.
20. Krichenko A, Benson LN, Burrows P, et al. Angiographic classifi cation of the isolated, persistently patent ductus arteriosus and implications for percutaneous catheter occlusion. Am J Cardiol. 1989;63:877–80.
21. Alwi M. Stenting the ductus arteriosus: case selection, technique and possible complications. Ann Pediatr Cardiol. 2008;1(1):38–45.
22. Silverman NH. Patent ductus arteriosus. In: Pediatric echocardiography. Baltimore: Lippincott Williams & Wilkins; 1993. p. 167–77.
23. Sohn DW, Kim YJ, Zo JH, et al. The value of contrast echocardiography in the diagnosis of patent ductus arteriosus with Eisenmenger's syndrome. J Am Soc Echocardiogr. 2001;14:57–9.
24. Mori K, Ando M, Takao A, et al. Distal type of aortopulmonary window: report of 4 cases. Br Heart J. 1978;40:681–9.
25. Balaji S, Burch M, Sullivan ID. Accuracy of cross-sectional echocardiography in diagnosis of aortopulmonary window. Am J Cardiol. 1991;67:650–3.
26. Bertolini A, Dalmonte P, Bava GL, et al. Aortopulmonary septal defects. A review of the literature and report of ten cases. J Cardiovasc Surg. 1994;35:207–13.
27. Gerbode F, Hultgren H, Melrose D, Osborn J. Syndrome of left ventricular–right atrial shunt: successful surgical repair of defect in five cases, with observation of bradycardia on closure. Ann Surg. 1958;148:433–46.
28. Riemenschneider TA, Moss AJ. Left ventricular-right atrial communication. Am J Cardiol. 1967;19:710–8.
29. Sakakibara S, Konno S. Congenital aneurysm of the sinus of Valsalva anatomy and classification. Am Heart J. 1962;63:405–24.
30. Saker E, Bahri GN, Montalbano MJ, Johal J, Graham RA, Tardieu GG, et al. Gerbode defect: a comprehensive review of its history, anatomy, embryology, pathophysiology, diagnosis, and treatment. J Saudi Heart Assoc. 2017;29(4):283–92.
31. Silbiger JJ, Kamran M, Handwerker S, Kumar N, Marcali M. The Gerbode defect: left ventricular to right atrial communication—anatomic, hemodynamic, and echocardiographic features. Echocardiography. 2009;26:993–8.
32. Levy MJ, Lillehei CW, Anderson RC. Aortico-left ventricular tunnel. Circulation. 1963;27:841–53.
33. Hovaguimian H, Cobanoglu A, Starr A. Aortico-left ventricular tunnel: a clinical review and new surgical classification. Ann Thorac Surg. 1988;45(1):106–12.
34. Fripp RR, Werner JC, Whitman V, et al. Pulsed Doppler and two-dimensional echocardiographic findings in aortico-left ventricular tunnel. J Am Coll Cardiol. 1984;4:1012–4.

35. Serino W, Andrade JL, Ross D, et al. Aorto-left ventricular communication after closure. Late postoperative problems. Br Heart J. 1983;49:501–6.
36. Lim DS, Mathern GP. Congenital anomalies of the coronary vessels and the aortic root. In: Allen HD, Driscoll DJ, Shaddy RE, Feltes TF, editors. Moss and Adams' heart disease in infants, children, and adolescents. 8th ed. Philadelphia: Lippincott Williams & Wilkins; 2013. p. 746–57.
37. Chu SH, Hung CR, How SS, Chang H, Wang SS, Tsai CH, Liau CS, Tseng CD, Tseng YZ, Lee YT, et al. Ruptured aneurysms of the sinus of valsalva in oriental patients. J Thorac Cardiovasc Surg. 1990;99(2):288–98.

5 Atrioventricular Septal Defect

Atrioventricular septal defects (AVSD) encompass a spectrum of anomalies characterized by a defective atrioventricular (AV) septum and anomalous atrioventricular (AV) valves. AVSD is also referred to as AV canal defect (AVCD) and Endocardial cushion defect (ECD).

Classification [1]

- **Partial AVSD**:
  - Characterized by ostium primum ASD without an inlet VSD.
  - The right and left AV valve annuli are distinct but focally contiguous.
  - The left AV valve invariably has a cleft (zone of apposition).
- **Transitional AVSD**:
  - A subtype of partial AVSD with a small inlet VSD.
  - The VSD is partially occluded by dense chordal attachments to the ventricular septum [1].
- **Complete AVSD**:
  - Ostium primum ASD is accompanied by a significant inlet VSD.
  - There is a common AV valve characterized by a single annulus and a single orifice.
- **Intermediate AVSD**:
  - A subtype of complete AVSD.
  - A tongue of tissue divides the common AV valve into distinct right and left orifices.
  - The AV valve has one common annulus and two orifices.
  - The inlet VSD is large (similar to other forms of complete AVSD).

Key point: Complete AVSDs demonstrate the physiology of both VSDs and ASDs; Partial AVSDs have the physiology of ASDs alone.

The following features are common to the entire spectrum of AVSD [1, 2]:

- The AV septum is absent or deficient; ostium primum ASD is invariably seen.
- Although a part of the inlet ventricular septum is always deficient, the defect may be sealed by valvular tissue in partial AVSD.
- The AV valves are displaced inferiorly, and the septal attachments of the valve leaflets are at the same level.
- Apical displacement of the left anterior AV valve leaflet results in a cleft (zone of apposition) in the left component of the AV valve; this cleft is directed towards the ventricular septum [3].
- The aortic valve is anteriorly displaced and is not wedged between the AV valve annuli.
- The LVOT is elongated (Gooseneck deformity): The distance from the left AV valve annulus to the left ventricular apex is less than

Supplementary Information The online version contains supplementary material available at https://doi.org/10.1007/978-981-96-0243-8_5.

A. Paul, *Manual of Echocardiography for Congenital Heart Diseases*,
https://doi.org/10.1007/978-981-96-0243-8_5

the distance from the aortic annulus to the apex.
- The LV papillary muscles are rotated in a counterclockwise fashion.
- AV valve abnormalities and LVOT obstruction are more frequent in two-orifice AV connections.

To summarize the confusing terminologies of the various types of AVSD:

- **Partial AVSD**: Two AV valve annuli + Two AV valve orifices + No inlet VSD
- **Transitional AVSD**: Two AV valve annuli + Two AV valve orifices + Restrictive inlet VSD
- **Intermediate AVSD**: Single AV valve annulus + Two AV valve orifices + Non-restrictive inlet VSD
- **Complete AVSD**: Single AV valve annulus + Single AV valve orifice + Non-restrictive inlet VSD

Key point: Although AV valve annular morphology may not be discernible with 2D echocardiography, the classification of the type of AVSD is possible on the basis of the number of AV valve orifices and the nature of the VSD.

5.1 Partial AVSD

The hallmark features of partial AVSD are (1) ostium primum ASD with no interventricular communication, and (2) two AV valve orifices with separate annuli.

5.1.1 Echocardiographic Diagnosis (Figs. 5.1 and 5.2)

- The internal cardiac crux and the ostium primum ASD can be visualized from the A4C view [4].
- Apical and subcostal four-chamber views allow reliable estimation of the size of the OP-ASD.
- The deficient inlet ventricular septum is sealed by valvular tissue and chords, attached to the crest of the ventricular septum [1].

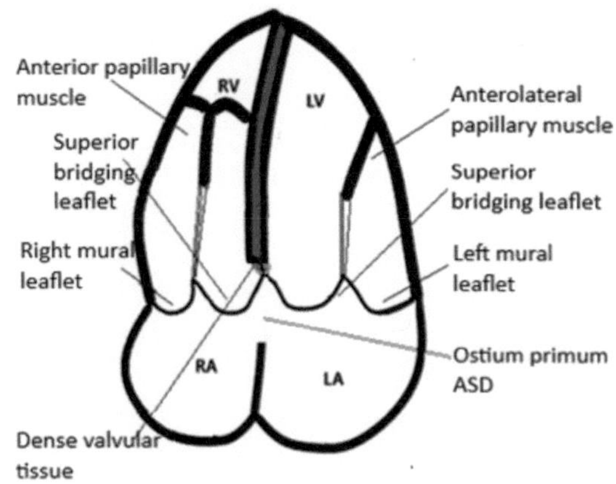

Fig. 5.1 Partial AVSD as seen from the apical view

- **Transitional form of partial AVSD** (Fig. 5.2e):
  - There is aneurysm formation at the site of the sealed inlet VSD, referred to as a '***tricuspid pouch***' [5].
  - Small shunts may occur through the tricuspid pouch, although the dense chordal attachments obstruct this flow to a great extent.

5.1.2 AV Valves

- The left AV valve is displaced towards the apex, bringing both the AV valve annuli at the same level, and effecting a focal contiguity between the annuli at the internal cardiac crux [1, 2].
- There is a 'common hinge point' of the common AV valve on the crest of the ventricular septum.
- The two separate AV valve orifices are equidistant from the cardiac apex.

Cleft AV valve: The cleft (zone of apposition) in the anterior leaflet of the left AV valve in partial AVSD is the commissure between the left superior (anterior) and left inferior (posterior) bridging leaflets [2].

- Best visualized from the parasternal and subcostal short-axis windows.
- The cleft in the anterior leaflet of the left AV valve is directed towards the ventricular sep-

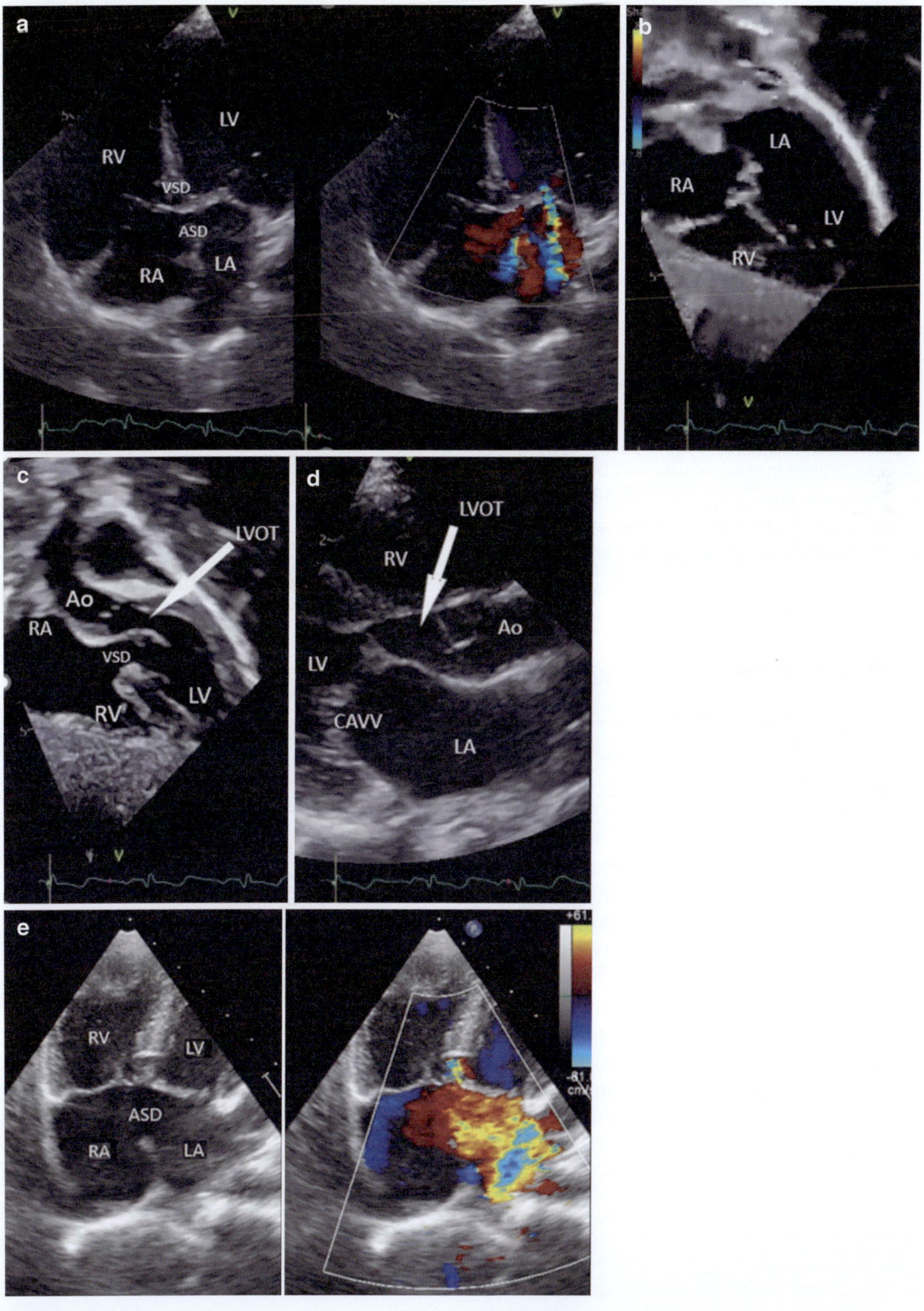

Fig. 5.2 Partial AVSD. (**a**) A4C view depicting partial AVSD with common AV valve regurgitation on colour Doppler. (**b**) Subcostal four-chamber view confirms the absence of inlet VSD. (**c, d**) Elongated LVOT (gooseneck deformity) is depicted from the subcostal LVOT view (**c**) and the PLAX view (**d**). (**e**) Transitional form of partial AVSD as seen from A4C view. Note the large OP-ASD and the turbulence at the site of restrictive inlet VSD

tum, along the anteroinferior rim of the septal defect [5].
- The left AV valve orifice appears triangular (as opposed to the normal elliptical orifice) owing to the cleft.
- The cleft left AV valve gradually becomes thickened to resemble a myxomatous MV.

AV valve regurgitation:

- Multiple jets of regurgitation may be present in an incompetent cleft AV valve.
- Closure of the ASD and repair of the cleft resolve LV-to-RA left AV valve regurgitation [1].
- LV-to-LA regurgitation usually points towards other intrinsic abnormalities of the valve [1].
- The regurgitant jets are visualized from apical, parasternal, and subcostal long-axis views.
- The degree of preoperative left AV valve regurgitation determines the degree of residual post-operative left AV valve regurgitation.

AV valve stenosis:

- ***Double-orifice left AV valve***:
  - A tongue of tissue may divide the left AV valve into two orifices [6].
  - The combined effective valve area of a double-orifice valve is less than the valve area of a single-orifice valve.
  - Post-operative stenosis is common.
  - Subcostal and parasternal short-axis views demonstrate the double-orifice valve.
- Parachute left AV valve is a rare cause of AV valve stenosis (discussed in Chap. 8).

Key point: The utility of spectral Doppler in evaluating the left AV valve stenosis in AVSD is limited because the ostium primum ASD will decompress the left atrium.

5.1.3 LVOT Obstruction

In the normal heart, the aortic valve is wedged between the mitral and tricuspid annuli. In AVSD, the aortic valve is displaced anteriorly, resulting in an elongated LVOT (Gooseneck deformity), which makes it prone to progressive subaortic obstruction [7]. The attachment of the superior bridging leaflet to the crest of the ventricular septum further elongates and narrows the LVOT. LVOTO is more frequent when two distinct AV valve orifices are present [1].

- The subaortic obstruction may be exacerbated by [2]:
  - Discrete subaortic fibromuscular ridges
  - Septal hypertrophy
  - Abnormal left AV valve chordal attachments
  - Abnormally oriented papillary muscles
- Progressive LVOT obstruction may develop after the initial repair of AVSD [1]

5.1.4 Associated Anomalies [2]

- Other shunt lesions: Ostium secundum ASD, Patent ductus arteriosus, Membranous VSD
- Persistent LSVC draining into the coronary sinus
- Pulmonary stenosis, tricuspid stenosis or atresia, hypoplastic right ventricle
- Pulmonary venous anomalies, Cor triatriatum
- Coarctation of the aorta, Hypoplastic left ventricle
- Tetralogy of Fallot, DORV, and pulmonary atresia (less common in partial AVSD)

5.2 Partial AVSD: Special Forms

5.2.1 Common Atrium

Common atrium is characterized by complete or near-total absence of the atrial septum [8].

- Common atrium commonly occurs in the setting of syndromes such as Ellis-van Creveld syndrome, asplenia, and polysplenia [9]
- Associated with TGA, DORV, anomalous pulmonary venous connections, and univentricular AV connection [2]

- Subcostal and apical four-chamber views aid in the diagnosis
- Care should be taken not to misinterpret a muscle bundle/band coursing through the atrium as the atrial septum

Key point: Common atrium, in the presence of two ventricles, is always associated with partial or complete AVSD.

5.2.2 Double Outlet Right Atrium

Double outlet right atrium (DORA) refers to the condition in which the RA drains into both ventricles [10].

- Characterized by malalignment of the atrial septum to the left of the AV junction (ventriculo-atrial malalignment) in the setting of a large ostium primum ASD [1]
- Results from subnormal leftward morphogenetic movement of the ventricular septum
- Always associated with AVSD—more commonly a partial AVSD
- The extreme form of this condition is associated with atresia of the left AV orifice
- Excessive leftward morphogenetic movement of the LV and IVS results in double outlet left atrium [1]
- Subcostal and apical four-chamber views aid in the diagnosis

5.3 Complete AVSD

Complete AVSD is characterized by (1) a large septal defect with interatrial and interventricular components, and (2) a common AV valve (single annulus and single orifice) that spans the entire septal defect.

The **common AV valve has five leaflets** [1, 2] (Fig. 5.3):

- Superior (anterior) bridging leaflet corresponds to the superior half of the anterior mitral leaflet
- Inferior (posterior) bridging leaflet, which overlies the inlet VSD, represents fusion of the septal tricuspid leaflet and the inferior half of the anterior mitral leaflet
- Two lateral leaflets—right (inferior) mural leaflet and left mural leaflet—correspond to the posterior tricuspid and posterior mitral leaflets, respectively
- Right anterior (anterosuperior) leaflet corresponds to the normal anterior tricuspid leaflet

The leaflets are prone to develop progressive regurgitation and myxomatous changes. The superior bridging leaflet may be completely or partially divided into right and left components (Rastelli types A and B, respectively—see below).

There are **five papillary muscles** beneath the five commissures:

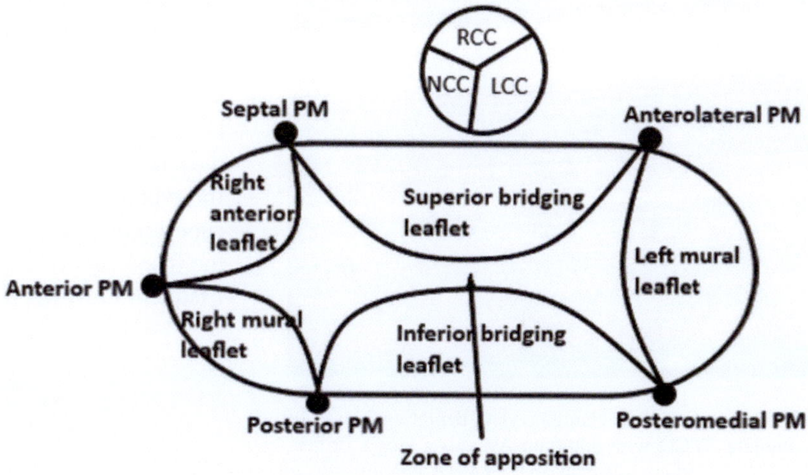

Fig. 5.3 Schematic diagram depicting the leaflets of the common AV valve, the commissures, and the corresponding papillary muscles (PM). The zone of apposition is shown between the superior and inferior bridging leaflets

- Anterolateral papillary muscle: corresponds to the superior bridging-left mural commissure
- Posteromedial papillary muscle: corresponds to the inferior bridging-left mural commissure
- Septal papillary muscle: corresponds to the superior bridging-right anterior commissure
- Anterior papillary muscle: corresponds to the right anterior-right mural commissure
- Posterior papillary muscle: corresponds to the inferior bridging-right mural commissure

Abnormal papillary muscle arrangement may contribute to progressive LVOT obstruction.

5.3.1 Echocardiographic Diagnosis

5.3.1.1 Septal Defects

Complete AVSD comprises of ostium primum ASD and inlet VSD. Apical and subcostal four-chamber views aid in visualizing the internal cardiac crux and the septal defects. The size and extent of the septal defects can be evaluated from these views (Fig. 5.4).

5.3.1.2 Common AV Valve

- The subcostal short axis *(en face)* view aids in determining AV valve morphology (Fig. 5.5)
  - This view is obtained by counterclockwise rotation from the subcostal four-chamber view until the AV valve leaflets appear *en face* [1].
  - Superior and inferior angulation of the probe facilitates cross-sectional inspection of all five valve leaflets.
- The following points need to be assessed in detail:
  - Commitment of the AV valve orifice to each ventricle (subcostal sagittal or A4C views—Fig. 5.2).
  - Degree of common AV valve regurgitation.
- A single left ventricular papillary muscle may occur in complete AVSD, reducing the effective valve orifice area (see 'Parachute mitral valve' in Chap. 8) [1].
- Relative leaflet hypoplasia may complicate AV valve repair.
- In a complete AVSD, when the common AV valve is divided into right and left orifices by a tongue of tissue that connects the two bridging leaflets, it is called an **intermediate AVSD**.

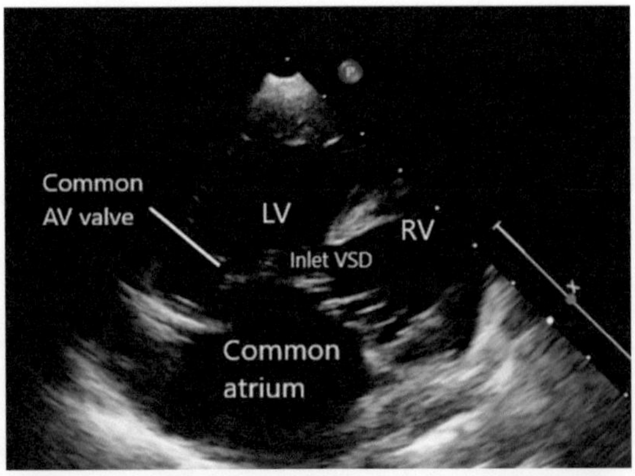

Fig. 5.4 Apical four-chamber view depicting a case of complete AVCD with common AV valve, single orifice, inlet VSD, and complete absence of atrial septum

5.3.2 Rastelli Classification

The variable anatomic relationship between the superior bridging leaflet and the ventricular septum forms the basis for the classification of complete AVSD described by Rastelli et al. [11]

Type A (most common):

- The superior bridging leaflet is completely divided at the level of the septum, and both the components are attached by chordae to the crest of the ventricular septum (Fig. 5.6).
- Multiple insertions of the chordae exist—to the septum or to a distinct medial papillary muscle.
- Interventricular communication beneath the superior bridging leaflet may be obliterated by the fused chordae.
- Frequent in Down syndrome.

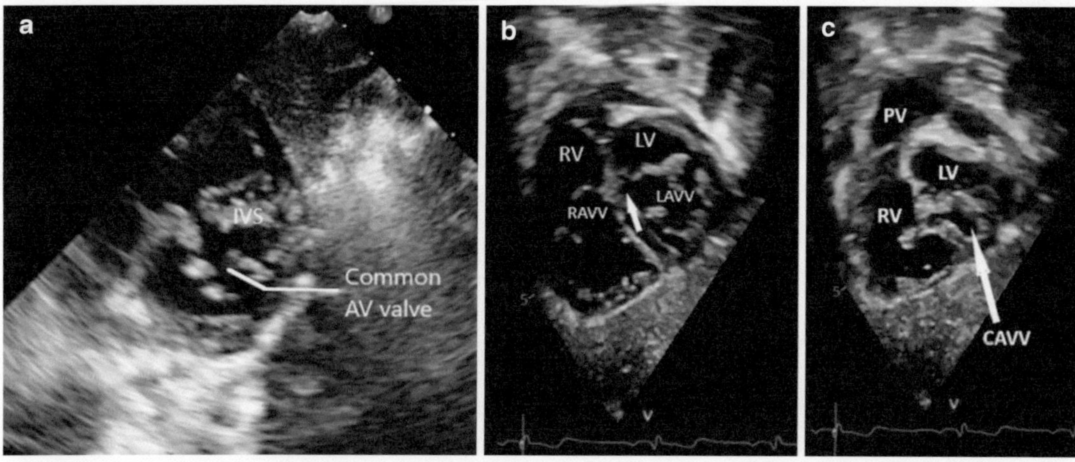

Fig. 5.5 (**a**) Subcostal en face view of the common AV valve in complete AVSD. (**b**) Intermediate AVSD—the common AV valve is divided into right AV valve (RAVV) and left AV valve (LAVV) orifices by a tongue of tissue that connects the superior and inferior bridging leaflets. The zone of apposition is denoted by an arrow. (**c**) Common AV valve (CAVV) is seen in complete AVSD from the subcostal RVOT view

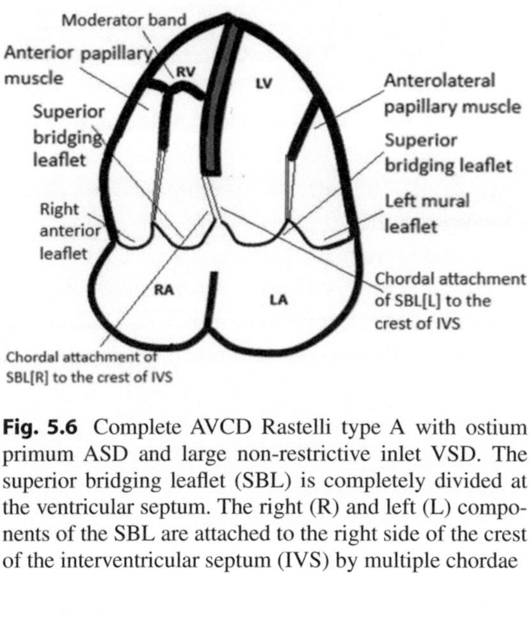

Fig. 5.6 Complete AVCD Rastelli type A with ostium primum ASD and large non-restrictive inlet VSD. The superior bridging leaflet (SBL) is completely divided at the ventricular septum. The right (R) and left (L) components of the SBL are attached to the right side of the crest of the interventricular septum (IVS) by multiple chordae

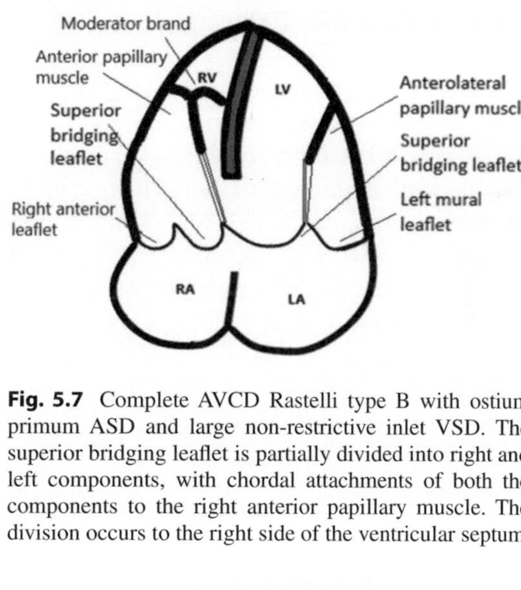

Fig. 5.7 Complete AVCD Rastelli type B with ostium primum ASD and large non-restrictive inlet VSD. The superior bridging leaflet is partially divided into right and left components, with chordal attachments of both the components to the right anterior papillary muscle. The division occurs to the right side of the ventricular septum

Type B (least common):

- The superior bridging leaflet is partially divided above and to the right of the ventricular septum (Fig. 5.7).
- Chordae attach to the RV papillary muscle on the septal surface or the moderator band.
- There is no direct chordal attachment between the superior bridging leaflet and the septum.
- Free interventricular communication exists beneath the leaflet.

Type C ('free floating'):

- The superior bridging leaflet is neither divided nor attached to the septum (Fig. 5.8)
- Chordae attach to the anterior papillary muscle located on the RV free wall
- Free interventricular communication exists beneath the leaflet
- Associated with complex anomalies such as tetralogy of Fallot (common in Down syndrome), DORV (common in asplenia), TGA, and heterotaxy syndromes

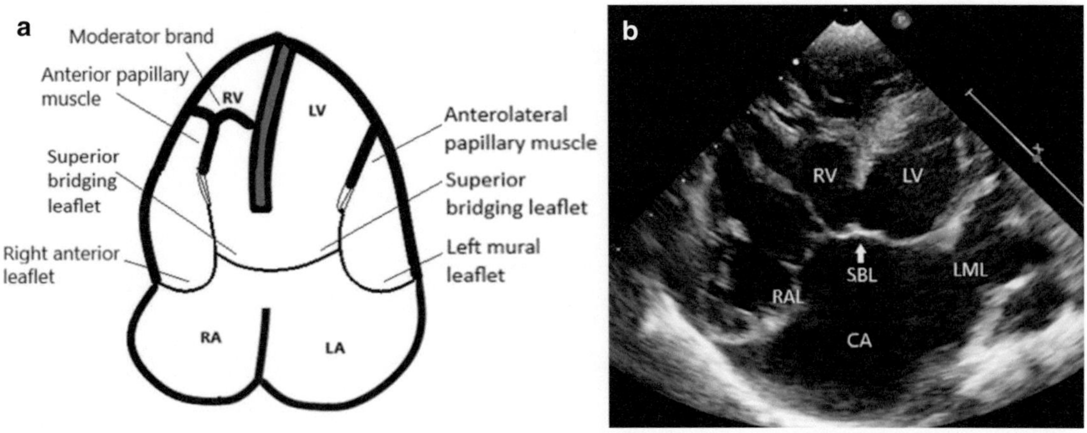

Fig. 5.8 (**a**) Complete AVCD Rastelli type C with ostium primum ASD and a large non-restrictive inlet VSD. The superior bridging leaflet is undivided and unattached to the septum ('free floating'). Laterally, the superior bridging leaflet attaches to the anterior papillary muscles of both ventricles. (**b**) Complete AVCD Rastelli type C with common atrium (CA), large inlet VSD, and floating superior bridging leaflet (SBL) with lateral chordal attachments to the anterior papillary muscles of each ventricle. *RAL* right anterior leaflet, *LML* left mural leaflet

5.3.3 Ventricular Balance

In AVSD, preferential commitment of the common AV valve to one ventricle compromises the flow into the other ventricle, resulting in relative hypoplasia of the latter. This phenomenon of 'ventricular imbalance' can occur in complete as well as partial AVSDs. If 75% or more of the annulus of a common AV valve is committed to one ventricular chamber owing to partial atresia or stenosis of a part of the common AV valve, it is considered to be a univentricular AV connection (common inlet ventricle).

Balanced AVSD (90%) refers to an AVSD in which the AV inlet is equally shared by both the ventricles.

Unbalanced AVSD (10%) [1]:

- One ventricle is typically hypoplastic, although the common AV valve is its usual size.
- The larger ventricle is termed the 'dominant' ventricle.
- Two-thirds of unbalanced AVSDs are RV dominant.
  - RV dominant AVSD: Associated with hypoplastic LV, severe coarctation of the aorta, and aortic arch anomalies.
  - LV dominant AVSD: Associated with hypoplastic RV and pulmonary valve stenosis or atresia, common in Down syndrome.

Determination of 'balance' with echocardiographic imaging is vital in deciding the surgical strategy. Biventricular repair is not possible in grossly unbalanced AVSD. When RV hypoplasia is modest, a so-called '1.5-ventricle' repair is performed by repairing the intracardiac shunts and unloading the RV by performing a bidirectional Glenn anastomosis [1].

Echocardiographic Assessment of 'Ventricular Balance'

Traditionally, the diagnosis of unbalanced AVSD was primarily based on ventricular size. However, 'imbalance' can be present even if the contralateral ventricle is not very small. On the other hand, left-to-right shunting of blood and RV volume overload may result in bowing of the ventricular septum towards the left side, creating a false impression of LV hypoplasia. Subsequent studies have focused on the derangement in the distribution of blood flow into the affected ventricle to determine balance.

- A hypoplastic appearance of the LV need not necessarily correlate with the severity of valve malalignment [1]. It may be the result of:
  - Underfilling due to preferential shunting of pulmonary venous blood across the ASD
  - Bowing of the IVS to the left due to RV volume overload owing to large left-to-right shunt
- Whether LV is apex-forming needs to be determined

Key point: Malalignment between the atrial and ventricular septa in the A4C view can be a clue to an unbalanced AVSD.

The various indices proposed to determine the feasibility of biventricular repair are:

1. **AV valve index (AVVI)** [12]
   (a) Measured from the subcostal sagittal (LAO) view:
   The circumference of the common AV valve orifice is traced during end diastole.
   A line is drawn from the tip of the infundibular septum to the crest of the muscular septum, thus dividing the common AV valve into left and right components (Fig. 5.9).
   AVVI is expressed as the ratio of the area of the smaller AV valve component of the divided common AV valve over the larger component.

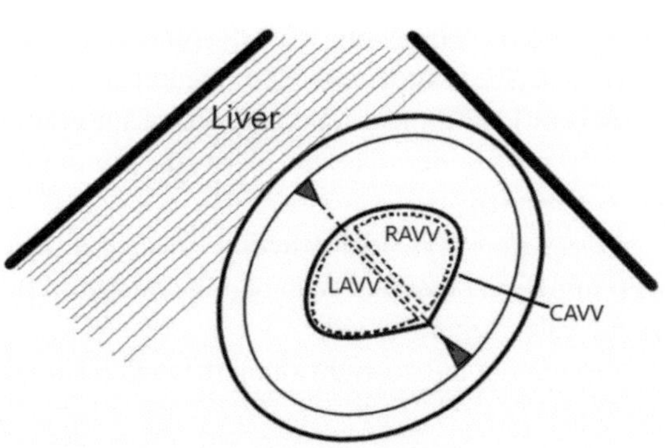

Fig. 5.9 Schematic diagram showing the calculation of AVVI from the subcostal LAO view

   (b) In an RV-dominant AVSD:
   AVVI is calculated as the **left component AVV area/right component AVV area**.
   This serves as an LV/RV area ratio.
   A balanced AVSD generally has an AVVI >0.67.
   An AVVI <0.67, in the presence of a large VSD, favours univentricular palliation.
2. **Modified AV valve index** [13]**:**
   (a) Modified AVVI is calculated as the ***left component AVV area/total AVV area***.
   (b) A value <0.4 implies an RV-dominant AVSD, and a value >0.6 defines an LV-dominant AVSD.
   (c) Modified AVVI value between 0.4 and 0.6 is safe for biventricular repair.
   (d) A value <0.19 warrants univentricular palliation, with variable outcomes for values in between.
3. **LV inflow index** [14]**:**
   (a) Obtained from A4C view using 2D imaging and colour Doppler.
   (b) AV valve annulus diameter is obtained by 2D imaging.
   (c) LV inflow diameter is measured at the level of papillary muscles using colour Doppler as secondary colour inflow diameter.
   (d) LV inflow index is calculated as: ***LV colour inflow diameter/2D AVV annulus diameter***.
   (e) LV inflow index <0.55 favours single ventricle palliation over biventricular repair.
4. **Ventricular cavity ratio (VCR)** [12]**:**
   (a) Obtained from A4C view.
   (b) Ventricular length is measured from the AV valve annulus to the apex for each ventricle.
   (c) Ventricular width is measured from the crest of the ventricular septum to the free wall of each ventricle.
   (d) VCR is calculated as: ***LV length × LV width/RV length × RV width***.

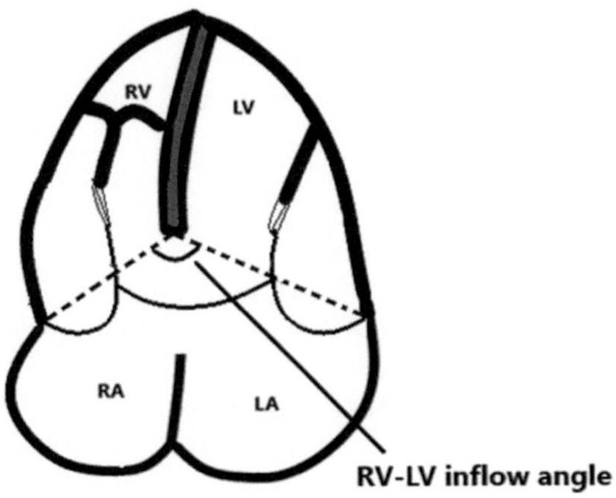

Fig. 5.10 Schematic diagram showing the measurement of RV/LV inflow angle from the A4C view

   (e) Value <0.46 is considered a risk factor for biventricular repair in RV-dominant AVSD.
5. **RV/LV inflow angle** [15]**:**
   (a) An indirect measure of left AV valve inflow.
   (b) In the A4C view, the angle of RV inflow to LV inflow is measured from the crest of the ventricular septum to each AV valve hinge point. The angle is derived at the crest of the ventricular septum (Fig. 5.10).
   (c) A steeper angulation between the RV and LV inflows favours RV-dominant unbalanced AVSD.
   (d) RV/LV inflow angle during systole ≤114° was found to correlate with the need for univentricular palliation.

5.3.4 Associated Anomalies

- Membranous VSD: The inlet VSD extends to the level of the membranous septum
- Ostium secundum ASD
- Patent ductus arteriosus
- Down's syndrome occurs in >50% of patients with complete AVSD and may be associated with tetralogy of Fallot
- LVOT obstruction is common in Rastelli type A complete AVSD; aortic coarctation needs to be ruled out
- Specific anomalies associated with Rastelli types A, B, and C have been described

Key point: Small PDAs are difficult to be appreciated due to aliasing of colour Doppler signals in the branch pulmonary arteries (owing to the increased pulmonary flow from AVSD).

5.3.5 Post-operative Assessment

Evaluation of a patient, who has undergone repair for AVSD, should focus on:

- Type of surgery performed
- Residual shunts
- Morphology of the AV valves
- Right and left AV valve stenosis or regurgitation
- LVOT obstruction
- LV function and dimensions
- Evidence of pulmonary hypertension:
  - Right ventricular systolic pressure: Calculated from the gradients across the VSD or right AV valve regurgitation (Residual VSD jet may alter the gradient across right AV valve regurgitation)
  - Ventricular septal flattening or bowing; RV size and function
  - PR gradient to assess pulmonary artery diastolic pressure

Echocardiographic parameters that may warrant reoperation in adults with previously repaired AVSD [1, 16]:

Significant regurgitation or stenosis of the AV valve, associated with an increase in LV dimensions and/or worsening LV dysfunction, may warrant repair or replacement of the left AV valve

Severe LVOT obstruction with a mean gradient >50 mmHg or a Doppler-derived peak gradient >70 mmHg

Less severe LVOT obstruction (mean gradient <50 mmHg) associated with significant mitral or aortic valve regurgitation

Significant left-to-right shunt through residual ASD or VSD

Points to Report

- Type of AVSD:
  - Complete/partial/transitional
  - Balanced/unbalanced: AV valve index, LV inflow index
- ASD: Size of primum ASD, direction of shunt, additional secundum ASDs (if any)
- VSD: Size of inlet VSD, direction of shunt, additional VSDs (if any)
- AV valve:
  - Morphology
  - Ventricular commitment
  - Severity of regurgitation/stenosis
  - Direction of regurgitation: LV-RA/LV-LA
- AV valve chordal anatomy: Straddling–Rastelli type
- Secondary effects:
  - LVOT obstruction
  - Pulmonary pressure—RVSP or mean PA pressure
- Associated lesions
- Pulmonary venous drainage
- LV and RV size and function
- **Post-repair:**
  - Residual ASD/VSD
  - LV and RV size and function
  - Left and right AV valve regurgitation/stenosis
  - RVSP or mean PA pressure
  - Residual/Progressive LVOTO

References

1. Cetta F, Minich LL, Maleszewski JJ, Dearani JA, Burkhart HM. Atrioventricular septal defects. In: Allen HD, Driscoll DJ, Shaddy RE, Feltes TF, editors. Moss and Adams' heart disease in infants, children, and adolescents. 8th ed. Philadelphia: Lippincott Williams & Wilkins; 2013. p. 691–712.
2. Perloff JK, Marelli AJ. Atrial septal defect: simple and complex. In: Clinical recognition of congenital heart disease. 6th ed. Philadephia: Saunders; 2012. p. 212–72.
3. di Segni E, Edwards JE. Cleft anterior leaflet of the mitral valve with intact septa: a study of 20 cases. Am J Cardiol. 1983;51:919–26.
4. Seward JB, Tajik AJ, Hagler DJ. Two-dimensional echocardiographic features of atrioventricular canal defect. In: Lundström N-R, editor. Pediatric echocardiography: cross sectional, M-Mode and Doppler. New York: Elsevier; 1980. p. 197–206.
5. Seward JB, Tajik AJ, Edwards WD, et al. Congenital heart disease. In: Two-dimensional echocardiographic atlas, vol. 1. New York: Springer; 1987.
6. Warnes CA, Somerville J. Double mitral valve orifice in atrioventricular defects. Br Heart J. 1983;49:59–64.
7. Reeder GS, Danielson GK, Seward JB, et al. Fixed subaortic stenosis in atrioventricular canal defect: a Doppler echocardiography study. J Am Coll Cardiol. 1992;20:386–94.
8. Titus JL, Rastelli GC. Anatomic features of persistent common atrioventricular canal. In: Feldt RH, McGoon DC, Ongley PA, et al., editors. Atrioventricular canal defects. Philadelphia: WB Saunders; 1976.
9. Peoples WM, Moller JH, Edwards JE. Polysplenia: a review of 146 cases. Pediatr Cardiol. 1983;4:129–37.
10. Corwin RD, Singh AK, Karlson KE, Double-outlet right atrium. A rare endocardial cushion defect. Am Heart J. 1983;106:1156–7.
11. Rastelli GC, Kirklin JW, Titus JL. Anatomic observations on complete form of persistent common atrioventricular canal with special reference to atrioventricular valves. Mayo Clin Proc. 1966;41:296–308.
12. Cohen MS, Jacobs ML, Weinberg PM, Rychik J. Morphometric analysis of unbalanced common atrioventricular canal using two-dimensional echocardiography. J Am Coll Cardiol. 1996;28(4):1017–23.
13. Jegatheeswaran A, Pizarro C, Caldarone CA, Cohen MS, Baffa JM, Gremmels DB, Mertens L, Morell VO, Williams WG, Blackstone EH, McCrindle BW, Overman DM. Echocardiographic definition and surgical decision-making in unbalanced atrioventricular septal defect: a Congenital Heart Surgeons' Society multiinstitutional study. Circulation. 2010;122(11):209–15.
14. Szwast AL, Marino BS, Rychik J, Gaynor JW, Spray TL, Cohen MS. Usefulness of left ventricular inflow index to predict successful biventricular repair in right-dominant unbalanced atrioventricular canal. Am J Cardiol. 2011;107(1):103–9.
15. Cohen MS, Jegatheeswaran A, Baffa JM, Gremmels DB, Overman DM, Caldarone CA, McCrindle BW, Mertens L. Echocardiographic features defining right dominant unbalanced atrioventricular septal defect: a multi-institutional Congenital Heart Surgeons' Society study. Circ Cardiovasc Imaging. 2013;6(4):508–13.
16. Warnes CA, et al. ACC/AHA 2008 guidelines for adults with congenital heart disease. J Am Coll Cardiol. 2008;52:1890–947.

6 Right Ventricular Inflow Anomalies

RV inflow anomalies are the congenital defects affecting the right AV valve apparatus and right atrium. The commonly encountered RV inflow anomalies are Ebstein's anomaly and tricuspid atresia. Other anomalies, such as tricuspid valve dysplasia and cor triatriatum dexter, are relatively rare.

The RV inflow is visualized from the apical and the subcostal four-chamber views, PSAX view, and the medially angulated PLAX (RV inflow) view.

6.1 Ebstein's Anomaly

Ebstein's anomaly is a complex congenital defect of the heart characterized by the following abnormalities [1–4]:

- The anterior tricuspid leaflet, which is usually attached to the true TV annulus, is redundant (sail-like), tethered, and fenestrated.
- Although the septal and posterior leaflets also originate from the true TV annulus in most cases, the adherence (tethering) of the proximal parts of these leaflets to the adjacent myocardium makes their origins appear apically displaced.
- The functional TV annulus appears apically displaced due to the tethering of leaflets (septal > posterior > anterior).
- The true TV annulus (i.e. the anatomic right AV junction) is dilated.
- The segment of RV between the true TV annulus and the apically displaced functional TV annulus is referred to as the 'atrialized portion of RV' (ARV).
- The septal and posterior (inferior) leaflets of the TV may sometimes be rudimentary or even absent.
- Incomplete coaptation of the leaflets and abnormal chordal attachments to the ventricular septum lead to varying degrees of TR.
- The atrialized portion of RV is dilated and thinned out, predisposing to thrombus formation.
- The functional portion of RV is diminished in size, which predisposes to RV failure.

Ebstein's anomaly has been classified echocardiographically as mild, moderate, or severe according to [5]:

- The extent of apical displacement of the TV leaflets
- Degree of TR
- Degree of RV dilation and dysfunction

Carpentier classification [6]

- Type A: The true RV volume is adequate.
- Type B: Large atrialized portion of the RV with freely mobile anterior leaflet.

Supplementary Information The online version contains supplementary material available at https://doi.org/10.1007/978-981-96-0243-8_6.

A. Paul, *Manual of Echocardiography for Congenital Heart Diseases*,
https://doi.org/10.1007/978-981-96-0243-8_6

- Type C: The mobility of the anterior tricuspid leaflet is severely restricted and may cause significant RVOTO.
- Type D: Near total atrialization of the RV except for a small infundibular component.

6.1.1 Echocardiographic Diagnosis (Fig. 6.1)

6.1.1.1 Apical Displacement of Tricuspid Valve

The commissure between the septal and posterior tricuspid leaflets is the most apically displaced point. Normally, the distance between this point and the septal attachment of the anterior mitral leaflet is <8 mm/m^2 body surface area [7]. The distance between the septal attachments of the septal tricuspid leaflet and AML is measured from the A4C view in either systole or diastole and divided by the body surface area.

- ***Apical displacement of the septal tricuspid leaflet (from the AML) by ≥8 mm/m^2 BSA is considered diagnostic of Ebstein's anomaly.***
- A displacement of more than 20 mm is also considered diagnostic in adults.

6.1.1.2 Delayed Closure of TV

A relative delay in the closure of the tricuspid valve (as compared to the mitral valve closure) is a characteristic feature of Ebstein's anomaly. This can be assessed by M-mode echocardiography. The dilated RV and the large redundant anterior tricuspid leaflet facilitate M-mode imaging of the anterior tricuspid leaflet and AML simultaneously [8].

- A ***delay of greater than 65 ms*** is considered diagnostic of Ebstein's anomaly.

6.1.2 Degree of Atrialization and RV Function

A greater degree of atrialization of the RV portends a worse prognosis in Ebstein's anomaly. The overall prognosis is poor if the area of the functional RV is ***less than one-third*** of the total RV area.

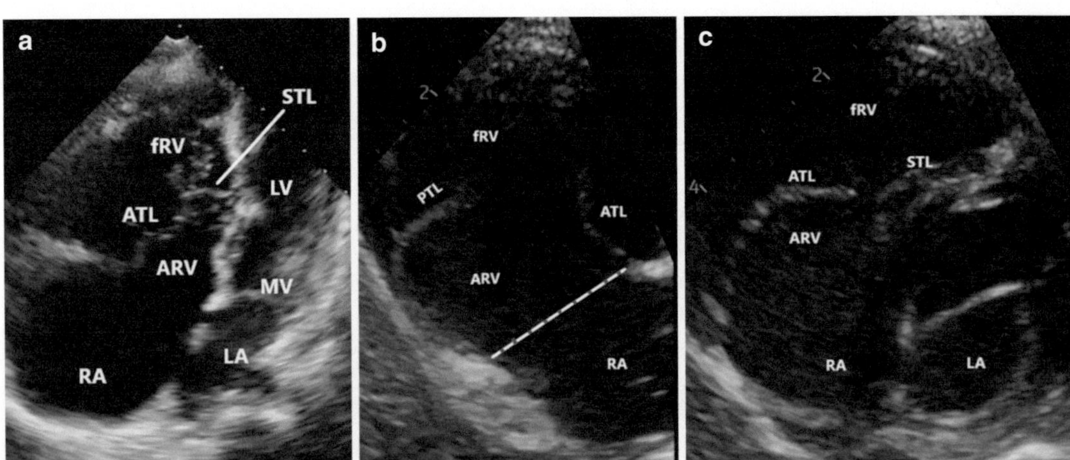

Fig. 6.1 Ebstein's anomaly. (**a**) A4C view depicting the apically displaced, tethered septal tricuspid leaflet (STL), redundant anterior tricuspid leaflet (ATL), atrialized right ventricle (ARV), and smaller functional right ventricle (fRV). (**b**) RV inflow view depicting tethered, apically displaced posterior tricuspid leaflet (PTL). The true tricuspid valve annulus (anatomic AV junction) is dilated (marked with a dotted line). (**c**) Modified PSAX view depicting redundant ATL and tethered STL

Celermajer's Echocardiographic Grading Score (GOSE Score) [9]

Great Ormond Street Echocardiography **(GOSE)** score is a quantitative index of the severity of Ebstein's anomaly (Fig. 6.2).

$$\text{GOSE score} = \left(\text{RA area} + \text{Area of atrialized RV}\right) / \left(\text{Area of functional RV} + \text{LA area} + \text{LV area}\right)$$

- Grade 1: score <0.5
- Grade 2: score of 0.5–0.99
- Grade 3: score of 1.0–1.49
- Grade 4: score ≥1.5

Neonates with the most severe GOSE scores (grades 3 and 4) have a very poor prognosis, with 100% mortality in those with cyanosis.

Key point: The area measurements for the GOSE score are made from the A4C view at end-diastole.

RV Size and Function

- The dimensions of the right ventricle, including the area of the atrialized RV (RV) and functional RV (fRV), need to be carefully documented.
- Indices of RV function, such as fractional area change (FAC) and the Tei index, are significant (discussed in Chap. 15), although geometric distortions of the ventricles make the assessments challenging.

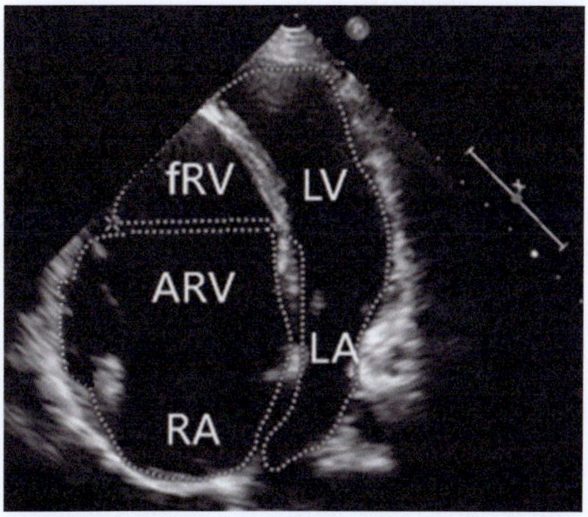

Fig. 6.2 Estimation of GOSE score in Ebstein's anomaly

6.1.3 RVOT Obstruction/Dilation

Significant RVOT obstruction, if present, leads to neonatal cyanosis. RVOT obstruction may result from [2, 5]:

- Pulmonary valve or branch PA stenosis
- Anatomic or functional pulmonary atresia
- Displacement of the redundant anterior TV leaflet into the RVOT

The level and severity of the obstruction, as well as the dimensions of the pulmonary arteries, need to be documented. Aneurysmal dilation of the RVOT is also observed in Ebstein's anomaly.

6.1.4 Tricuspid Valve Anatomy and Function

The morphology of the TV in Ebstein's anomaly is variable. Leaflet malformation, dysplasia, and tethering may lead to regurgitation and rarely stenosis. Shortened chordae, underdeveloped papillary muscles, thickening and rolling of the leaflets, and multiple orifices may be present [4, 5, 7].

- The abnormal anatomy and exaggerated motion of the anterior tricuspid leaflet can be appreciated from the A4C view and RV inflow view. The attachment of the anterior leaflet to the conal papillary muscle (Lancisi), and the degree of obstruction caused by its displacement into the RVOT are ascertained in the subcostal coronal view [2].

- Abnormal chordal attachments between the septal tricuspid leaflet and the ventricular septum can be appreciated from the A4C view.
- The posterior tricuspid leaflet, which is sometimes more severely affected than the septal leaflet, is best visualized from the RV inflow view and the PSAX view (if the tricuspid annulus is dilated).

6.1.4.1 The Extent of Leaflet Tethering

The extent of leaflet tethering should be ascertained from multiple views.

- Demonstration of at least three accessory attachments between the tricuspid leaflet and the ventricular wall confirms tethering of the leaflet [10].
- Valve replacement (rather than repair) may be required in cases of significant tethering, characterized by:
  - Absence of free leading edge.
  - <50% delamination of the anterior tricuspid leaflet.

6.1.4.2 Tricuspid Regurgitation

Tricuspid regurgitation is the most common functional abnormality of the TV encountered in Ebstein's anomaly, the severity of which is a significant prognostic factor. The various points to be assessed during echocardiography are:

- Severity of TR: Vena contracta width >0.7 cm is indicative of severe TR. Multiple jets of TR may make the assessment of severity difficult.
- Site of origin/mechanism of TR: Non-coaptation of leaflets; abnormal subvalvular apparatus; dilated tricuspid annulus.
- The feasibility of valve repair needs to be assessed based on the extent of tethering (see above) and the magnitude of deformity/dysplasia of the leaflets [5].
- The size of the TV annulus (Z-score) should be documented from the A4C view.

Key point: Tricuspid regurgitation in Ebstein's anomaly is typically a low-pressure TR.

6.1.4.3 Tricuspid Stenosis

- The normal maximal velocity of blood flow across the tricuspid valve is <0.8 m/s.
- Peak and mean diastolic gradients across the tricuspid valve should be documented.

6.1.5 Associated Cardiac Defects [11, 12]

- ASD/PFO (80–90%): Right-to-left shunting through an interatrial shunt is common owing to increased right atrial pressure, thereby raising the risk of paradoxical embolism.
- VSD, PDA
- Anomalies involving the left side of the heart [11]:
  - Mitral valve prolapse; accessory mitral valve tissue
  - Bicuspid or atretic aortic valves; subaortic stenosis; coarctation of the aorta
  - Left ventricular dysplasia resembling noncompaction
  - Anomalous muscle bands in the LV
  - Diastolic and systolic dysfunction

6.1.6 Post-Repair Evaluation

The assessment of a post-surgical case of Ebstein's anomaly includes:

- Residual TR and its severity
- Pressure gradient across the native or prosthetic tricuspid valve
- RV size and function
- RVOT obstruction
- Residual shunts
- LV function

Congenitally corrected transposition of the great arteries (cc-TGA) and Ebstein's anomaly: Up to 50% of patients with cc-TGA have an abnormal systemic AV valve (morphologic TV) with features characteristic of Ebstein's anomaly [13].

The distinguishing features are [14]:

- The Ebsteinoid displacement of the TV into the morphologic RV in cc-TGA ***does not result in atrialization of the RV free wall.***
- The morphologic RV (systemic ventricle) in cc-TGA is usually not dilated.
- Ebsteinoid malformation of the TV in cc-TGA is associated with a ***high-pressure TR,*** in contrast to the low-pressure TR of classic Ebstein's anomaly.

Points to Report

- Tricuspid valve:
  - TV leaflet anatomy and displacement (indexed to BSA)
  - Extent of leaflet tethering
  - Tricuspid valve Z score
  - TR severity and diastolic gradient across the TV
- Right ventricle:
  - Degree of atrialization of RV–GOSE score
  - Size of functional RV
  - RV function
  - RVOT obstruction
- Shunting across interatrial septum
- LV size and function
- Associated defects

6.2 Congenital Tricuspid Valve Regurgitation

Congenital TR includes a heterogeneous group of anatomical defects that lead to tricuspid regurgitation in neonates without Ebstein's anomaly. Severe congenital TR is a cause of massive cardiomegaly in the newborn. The various causes of congenital TR include:

Primary valve abnormalities [15, 16]:

- Congenital absence of chordae
- TV dysplasia without apical displacement of TV annulus

Secondary regurgitation in association with other anomalies such as [17, 18]:

- cc-TGA
- AV septal defects
- Congenital coronary arterial fistula with secondary RV enlargement
- Uhl's anomaly—complete or partial absence of the myocardial layer of the RV

Echocardiographic evaluation is similar to the evaluation of TR in Ebstein's anomaly, with a special focus on deciphering the aetiology of TR.

6.3 Tricuspid Atresia

Tricuspid atresia is an RV inflow anomaly in which the tricuspid valve and the inlet portion of the RV are completely atretic, resulting in no direct communication between the RA and RV. The RV is composed largely of the infundibular portion and an incompletely formed trabecular portion. Since there is no inflow into the RV from the RA, there is an obligatory right-to-left shunt at the atrial level (an ASD or a large PFO) to maintain circulation [19, 20]. The degree of RV hypoplasia is determined by the amount of inflow into the RV through a VSD (usually perimembranous; less commonly muscular or inlet VSD).

The physiology of tricuspid atresia is determined by the size of the VSD [19, 20]:

- If a large VSD is present:
  - The trabecular portion of the RV may be better developed, resulting in a larger RV cavity.
- If no VSD is present:
  - The RV is either rudimentary or completely absent.
  - The pulmonary valve will also be atretic.
  - Pulmonary blood flow is supplied by the PDA.

Modified Kuhne Classification [21]

Type I: Tricuspid atresia with normally related great arteries (70–80%)

(a) Intact ventricular septum with pulmonary atresia
(b) Small VSD and pulmonary stenosis

(c) Large VSD without pulmonary stenosis

Type II: Tricuspid atresia with D-transposition of great arteries (12–25%)

(a) VSD with pulmonary atresia

(b) VSD with pulmonary stenosis

(c) VSD without pulmonary stenosis

Type III: Tricuspid atresia with more complex associated lesions such as L-transposition or malposition of great arteries (3–6%). Truncus arteriosus and/or atrioventricular septal defect may be associated.

6.3.1 Echocardiographic Diagnosis and Classification

6.3.1.1 Absent Tricuspid Valve

- An imperforate linear echo density at the right AV junction (Fig. 6.3)
- No demonstrable flow from the RA to the RV

6.3.1.2 Size and Adequacy of Interatrial and Interventricular Shunts

Interatrial shunt: A right-to-left interatrial shunt is obligatory for the systemic venous blood received by the RA to re-enter the circulation. The size of this interatrial defect (ASD/PFO) needs to be estimated. Subcostal coronal and sagittal views are helpful in this regard. If there is restriction to this interatrial blood flow, balloon atrial septostomy may be necessary.

Interventricular shunt: The size of the VSD, which is a key determinant of the physiologic phenotype, should be documented. In patients with associated TGA, where the RV forms a subaortic chamber, there may be restriction to blood flow into the systemic circulation if the VSD is small [19, 20].

6.3.1.3 Great Arterial Relationship and Pulmonary Blood Flow

Transposition/malposition of the great arteries occurs in up to 30% of cases of tricuspid atresia. This may be ascertained from the parasternal and the subcostal views (discussed in Chap. 11).

RVOT obstruction in tricuspid atresia, if present, varies from mild degrees of pulmonary stenosis to complete atresia of the pulmonary valve. The severity of RVOTO needs to be assessed from PSAX and RV outflow views (discussed in Chap. 7).

6.3.2 Associated Defects

Common in those with abnormally related great arteries [20].

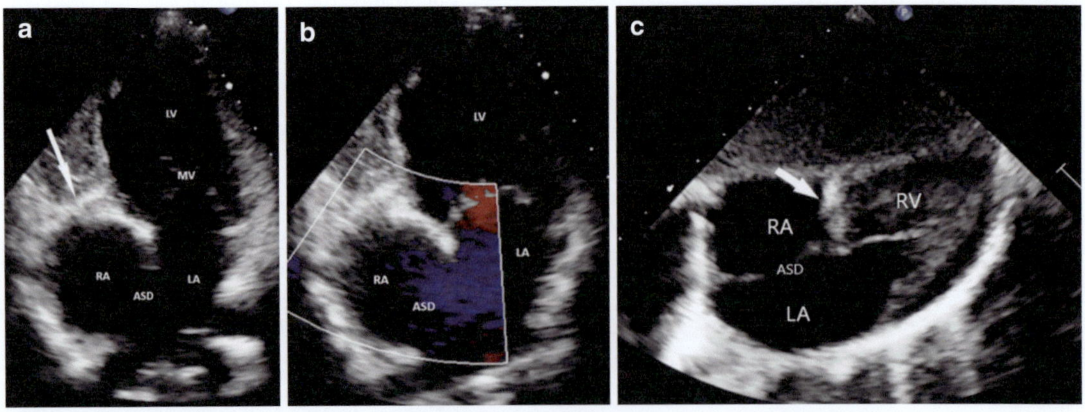

Fig. 6.3 Tricuspid atresia. (**a**) Apical four-chamber view depicting the imperforate linear echo density (denoted by an arrow) at the site of the tricuspid valve. (**b**) Obligatory interatrial shunt from right to left is seen by colour flow Doppler. (**c**) Subcostal coronal view depicting the atretic tricuspid valve (arrow) and ASD

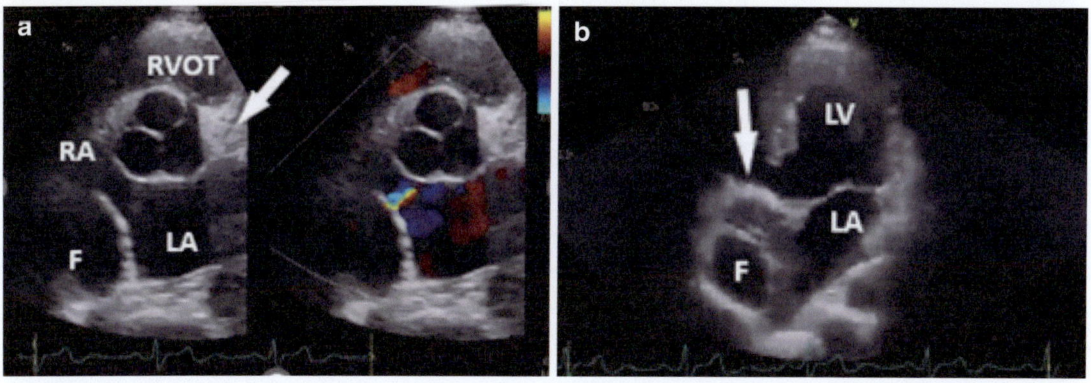

Fig. 6.4 Tricuspid atresia following fenestrated lateral tunnel Fontan procedure and MPA ligation. (**a**) The ligated MPA is denoted by the arrow. Turbulent flow across the fenestration is seen. (**b**) The atretic tricuspid valve is marked by the arrow. *F* fontan circuit

- Coarctation of aorta (8%)
- Right aortic arch
- Persistent left superior vena cava (LSVC)
- Patent ductus arteriosus
- Anomalous origin of the coronary artery from the pulmonary artery
- Juxtaposition of atrial appendages
- Mitral regurgitation
- LV dysfunction: Chronic volume overload and hypoxaemia may rarely lead to LV dysfunction.

6.3.3 Evaluation of Post-surgical Patients

- **Palliative procedures**
  - Initial surgical palliation may be done by aortopulmonary shunt (in case of inadequate pulmonary blood flow) or PA banding (in case of excessive pulmonary blood flow) (Fig. 6.4a).
  - PA pressure should be estimated by determining the ***pressure gradient through the AP shunt or across the PA band.***
- **Modified Fontan procedure** (discussed in Chap. 12) (Fig. 6.4b): The major points to be assessed after a Fontan procedure are [20]:
  - Ventricular dysfunction
  - Mitral valve insufficiency
  - Obstruction to RA-PA flow
  - Presence of right atrial thrombi due to sluggish flow through the right side of the heart

References

1. Perloff JK, Marelli AJ. Ebstein's anomaly of the tricuspid valve. In: Clinical recognition of congenital heart disease. 6th ed. Philadephia: Saunders; 2012. p. 176–95.
2. Said SM, Dearani JA. Ebstein's anomaly, congenital tricuspid valve regurgitation, and dysplasia. In: Allen HD, Driscoll DJ, Shaddy RE, Feltes TF, editors. Moss and Adams' heart disease in infants, children, and adolescents. 8th ed. Philadelphia: Lippincott Williams & Wilkins; 2013. p. 889–912.
3. Edwards WD. Embryology and pathologic features of Ebstein's anomaly. Prog Pediatr Cardiol. 1993;2:5–15.
4. Dearani JA, Danielson GK. Ebstein's anomaly of the tricuspid valve. In: Mavroudis C, Backer CL, editors. Pediatric cardiac surgery. 3rd ed. Philadelphia: Mosby; 2003. p. 524–36.
5. Attenhofer Jost CH, Connolly HM, Dearani JA, et al. Ebstein's anomaly. Circulation. 2007;115:277–85.
6. Carpentier A, Chauvaud S, Mace L, et al. A new reconstructive operationfor Ebstein's anomaly of the tricuspid valve. J Thorac Cardiovasc Surg. 1988;96:92–101.
7. Anderson KR, Zuberbuhler JR, Anderson RH, et al. Morphologic spectrum of Ebstein's anomaly of the heart: a review. Mayo Clin Proc. 1979;54:174–80.
8. Daniel W, Rathsack P, Walpurger G, et al. Value of M-mode echocardiography for non-invasive diagnosis of Ebstein's. Br Heart J. 1980;43:38–44.

9. Celermajer DS, Bull C, Till JA, et al. Ebstein's anomaly: presentation and outcome from fetus to adult. J Am Coll Cardiol. 1994;23:170–6.
10. Giuliani ER, Fuster V, Brandenburg RO, et al. Ebstein's anomaly: the clinical features and natural history of Ebstein's anomaly of the tricuspid valve. Mayo Clin Proc. 1979;54:163–73.
11. Attenhofer Jost CH, Connolly HM, O'Leary PW, et al. Left heart lesions inpatients with Ebstein anomaly. Mayo Clin Proc. 2005;80:361–8.
12. Sumner RG, Jacoby WJ Jr, Tucker DH. Ebstein's anomaly associated with cardiomyopathyand pulmonary hypertension. Circulation. 1964;30:578–87.
13. van Son JA, Danielson GK, Huhta JC, et al. Late results of systemic atrioventricularvalve replacement in corrected transposition. J Thorac Cardiovasc Surg. 1995;109:642–52.
14. Said SM, Burkhart HM, Schaff HV, et al. Congenitally corrected transposition of great arteries: surgical options for the failing right ventricle and/or severe tricuspid regurgitation. World J Pediatr Congenit Heart Surg. 2011;2:64–79.
15. Shikata F, Nagashima M, Nishimura K, et al. Repair of congenitally absentchordae in a tricuspid valve leaflet with hypoplastic papillary muscle usingartificial chordae. J Card Surg. 2010;25:737–9.
16. Mohan JC, Passey R, Arora R. Echocardiographic spectrum of congenitallyunguarded tricuspid valve orifice and patent right ventricular outflow tract. Int J Cardiol. 2000;74:153–7.
17. Ando M, Takahashi Y. Variations of atrioventricular septal defects predisposing to regurgitation and stenosis. Ann Thorac Surg. 2010;90:614–21.
18. Vaidyanathan K, Agarwal R, Johari R, et al. Isolated congenital pulmonaryregurgitation with right ventricular outflow tract aneurysm-a rarevariant of Uhl's anomaly. J Card Surg. 2010;25:415–7.
19. Perloff JK, Marelli AJ. Tricuspid atresia. In: Clinical recognition of congenital heart disease. 6th ed. Philadephia: Saunders; 2012. p. 439–53.
20. Epstein ML. Tricuspid atresia, stenosis, regurgitationand Uhl's anomaly. In: Allen HD, Driscoll DJ, Shaddy RE, Feltes TF, editors. Moss and Adams' heart disease in infants, children, and adolescents. 8th ed. Philadelphia: Lippincott Williams & Wilkins; 2013. p. 877–88.
21. Tandon R, Edwards JE. Tricuspid atresia: a re-evaluation and classification. J Thorac Cardiovasc Surg. 1974;67:530–42.

7 Right Ventricular Outflow Anomalies

7.1 Right Ventricular Outflow Tract Obstruction

RVOTO can occur at multiple levels: subvalvular (proximal to the valve), valvular, and/or supravalvular (distal to the valve). The morphological variants of RVOTO at different anatomic levels can be categorized as [1]:

- **Subvalvular (Intracavitary)**: Obstruction proximal to the pulmonary valve
  - Infundibular
  - Subinfundibular (Double Chambered Right Ventricle (DCRV))
- **Valvular**: Intrinsic abnormality of the valve
  - Typical mobile dome-shaped
  - Dysplastic
  - Bicuspid (associated with Tetralogy of Fallot)
- **Supravalvular**: Obstruction distal to the pulmonary valve
  - Discrete membrane-like lesions
  - Long tubular stenoses
  - Tubular hypoplasia

7.1.1 Echocardiography: General Considerations

The PSAX view at the aortic valve level and the subcostal sagittal/coronal views are ideal for assessing the complex geometry of the RVOT. Orienting the ultrasound beam parallel to the jet of blood requires appropriate use of all available windows.

- Multiple levels of obstruction may be present in the RVOT and PAs. Exact localization of the site of stenosis with PWD becomes difficult when narrowing occurs at more than one level.
- Colour Doppler can aid in identifying the site(s) of obstruction by spotting the origin of the mosaic colour pattern. This also enables parallel alignment of the ultrasound beam with the jet of blood for accurate estimation of the peak instantaneous gradient (PIG) by CWD.
- The spectral Doppler profile across the RVOT can differentiate dynamic subvalvular obstruction from fixed obstruction at any level. The late peaking Doppler signal of dynamic subvalvular stenosis ('lobster claw' shape) is distinct from the mid-systolic peaking of fixed obstruction at any level (Fig. 7.1).

Supplementary Information The online version contains supplementary material available at https://doi.org/10.1007/978-981-96-0243-8_7.

A. Paul, *Manual of Echocardiography for Congenital Heart Diseases*,
https://doi.org/10.1007/978-981-96-0243-8_7

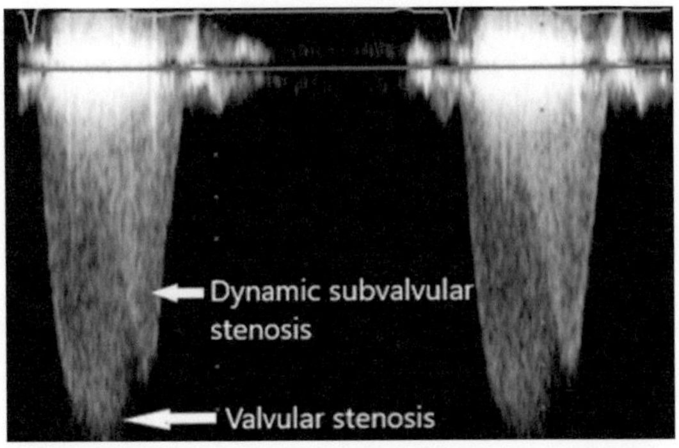

Fig. 7.1 Late-peaking CWD tracing of dynamic subvalvular stenosis is superimposed on the symmetric tracing of valvular PS with mid-systolic peaking. Dynamic subvalvular obstruction can result from RV hypertrophy secondary to valvular PS

Table 7.1 Grading of severity of RVOTO

Severity of RVOTO	Mild	Moderate	Severe
Peak flow velocity across RVOT	<3 m/s	3–4 m/s	>4 m/s
Peak pressure gradient across RVOT	<36 mmHg	36–64 mmHg	>64 mmHg
RV pressure/LV pressure ratio	<0.5	0.5–0.75	>0.75

Key point: In the setting of RVOTO, Doppler-derived pressure gradients across the RVOT do not reflect PASP.

7.1.2 Severity of RVOTO (Table 7.1)

In patients with normal cardiac output, the severity of RVOTO at any level is graded as shown in Table 7.1.

Better alignment of the ultrasound beam with the turbulent jet may be facilitated by:

- Visualization of the mosaic pattern by colour Doppler
- Use of a lower interspace for superiorly directed jets
- Subcostal approach in children
- Subcostal or suprasternal approach when there is anterior displacement of the valve due to dilated PA (as in post-stenotic dilation of valvular PS)

Key point: If significant subvalvular stenosis coexists, the peak flow velocity proximal to the valvular obstruction must be taken into account when estimating the pressure drop across the valve using Bernoulli's equation (discussed in Chap. 1).

7.1.3 Measurement of RVOT and PA Dimensions [2]

Reduction in pulmonary arterial blood flow secondary to RVOTO can affect the development of PAs. Hence, evaluation of any form of RVOTO should include an assessment of the size of the PAs as well.

- **Infundibulum:**
  - Measured in PLAX (proximal RVOT) or PSAX (distal RVOT) views during mid-systole.
  - Largest diameters are measured from inner edge to inner edge.
- **Pulmonary valve annulus:**
  - Measured in PSAX view or RV outflow view during mid-systole.
  - Balloon size for valvuloplasty is determined from the size of the PV annulus.
- **MPA, RPA, LPA:**
  - MPA can be visualized from PSAX view, parasternal RV outflow view, and subcostal views. The measurement is taken midway between PV and PA bifurcation.
  - Branch PAs are measured at the level of their origin. RPA can be visualized from subcostal coronal view, high left PSAX view, right PSAX view, and suprasternal long-axis and short-axis views. The LPA can be visualized from high left PSAX view and suprasternal long-axis view with a tilt to the left.
  - Supravalvular PA stenosis is reported at the smallest dimension.

7.1.4 Site of Obstruction

7.1.4.1 Subvalvular Obstruction

Subvalvular obstruction may be infundibular or subinfundibular.

Infundibular obstruction may be due to:

- Anterior malalignment of the infundibular septum or
- Infundibular hypertrophy secondary to severe valvular PS

Subinfundibular stenosis:

A classic example of subinfundibular stenosis is **Double-Chambered Right Ventricle (DCRV)** (Fig. 7.2). In this condition, anomalous hypertrophied muscular bands extend from the anterior wall of the RV to the ventricular septum, dividing the RV cavity into a proximal high-pressure chamber (the sinus portion of the RV) and a distal low-pressure chamber (infundibulum) [3].

DCRV is almost always associated with a VSD, most frequently in the perimembranous septum (occasionally a subarterial VSD) [4]. Associated discrete subaortic stenosis may be present. In adults diagnosed with DCRV (rare), the perimembranous VSD demonstrates minimal shunt flow because the VSD connects to the high-pressure inflow part of the RV [5].

Key point: The anomalous bands of DCRV can be distinguished from the moderator band by the site of septal attachment. Anomalous bands attach just inferior to the septal leaflet of the TV, whereas the septal attachment of the moderator band is to the apical third of the ventricular septum.

As discussed earlier, dynamic subvalvular stenosis is characterized by the 'lobster claw' appearance of spectral Doppler tracing. M-mode echo may demonstrate abnormal fluttering of the pulmonary valve in subvalvular stenosis.

7.1.4.2 Valvular Stenosis

Isolated pulmonary valve stenosis is found in 80–90% of all patients with RVOTO.

Typical dome shaped valvular PS is characterized by thickened and fused valve cusps, an incompletely formed raphe, and a small orifice. Post-stenotic dilation of the MPA may sometimes extend to the proximal LPA [1, 6].

Dysplastic pulmonary valve is characterized by myxomatous thickening of the valve leaflets, a hypoplastic PV annulus, and supra-annular narrowing of the proximal MPA [1, 7]. It is frequently seen in association with Noonan syndrome.

Imaging features (Fig. 7.3):

- Restricted mobility and systolic doming of the valve cusps may be visualized from the standard and high PSAX views, the parasternal RV outflow view, and the subcostal sagittal view.

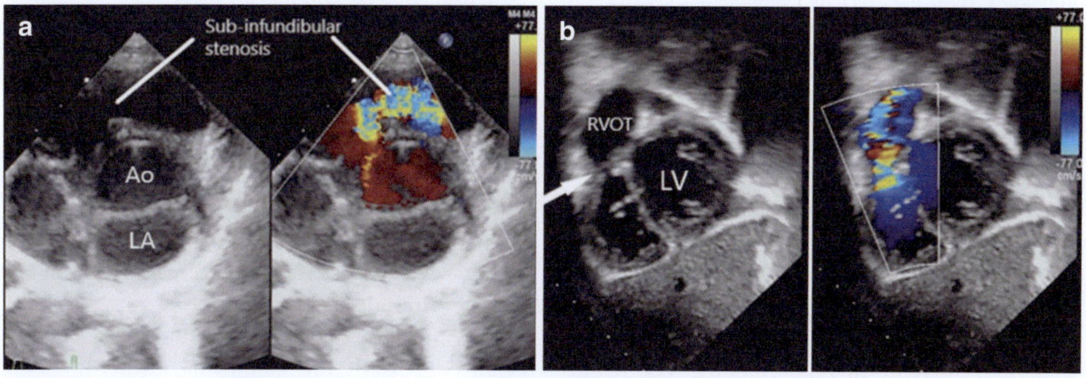

Fig. 7.2 Double-chambered right ventricle. (**a**) PSAX view showing turbulent flow across the subinfundibular stenosis. Also seen is a perimembranous VSD (11 o'clock position). (**b**) Subcostal sagittal view depicting narrowing and turbulence at the subinfundibular region

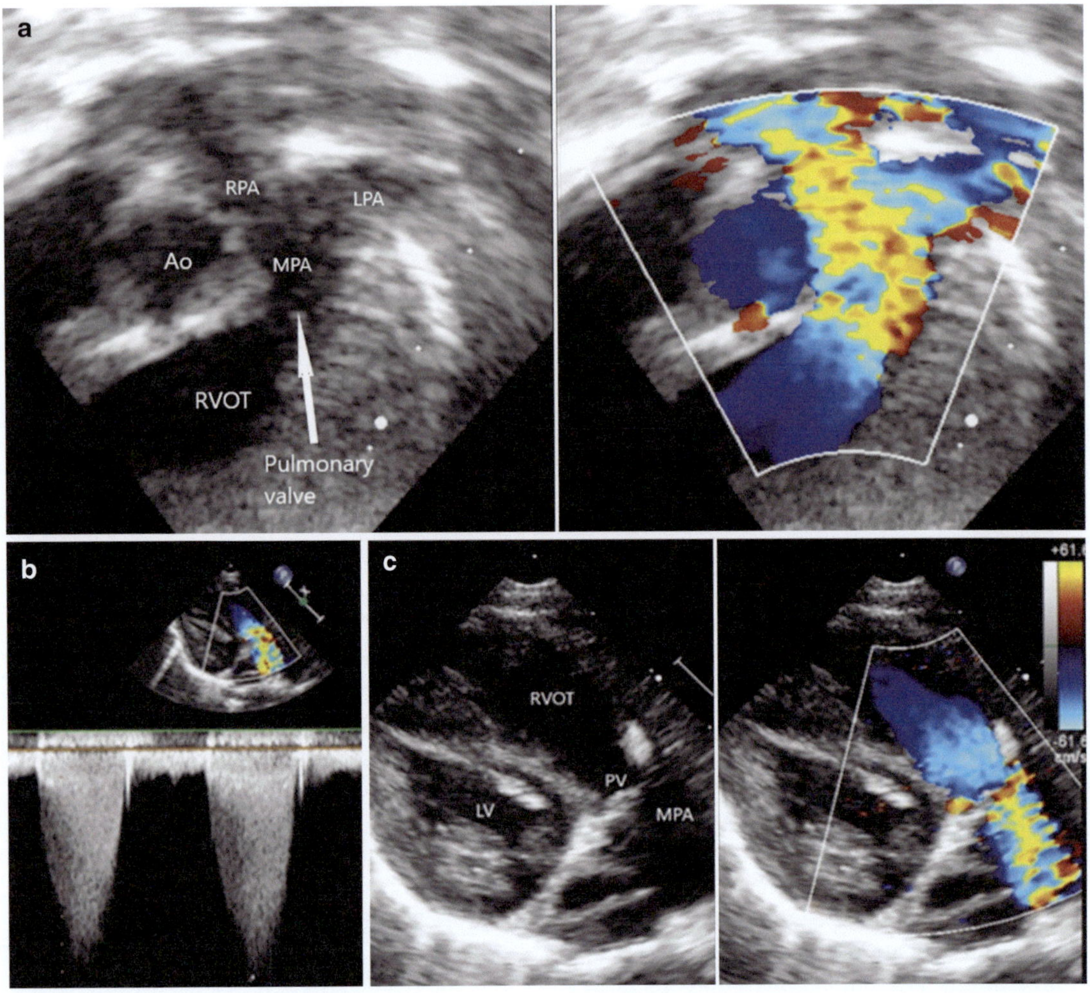

Fig. 7.3 Valvular pulmonary stenosis. (**a**) PSAX view depicting a stenotic pulmonary valve and flow turbulence originating from the level of the pulmonary valve. (**b**) CWD tracing from the pulmonary artery depicting the symmetric contour and mid-systolic peaking of valvular PS. (**c**) RV outflow view depicting a domed pulmonary valve with turbulent flow into the MPA

- **RV hypertrophy** with trabecular prominence is seen in severe pulmonary stenosis. The RV cavity is usually normal in size, except in newborns with critical PS, where the RV is hypoplastic.
- Diffuse infundibular hypertrophy often causes ***dynamic subvalvular obstruction,*** with late systolic peaking of flow velocity. The resulting 'lobster claw' pattern of spectral Doppler tracing is characteristic (Fig. 7.1).
- The tricuspid valve may be thickened and regurgitant.
- The RA may appear dilated owing to raised RV filling pressures.

7.1.4.3 Supravalvular Stenosis (Peripheral PS or PA Stenosis)

Supravalvular pulmonary stenosis can be at multiple levels, involving the MPA as well as several smaller peripheral PA branches, especially in the setting of pulmonary atresia.

It commonly occurs in the setting of Rubella syndrome, Williams syndrome, Allegille syndrome, and secondary to surgical procedures, such as previous Blalock-Taussig shunt.

Imaging features (Fig. 7.4):

- The most frequent site of supravalvular PS is the distal MPA, close to the bifurcation.

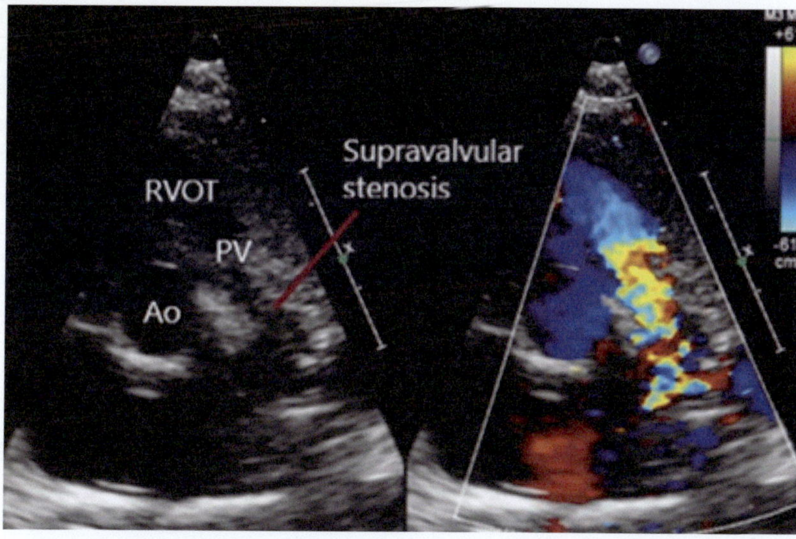

Fig. 7.4 PSAX view depicting valvular and supravalvular stenosis. Note the narrowing and turbulence at valvular as well as supravalvular levels

- Apart from pulmonary stenosis, ***turbulent flow in the PAs can also result from*** PDA, subpulmonic VSD, and AP window.
- Doppler gradients may underestimate pulmonary stenosis in the following situations [8]:
  - Unilateral PA stenosis with abnormal flow distribution in the lungs, away from the stenotic branch
  - Low pulmonary flow situations such as the Glenn shunt and Fontan circulation
- The clues to the possibility of distal branch PA stenosis, which cannot be visualized by echocardiography, are:
  - Unexplained right ventricular hypertrophy
  - Pulsatile proximal pulmonary artery

Key point: In general, PA stenosis is considered significant when there is a peak Doppler pressure gradient of >20 mmHg.

7.1.5 Post-Repair Evaluation

Indications for repair of PS include severe PS, RV dysfunction, unexplained heart failure, exercise intolerance, and cyanosis due to right-to-left shunt.

Balloon valvuloplasty is the preferred mode of intervention for valvular PS, wherever feasible.

- Balloon size is determined as **1.2–1.5 times the size of the PV annulus**.
- PR occurs in 10–40% after balloon dilatation.

Surgical intervention is reserved for patients with subvalvular or supravalvular PS, and valvular PS with a dysplastic valve and hypoplastic annulus (as in Noonan's syndrome).

Post-repair evaluation should focus on the following aspects, in addition to the factors evaluated in the initial echocardiographic study:

- Residual PS (at various levels)
  Severity of PR (discussed in Chap. 10)
  RV function

Points to Report

- Site and morphology of RVOTO
  - Subvalvular (subinfundibular/infundibular)
  - Valvular
  - Supravalvular
  - Multi-level stenosis
- Severity of RVOTO: Mild/Moderate/Severe
- Dimensions: Subvalvular RVOT, PV annulus, MPA, RPA, LPA
- RVSP (assessed from TR gradient)
- RV function and size
- Associated anomalies
- Post-repair: Residual PS, PR severity, RV function

7.2 Pulmonary Atresia with Intact Ventricular Septum

Pulmonary atresia with intact IVS (PA-IVS) is characterized by a totally imperforate RVOT. The morphologic variants of atretic RVOT are [9]:

- Membranous (valvular) atresia: Imperforate pulmonary valve and well-formed infundibulum
- Muscular (infundibular) atresia: Primitive pulmonary valve, hypoplastic RV, and a severely narrowed or atretic infundibulum

Based on the presence or absence of the three portions of the RV—inlet, trabecular, and infundibular portions, PA-IVS has been classified as [10]:

- Tripartite type: All three portions are present, and the RV is almost normal in size.
- Bipartite type: The inlet and infundibular portions are present, but the trabecular portion is obliterated.
- Monopartite type: Only the inlet portion is present, and the RV is diminutive.

A rare type of pulmonary atresia, associated with tricuspid insufficiency, less thickened but dilated RV cavity, and dilated RA, has also been described.

7.2.1 Echocardiographic Diagnosis

Pulmonary atresia is echocardiographically characterized by the absence of outflow of blood from the RV into the pulmonary artery.

- The membranous form of pulmonary atresia needs to be distinguished from severe forms of PV stenosis, where a small jet of forward flow may be masked by ductal flow.
- Detection of any small jet of forward flow across the pulmonary valve rules out the diagnosis of pulmonary atresia.

Functional pulmonary atresia is the lack of demonstrable forward flow across a morphologically normal PV [11]. This can result from high PA pressure in the setting of poor RV function or very severe TR. In patients on positive pressure ventilation, Doppler echocardiography may depict forward flow during transient opening of the PV.

Key point: PDA jet may sometimes regurgitate across the PV during systole. When there is no demonstrable forward flow across the PV, detection of this systolic regurgitant jet proves the patency of the pulmonary valve and rules out anatomic pulmonary atresia.

7.2.2 Pulmonary Blood Supply

Although there is no blood flow across the PV in pulmonary atresia, the main pulmonary trunk is almost always present, and PAs are usually confluent, with a left-sided PDA supplying the blood to the PAs. When the PAs are non-confluent, bilateral PDA or major aortopulmonary collateral arteries (MAPCAs) feed the pulmonary arterial branches [12]. The PDA in pulmonary atresia usually connects to the aorta at an acute angle inferiorly (hence the term, 'vertical' ductus).

High parasternal short-axis and suprasternal views are useful for delineating the source of blood supply to the pulmonary tree.

7.2.3 Adequacy of Interatrial Communication

Obligatory interatrial shunting can occur via ASD, PFO, or unroofed coronary sinus. The adequacy of this shunt should be ascertained from the subcostal coronal and sagittal views.

When the ASD is restrictive, herniation of aneurysmal septum primum through the mitral valve may be seen [11].

7.2.4 Size and Morphology of the Tricuspid Valve

Tricuspid valve abnormalities in PA-IVS range from severe stenosis to severe regurgitation [13].

Severe stenosis of the tricuspid valve usually occurs in the setting of hypoplastic RV and RV-dependent coronary circulation. The valve apparatus is usually abnormal, and the tricuspid annulus is hypoplastic.

Severe tricuspid regurgitation usually occurs in the setting of dilated RV. Ebsteinoid malformation and dysplasia of the TV are common in this setting.

Size of the tricuspid valve annulus indexed to body surface area needs to be documented. Tricuspid diameter z-score correlates with the size of the RV cavity, and thus determines the feasibility of biventricular repair.

7.2.5 Coronary Circulation

The coronary artery abnormalities in PA-IVS need detailed evaluation with angiocardiography. However, echocardiography can detect certain abnormalities of coronary circulation in PA-IVS, including [11, 14]:

- Single coronary artery from the aorta or pulmonary trunk
- Absence of proximal aortic connection of a coronary artery
- Coronary arterial stenosis or interruption
- Coronary-cameral fistula to the RV

Ventriculo-coronary artery connections: If a proximal coronary artery is obstructed, coronary circulation is maintained by retrograde flow of desaturated RV blood into the coronary tree through ventriculo-coronary artery connections (**RV-dependent coronary circulation**). These connections occur only in the setting of a high-pressure hypoplastic RV.

Echocardiographic findings suggestive of the presence of ventriculo-coronary connections and RV-dependent coronary circulation include [14]:

- Absence of normal origin of a coronary artery
- Dilated or tortuous proximal coronary arteries
- Retrograde flow in a proximal coronary artery demonstrated by colour Doppler
- Low tricuspid valve diameter Z score and hypoplastic RV

Key point: A negative tricuspid valve diameter Z score (i.e. a small tricuspid valve) is highly predictive of the presence of ventriculo-coronary artery connections.

7.2.6 Associated Anomalies

- Left ventricular hypertrophy
- LVOTO resulting from a bulging IVS
- Aortic valve stenosis

7.2.7 Evaluation of Patients After Surgical Repair

The primary goal of surgical repair in PA-IVS is to allow RV decompression by performing RVOT reconstruction—surgical valvotomy, RVOT patch repair, or balloon valvuloplasty—before considering biventricular repair. Univentricular palliation is considered in cases with complex coronary anatomy and/or predominant RV-dependent coronary circulation [11].

Aorto-pulmonary shunt, PDA stenting, and/or pulmonary valve replacement may be needed at different stages of treatment, depending on the anatomy. Fistula occlusion with RV decompression is done for significant coronary-cameral fistulae, and RV decompression alone is done for ventriculo-coronary arterial connections [11].

Echocardiographic evaluation of a post-surgical patient depends on the type of procedure performed. (Post-surgical evaluation of various surgical procedures is discussed in Chaps. 10 and 12).

Points to Report

- Morphology:
  - Atresia: Valvular (membranous)/ Infundibular (muscular)
  - RV: Normal/Hypoplastic
- Pulmonary circulation:
  - Pulmonary arteries: Confluent/ Non-confluent

- PA dimensions
  - Source of blood supply: PDA/MAPCAs
- Tricuspid valve:
  - Annulus diameter: z score
  - Function: stenosis/regurgitation
- Adequacy of interatrial communication
- Coronary circulation:
  - Ventriculo-coronary artery connections (RV dependence)
  - Coronary artery anomalies, such as coronary-cameral fistulae
- RV and LV function
- Associated anomalies

7.3 Pulmonary Atresia with VSD

Pulmonary valve atresia with ventricular septal defect (PA–VSD) is the most severe morphologic form of tetralogy of Fallot (TOF), with the following physiologic differences [15]:

- Since the RVOT ends blindly, the entire RV stroke volume is directed to the aorta via the VSD.
- The pulmonary arterial system is more complex in PA–VSD, with well-developed systemic collateral vessels supplying the lungs totally or partially.

Based on the source of pulmonary blood supply, PA–VSD has been classified as [16]:

- Type A: Unifocal confluent native pulmonary arteries supplied by PDA; no MAPCAs
- Type B: Native pulmonary arteries and MAPCAs co-exist, supplying various bronchopulmonary segments
- Type C: Non-confluent or absent pulmonary arteries; the entire pulmonary blood supply is by MAPCAs

Somerville classification of pulmonary arterial anatomy in PA–VSD [17]:

- Type 1: Atresia of pulmonary valve with confluent PAs
- Type 2: Atresia of pulmonary valve and MPA
- Type 3: Atresia of pulmonary valve, MPA, and one main branch
- Type 4: Atresia of pulmonary valve, MPA, and both main branches

7.3.1 Echocardiographic Diagnosis

PA–VSD is characterized by an anteriorly malpositioned infundibular septum, which is fused with the RV free wall. The aorta overrides the resultant malalignment VSD, and RV stroke volume drains into the overriding aorta through the VSD [15].

PLAX view depicts a large aortic valve that overrides the VSD (Fig. 7.5a). The presence of an obliterated RVOT and a single large vessel draining both the ventricles raises the suspicion of truncus arteriosus (Fig. 7.5b, c). However, truncus arteriosus can be distinguished by the origin of pulmonary arteries directly from the posterolateral aspect of the truncal root prior to the arch.

7.3.2 Central Pulmonary Arterial Confluence

The size and confluence of pulmonary arteries and the adequacy of pulmonary blood flow are important determinants of the surgical strategy for the repair of PA–VSD. Demonstration of confluent PAs (i.e. branch PAs arising from the MPA) is an essential prerequisite for a palliative systemic-to-PA shunt procedure.

Suprasternal and high parasternal (right and left) windows are useful for the assessment of the PA confluence. The size of the MPA and branch PAs, if present, should be documented.

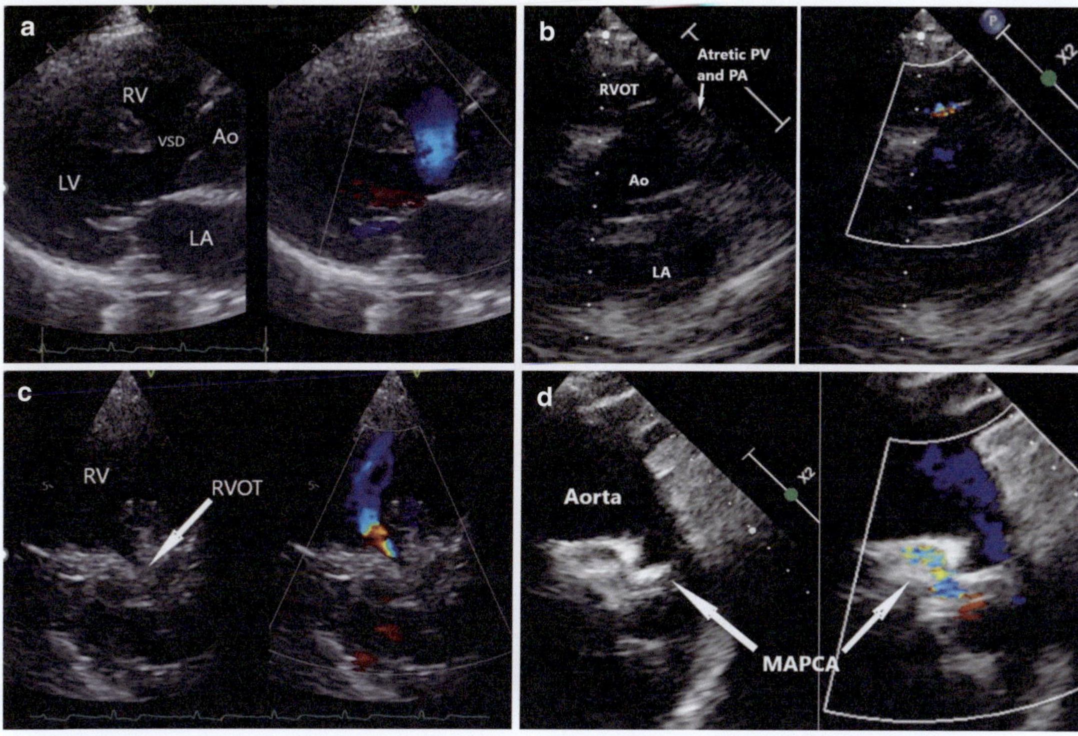

Fig. 7.5 Pulmonary atresia with VSD. (**a**) PLAX view depicting a non-restrictive subaortic VSD with a right-to-left shunt, a large aortic valve overriding the VSD, and a hypertrophic RV. (**b**) PSAX view depicting the atretic pulmonary valve (PV) and artery (PA), with no significant RV outflow in colour flow Doppler. (**c**) Apical view with anterior angulation depicting the atretic pulmonary valve and artery with no flow. (**d**) Suprasternal view depicting a tortuous major aorto-pulmonary collateral artery (MAPCA)

7.3.3 Aortopulmonary Collateral Arteries

- Major aortopulmonary collateral arteries (MAPCAs) are vessels arising from the aorta or its branches and connecting distally to the pulmonary vasculature, in order to provide pulmonary blood flow [18].
- MAPCAs, usually 1–6 in number, arise from [18, 19]:
  - The anterior aspect of the descending thoracic aorta or aortic arch (type II: most common)
  - Branches of the aorta, such as the subclavian artery or its branches, coeliac artery (type III)
  - The bronchial artery (type I)
  - Coronary arteries (very rare)
- Detection of MAPCAs and delineation of their origin (to the maximum extent possible) are important aspects of the evaluation of PA–VSD. Suprasternal views are particularly useful in this regard (Fig. 7.5d).

Key point: Ductus arteriosus can be distinguished from MAPCAs by their morphology—ductus is less tortuous and non-branching, whereas MAPCAs usually branch before joining the PA and are tortuous.

7.3.4 Associated Malformations

ASD, muscular VSDs, coronary artery abnormalities (refer to the section on pulmonary atresia with intact ventricular septum).

7.3.5 Evaluation of Patients After Surgical Repair

Complete repair (with RVOT reconstruction and VSD closure) is done if the PA anatomy is amenable. The various surgical palliative options in cyanotic infants to improve the pulmonary blood flow include an aortopulmonary shunt and/or unifocalization of MAPCAs. Unifocalization involves transplantation of the proximal end of the MAPCAs from the aorta to the native or reconstructed PA [15].

Interventional options in PA–VSD include balloon valvuloplasty, PDA stenting, and angioplasty to MAPCAs.

RV–PA conduit may become necessary in the presence of:

- Symptomatic moderate-to-severe pulmonary regurgitation
- Diminished RV function and/or progressive RV dilation
- RVSP greater than 80 mmHg
- Progressive tricuspid regurgitation

Long-term prognosis after surgery is determined by the presence of:

- RV–PA conduit obstruction/degeneration
- Residual PA branch stenosis
- Residual VSD
- Aortic root dilatation
- RV volume and pressure overload, and RV dysfunction

Echocardiographic evaluation of a post-surgical patient should focus on the above elements. (Post-surgical evaluation of major surgical procedures is discussed in Chaps. 10 and 12).

Points to Report

- Morphology of atretic segment: Short/long
- Pulmonary circulation:
  - Pulmonary arteries: Confluent/Non-confluent
  - PA dimensions
  - Source of blood supply: PDA/MAPCAs
- MAPCAs: Size, number, origin
- VSD: Location and size
- RV and LV size and function
- Associated anomalies
- Post-surgical evaluation:
  - RV–PA conduit function
  - Residual VSD
  - RVSP and RV function
  - Progressive PR, TR
  - Aortic root dilatation

7.4 Double Outlet Right Ventricle

DORV is a conotruncal anomaly, in which 50% or more of both great arteries arise from the right ventricle (RV). One of the hallmark features of DORV is the presence of a bilateral conus. The degree of conal development on either side and the consequent rotation of the great arteries determine the relative position of the VSD and the degree of commitment of the great arteries to either ventricle [20, 21]. In a rarer variant of DORV with a subaortic VSD, both great arteries arise completely from the RV (200% DORV).

The various physiological categories of DORV with their key morphologic features are as follows [20, 21]:

1. **'Tetralogy type' and DORV with subaortic VSD with PS (40%):**
   (a) Normal subpulmonary conus and minimal subaortic conus.
   (b) There is a loss of aorto–mitral continuity.
   (c) Antero-superior position of the pulmonary valve.
   (d) Aorta overrides the interventricular septum.
2. **'Transposition type', 'Taussig-Bing anomaly', and DORV with subpulmonic VSD (20%):**
   (a) Predominantly subaortic conus with a small subpulmonary conus.
   (b) There is a loss of pulmonary–mitral continuity.
   (c) Aorta is positioned anteriorly and superiorly, such that the aortic position is more rightward relative to the PA.
3. **'VSD type' and DORV with subaortic VSD without PS (15%):**
   (a) Bilateral conus with more subpulmonary conus than subaortic conus.
   (b) There is a loss of mitral–aortic continuity.
   (c) Pulmonary valve is anterior, but the aorta is rotated rightward.
   (d) There is significant left-to-right shunting and increased pulmonary blood flow.
4. **DORV with 'non-committed VSD' (10%):**
   (a) Bilateral conus with equally significant subaortic and subpulmonary tissue.
   (b) VSD, which is not directly related to either great vessel, opens primarily into the RV cavity.
   (c) Tricuspid valve leaflets may be interposed between the VSD and the great arteries, posing a challenge to VSD closure.
5. **DORV with 'doubly committed VSD' (10%):**
   (a) Aortic and pulmonary valve leaflets are separated by a fibrous conal septum.
   (b) Both great vessels override the VSD.
   (c) The pathophysiology is predominantly left-to-right shunting with increased pulmonary blood flow.

7.4.1 Echocardiographic Diagnosis

The cardinal features of DORV can be clearly demonstrated by echocardiography (Fig. 7.6):

- Both great arteries arise from the RV and run a parallel proximal course
- Mitral-semilunar discontinuity (in other words, absence of normal aortic-mitral continuity)

In the PLAX view, mitral-aortic discontinuity is established by recognizing conus tissue beneath the aortic valve, separating it from the mitral valve.

The A4C view demonstrates the degree of aortic override over the crest of the interventricular septum.

Key point: Aortic override of more than 50% and aortic-mitral discontinuity distinguish DORV from tetralogy of Fallot.

7.4.2 Ventricular Septal Defect

- The typical VSD in DORV is a perimembranous VSD extending into the muscular septum, bordered superiorly by the conus.
- The relative position of the VSD with the great arteries (subaortic vs subpulmonary vs doubly committed vs non-committed) determines the commitment of the great arteries to the ventricles as well as the direction of streaming of blood from the RV [20].

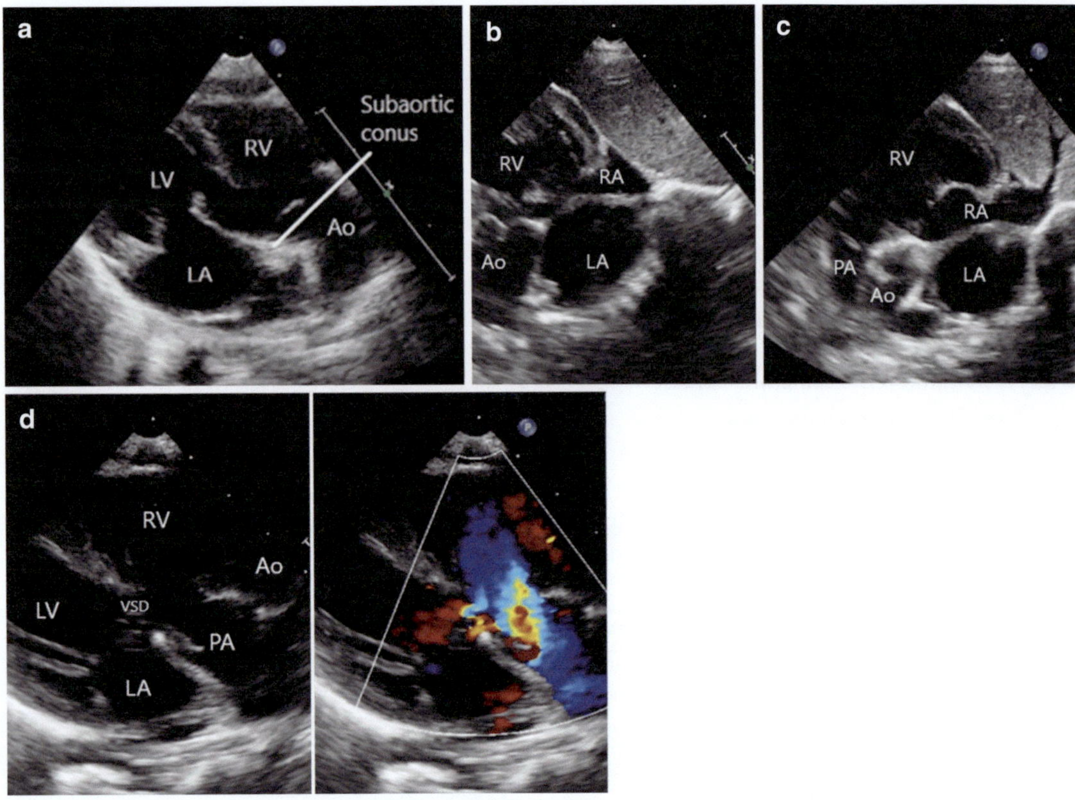

Fig. 7.6 Double outlet right ventricle. (**a**) PLAX view: Normal aortic-mitral continuity is disrupted by subaortic conus; aortic override of the VSD is evidently more than 50%, resulting in commitment of the aortic valve annulus to the RV. (**b**) Subcostal sagittal view depicting the aorta originating from the RV. (**c**) Subcostal sagittal view depicting both great arteries arising from the RV and running a proximal parallel course. (**d**) DORV with anterior aorta and subpulmonic VSD (Taussig-Bing anomaly)

- PLAX, PSAX, subcostal, and A5C views aid in determining the location of the VSD (may refer to the chapter on shunt lesions).

7.4.3 Pulmonary Stenosis

- Pulmonary stenosis—valvular or subvalvular—is present in 50% of DORV cases.
- Subcostal views demonstrate the narrow subpulmonic tunnel in subvalvular stenosis.

7.4.4 Feasibility of Repair

The size of the ventricles and AV valves determines the strategy for DORV repair.

- If the ventricles and the AV valves are adequate in size, biventricular repair is possible.
- If the tricuspid valve or RV is small, 'one and a half ventricle' repair is done.
- Chordal attachments of each AV valve must be noted in relation to the VSD.

One and a half ventricle repair:

In 'one and a half ventricle' repair, the RV receives only the systemic venous blood from the IVC; blood from the SVC reaches the lungs through a cavopulmonary anastomosis such as a bidirectional Glenn shunt.

Feasibility of LV-to-aorta baffle (biventricular repair): Is the VSD routable to the aorta?

The tricuspid valve and the pulmonary valve should be adequately spaced out from each other, and the edge of the VSD must be adequately close to the aortic valve for an LV-to-aorta baffle to be feasible.

IVS-aortic valve distance is an echocardiographic index of the routability of the VSD to the aorta. The distance between the crest of the interventricular septum and the aortic valve (IVS–AV length), measured from the PLAX view, roughly equals the length of intraventricular rerouting. The ratio of IVS-AV length to LV diastolic dimension of less than 0.8 has been found to be associated with lesser surgical mortality and LVOTO following biventricular repair [22].

Key point: Patients with subaortic or subpulmonic VSD can have biventricular repair. Remote or non-committed VSDs are usually not suitable for biventricular repair.

7.4.5 Associated Anomalies [20, 23]

- Ostium secundum ASD (25%), AV canal defects, PDA, additional muscular VSDs.
- Subaortic stenosis, mitral atresia.
- Right aortic arch; coarctation, hypoplasia, or interruption of the aortic arch (10%).
- LSVC to the coronary sinus or left atrium.
- AV discordance; univentricular AV connection; heterotaxy.
- Coronary artery anomalies (10%):
  - Anomalous origin of the anterior descending coronary artery from the RCA is the commonest coronary artery anomaly.
  - Vital in determining the feasibility of conduit placement or coronary transfer during repair.

Points to Report

- Degree of great arterial commitment to RV
- VSD: Site, severity
- Pulmonary stenosis: Site, severity
- Ventricular size and function
- AV valves: Size, chordal attachments
- Routability of VSD to aorta: IVS-AV length and its ratio with LV diastolic dimension
- Associated anomalies: ASD, PDA, LVOTO, coronary artery anomaly

References

1. Perloff JK, Marelli AJ. Congenital pulmonary stenosis. In: Clinical recognition of congenital heart disease. 6th ed. Philadephia: Saunders; 2012. p. 147–68.
2. Lopez L, Colan SD, Frommelt PC, Ensing GJ, Kendall K, Younoszai AK, et al. Recommendations for quantification methods during the performance of a pediatric echocardiogram: a report from the pediatric measurements writing group of the American Society of Echocardiography Pediatric and Congenital Heart Disease Council. J Am Soc Echocardiogr. 2010;23:465–95.
3. Prieto LR, Latson LA. Pulmonary stenosis. In: Allen HD, Driscoll DJ, Shaddy RE, Feltes TF, editors. Moss and Adams' heart disease in infants, children, and adolescents. 8th ed. Philadelphia: Lippincott Williams & Wilkins; 2013. p. 913–38.
4. Hachiro Y, Takagi N, Koyanagi T, et al. Repair of double-chamberedright ventricle: surgical results and long-term follow-up. Ann Thorac Surg. 2001;72:1520–2.
5. McElhinney DB, Chatterjee KM, Reddy VM. Double-chambered right ventricle presenting in adulthood. Ann Thorac Surg. 2000;70:124–7.
6. Edwards J. Congenital malformations of the heart and great vessels. In: Gould S, editor. Pathology of the heart. Springfield: Charles C. Thomas; 1953.
7. Koretzky ED, Moller JH, Korns ME, et al. Congenital pulmonary stenosisresulting from dysplasia of valve. Circulation. 1969;40:43–53.
8. Park MK. Obstructive lesions. In: 5th, editor. Park's pediatric cardiology for practitioners. Philadephia: Saunders; 2016. p. 116–32.
9. Perloff JK, Marelli AJ. Pulmonary atresia with intact ventricular septum. In: Clinical recognition of congenital heart disease. 6th ed. Philadephia: Saunders; 2012. p. 429–38.
10. Bull C, de Leval MR, Mercanti C, Macartney FJ, Anderson RH. Pulmonary atresia and intact ventricular septum: a revised classification. Circulation. 1982;66(2):266–72.
11. Nykanen DG. Pulmonary atresia and intact ventricular septum. In: Allen HD, Driscoll DJ, Shaddy RE, Feltes TF, editors. Moss and Adams' heart disease in infants, children, and adolescents. 8th ed. Philadelphia: Lippincott Williams & Wilkins; 2013. p. 939–58.
12. Luciani GB, Swilley S, Starnes VA. Pulmonary atresia, intact ventricularseptum, and major aortopulmonary collaterals: morphogenetic and surgicalimplications. J Thorac Cardiovasc Surg. 1995;110:853–4.
13. Choi YH, Seo JW, Choi JY, et al. Morphology of tricuspid valve in pulmonaryatresia with intact ventricular septum. Pediatr Cardiol. 1998;19:381–9.
14. Satou GM, Perry SB, Gauvreau K, et al. Echocardiographic predictors ofcoronary artery pathology in pulmonary atresia with intact ventricular septum. Am J Cardiol. 2000;85:1319–24.
15. O'Leary PW, Edwards WD, Julsrud PR, Burkhart HMD. Pulmonary atresia and ventricular septal

defect. In: Allen HD, Driscoll DJ, Shaddy RE, Feltes TF, editors. Moss and Adams' heart disease in infants, children, and adolescents. 8th ed. Philadelphia: Lippincott Williams & Wilkins; 2013. p. 959–68.
16. Tchervenkov CI, Roy N. Congenital heart surgery nomenclature and database project: pulmonary atresia–ventricular septal defect. Ann Thorac Surg. 2000;69(4):97–105.
17. Somerville J. Management of pulmonary atresia. Br Heart J. 1970;32(5):641–51.
18. Wernovsky G, Anderson RH, Kumar K, et al. Anderson's pediatric cardiology. 4th ed. Philadelphia: Elsevier; 2019. p. 660–3.
19. Perloff JK, Marelli AJ. Ventricular septal defect with pulmonary stenosis. In: Clinical recognition of congenital heart disease. 6th ed. Philadephia: Saunders; 2012. p. 316–47.
20. Wright GE, Maeda K, Silverman NH, Hanley FL, Roth SJ. Double outlet right ventricle. In: Allen HD, Driscoll DJ, Shaddy RE, Feltes TF, editors. Moss and Adams' heart disease in infants, children, and adolescents. 8th ed. Philadelphia: Lippincott Williams & Wilkins; 2013. p. 1161–74.
21. Sridaromont S, Ritter DG, Feldt RH, et al. Double-outlet right ventricle. Anatomic and angiocardiographic correlations. Mayo Clin Proc. 1978;53:555–77.
22. Fujii Y, Kotani Y, Takagaki M, Arai S, Kasahara S, Otsuki S, Sano S. The impact of the length between the top of the interventricular septum and the aortic valve on the indications for a biventricular repair in patients with a transposition of the great arteries or a double outlet right ventricle. Interact Cardiovasc Thorac Surg. 2010;10(6):900–5; discussion 905.
23. Sridaromont S, Feldt RH, Ritter DG, et al. Double outlet right ventricle: hemodynamic and anatomic correlations. Am J Cardiol. 1976;38:85–94.

8 Anomalies of Pulmonary Veins, Left Atrium, and Left Ventricular Inflow

8.1 Anomalies of Pulmonary Veins

Anomalies of pulmonary veins fall into the following broad categories:

Abnormal pulmonary venous connections and/or drainage

- **TAPVC**: All the pulmonary veins connect anomalously to one or more of the systemic veins.
- **PAPVC**: One or more of the pulmonary veins connect anomalously to the systemic veins.
- **PAPVD/TAPVD**: In partial or total anomalous pulmonary venous drainage, all pulmonary veins are connected normally to the left atrium. However, there is anomalous drainage of pulmonary venous return into the RA or its tributaries, such as in a common atrium, malposition of septum primum, or sinus venosus ASD.

Pulmonary venous obstruction

- Stenosis in one or more of the pulmonary veins results in obstruction to pulmonary venous return.

Variation in the number of pulmonary veins [1, 2]

- Normally, there are two right and two left pulmonary veins.
- In visceral heterotaxy with asplenia, there may only be a single pulmonary vein on the left side.
- All pulmonary veins may seldom join a common pulmonary vein that drains into the LA.
- The pulmonary veins may sometimes be more than four in number.

8.1.1 Partial Anomalous Pulmonary Venous Connection

In PAPVC, one or more of the pulmonary veins connect anomalously to any of the systemic veins or their tributaries [2]:

- Left-sided pulmonary veins tend to connect to derivatives of the left cardinal system (i.e. the CS or the left innominate vein (LIV)).
- Right-sided pulmonary veins usually connect to derivatives of the right cardinal system (i.e. the SVC or the IVC).

Among all the possible venous connections in PAPVC, the most commonly observed types are:

- Left pulmonary veins connecting to the LIV
- Pulmonary veins from the right lung connecting to the IVC (called **Scimitar syndrome** in the presence of associated right lung hypoplasia—see below)

Supplementary Information The online version contains supplementary material available at https://doi.org/10.1007/978-981-96-0243-8_8.

A. Paul, *Manual of Echocardiography for Congenital Heart Diseases*,
https://doi.org/10.1007/978-981-96-0243-8_8

The other uncommon sites of attachment of left pulmonary veins are the CS, IVC, right SVC, left subclavian vein, or azygous vein.

8.1.1.1 Echocardiographic Diagnosis

The two vital echocardiographic clues that should prompt a thorough search for anomalous pulmonary venous connections are:

- ***Presence of RV volume overload in the absence of an ASD*** (or other pre-tricuspid shunts).
- ***Presence of an abnormally dilated systemic vein(s) with evidence of increased flow by colour Doppler***. For example, an anomalous connection of a left pulmonary vein into the LIV results in dilation of the LIV and SVC with increased flow (Fig. 8.1).

The size, course, connection, and drainage of each pulmonary vein need to be clearly defined before ruling out the possibility of an anomalous pulmonary venous connection (discussed in Chap. 3) [3]. 2-D imaging should be complemented with colour Doppler flow mapping.

- PAPVC to the SVC is best imaged from the suprasternal views or high parasternal long-axis view of the SVC.
- PAPVC to the LIV is best imaged from suprasternal (or high parasternal) short- and long-axis views.
- PAPVC to the CS is best imaged from the subcostal, apical, and parasternal windows by scanning the posterior left AV groove.

8.1.1.2 Scimitar Syndrome

This refers to the association of: (**a**) anomalous connection of one or more of the right pulmonary veins to the IVC (rarely to the RA or portal vein) and (**b**) right lung hypoplasia [4].

The anomalous connection can be imaged from the subcostal view. The other associated abnormalities are [2, 4]:

- Mesocardia or dextrocardia (70%)
- RPA is smaller in size as compared to LPA
- Aortopulmonary collaterals (demonstrated by colour Doppler)

8.1.2 Total Anomalous Pulmonary Venous Connection

In TAPVC, all the pulmonary veins are anomalously connected to one or more of the systemic veins. Commonly, all the pulmonary veins drain into a common confluence, which in turn drains into a systemic vein.

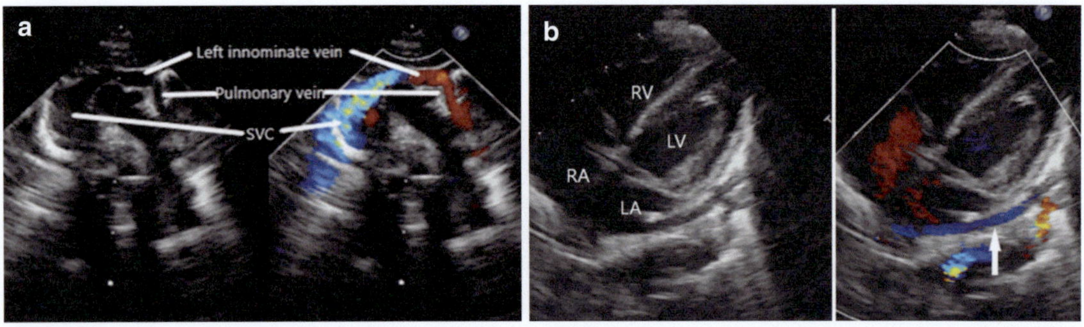

Fig. 8.1 PAPVC with anomalous drainage of the left pulmonary vein. (**a**) The anomalous drainage of the pulmonary vein into the left innominate vein is appreciated from the suprasternal coronal view. Note the increased flow in the LIV and SVC demonstrated by colour Doppler. (**b**) The left pulmonary vein (arrow) is seen coursing posterior to the left atrium without draining into it

Based on the site of drainage of the pulmonary venous confluence, TAPVC is classified by Craig et al. as follows [5]:

1. **Supracardiac TAPVC** (50%): The pulmonary venous confluence (PVC), which is located posterior to the LA (also called **the posterior chamber**), drains into one of the following:
   (a) RSVC or Azygous vein: A venous channel originating from the right side of the PVC ascends anterior to the hilum of the right lung and enters the posterior aspect of the right SVC (or, rarely, the azygous vein).
   (b) Left innominate vein: An anomalous vein (known as **the vertical vein**) originating from the left side of the confluence passes anterior to the LPA and the aortic arch and drains into the LIV close to its origin. The LIV drains into the right SVC normally.
2. **Cardiac TAPVC** (20%): The PVC drains into the CS in the region of the AV groove. The dilated CS, in turn, drains into the RA as usual.
3. **Infracardiac (infradiaphragmatic) TAPVC** (20%): A vertical vein descends from the PVC anterior to the oesophagus and drains into one of the following:
   (a) Portal vein confluence (most common)
   (b) Ductus venosus
   (c) One of the hepatic veins
   (d) Inferior vena cava

   Obstruction of pulmonary venous drainage is common in infradiaphragmatic TAPVC. Obstruction to the flow can be due to intrinsic narrowing or extrinsic compression of the venous channel [2].
   (a) The vertical vein to LIV may be compressed between the LPA and left main bronchus.
   (b) The venous channel to the SVC may be compressed between the RPA and trachea.
4. **Mixed type** (10%): Anomalous connection occurs at two or more levels. In the most common type, left pulmonary veins drain into the LIV, and right pulmonary veins into the CS.

Key point: The presence of an interatrial communication (ASD or PFO) with right-to-left shunt is obligatory in TAPVC. A restrictive interatrial communication resembles obstructed TAPVC clinically.

8.1.2.1 Echocardiographic Diagnosis

The echocardiographic diagnosis of TAPVC requires a high index of suspicion. The features that should alert the echocardiographer towards the possibility of TAPVC are:

- **Signs of RV volume overload** (in the absence of a left-to-right shunt):
  – RA, RV, and pulmonary arteries appear dilated
  – Bowing of the IAS and IVS towards the left, with paradoxical movement of the IVS
- **Signs of LA/LV underfilling:**
  – The LV appears to be compressed by the RV, with a reduction in LV volume
  – The LA appears small and underfilled
- **Right-to-left shunting** of blood across the interatrial communication—ASD/PFO
- **Absence of pulmonary venous inflow** into the LA, with the inability to image the pulmonary veins entering the LA

The presence of all the aforementioned features in a patient confirms the diagnosis of TAPVC.

8.1.2.2 Delineation of Pulmonary Venous Drainage

Once the diagnosis is made, the anatomy of pulmonary venous drainage needs to be delineated to the maximum possible extent. The three elements to be characterized are (**a**) the pulmonary veins, (**b**) the pulmonary venous confluence, and (**c**) the pulmonary venous channel that drains into the systemic vein.

1. **Pulmonary veins:** The course, diameter, connection, and drainage of each pulmonary vein

need to be determined separately, so as to confirm that all four pulmonary veins drain into the pulmonary venous confluence (imaging of pulmonary veins is discussed in Chap. 3). Pulmonary vein hypoplasia and/or stenosis portend a poor prognosis.

2. **Pulmonary venous confluence (Posterior chamber):**
   (a) The posterior chamber appears as an echo-free space behind the LA in supra- and infracardiac TAPVC (Fig. 8.2).
   (b) The size and orientation (horizontal vs vertical) of the PVC and its anatomic relationship with the LA should be clearly delineated.
   (c) The subcostal, parasternal, and suprasternal windows are useful in imaging the PVC.
3. **Common pulmonary venous channel:**

   Supracardiac TAPVC: The venous channel in supracardiac TAPVC (vertical vein) is best imaged from the suprasternal or high parasternal windows (Fig. 8.3). In the suprasternal short-axis view, colour Doppler imaging shows the vertical venous flow as a low-velocity jet directed away from the heart and draining into the dilated LIV.

   Cardiac TAPVC: Parasternal and apical views are useful in delineating the drainage of the venous channel into the RA through the CS.

   Infradiaphragmatic TAPVC: The anomalous pulmonary venous channel is visualized from the subcostal coronal and sagittal views. It is often seen as a distinct chamber, the connections of which can be delineated by tracing it in the transverse and vertical directions. In the transverse plane, the channel is located in an intermediate position between the descending aorta (which lies posteriorly and to the left) and the IVC (which is placed anteriorly and to the right) [2, 3]. Visualization of the vein may be difficult if it is compressed by a congested liver.

Doppler interrogation aids in differentiating the flow characteristics among the various abdominal vessels:

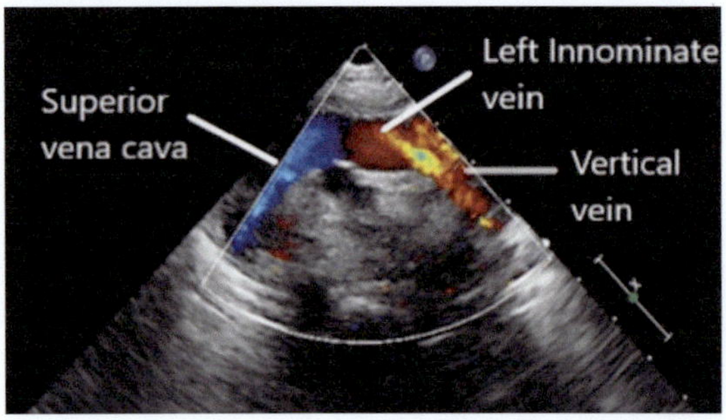

Fig. 8.3 TAPVC. Suprasternal view depicting the vertical vein draining into the left innominate vein. In contrast to PAPVC depicted in Fig. 8.1, it can be seen that the LIV and SVC appear more dilated owing to the larger volume of pulmonary venous blood drained by the vertical vein

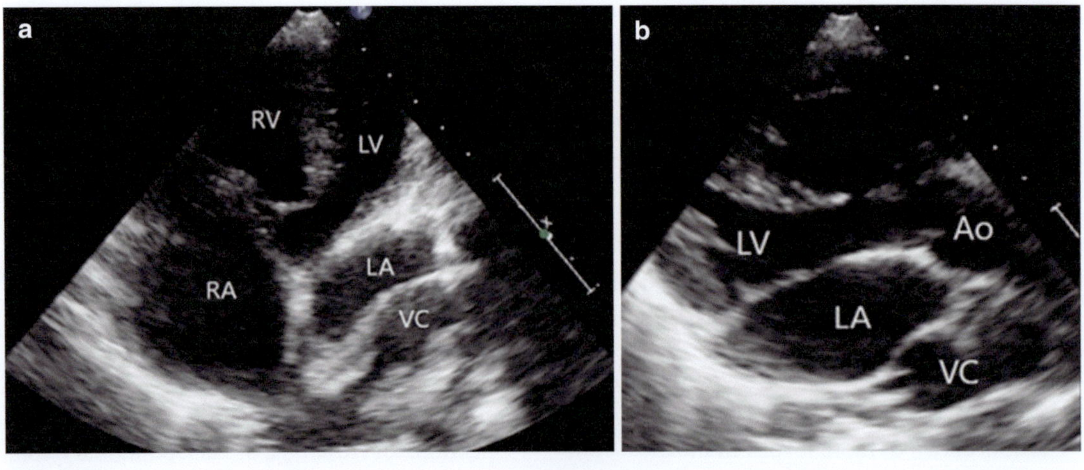

Fig. 8.2 Supracardiac TAPVC with pulmonary venous confluence (PVC) posterior to the left atrium (LA), as seen from the apical view (**a**) and PLAX view (**b**)

- Flow in the descending aorta is in a direction away from the heart and has a systolic laminar profile.
- Flow in the IVC is towards the heart with a nearly continuous, phasic profile.
- Flow in the common pulmonary venous channel resembles the venous flow pattern of the IVC, but it is directed away from the heart towards the abdomen.

Key point: The flow in anomalous pulmonary venous channels is directed away from the heart. Detection of any venous flow directed away from the heart should flag the possibility of anomalous pulmonary venous drainage.

8.1.2.3 Obstruction to Pulmonary Venous Flow

Although obstruction to pulmonary venous drainage is more common in infradiaphragmatic TAPVC, it can occur in supracardiac TAPVC as well. The most common site of obstruction is the junction of the pulmonary venous channel with the systemic vein. The course of the pulmonary venous channel in relation to the branch pulmonary arteries and the bronchi determines the risk of obstruction [6].

Pulmonary venous flow in an unobstructed vessel is characterized by a low-velocity, phasic laminar pattern, with brief flow reversal during atrial systole. The echocardiographic features suggestive of obstructed pulmonary venous drainage are:

- Luminal narrowing associated with turbulence and flow acceleration
- Dilation of the pulmonary venous channel proximal to the obstruction
- High-velocity, continuous, non-phasic venous flow profile in spectral Doppler

8.1.2.4 Interatrial Communication

As mentioned earlier, a right-to-left interatrial shunt is obligatory in TAPVC. TAPVC with a restrictive interatrial shunt is physiologically similar to an obstructed TAPVC, with reduced systemic output.

8.1.2.5 Exclude Additional Structural Cardiac Anomalies [6]

- TAPVC to the RA occurs in visceral heterotaxy syndromes, more commonly with polysplenia than asplenia. Hence, anomalies such as common AV canal, conotruncal anomalies, pulmonary or mitral atresia, and systemic venous anomalies may be associated.
- TAPVC may be associated with TGA, TOF, single ventricle, truncus arteriosus, tricuspid/pulmonary atresia, HLHS, coarctation of the aorta, vascular sling, and VSD.

Points to Report

- Extent of anomalous pulmonary venous connection—Partial/Total
- Course and drainage of each pulmonary vein
- Pulmonary venous confluence
  - Location and relationship with the LA
  - Size
  - Orientation: horizontal/vertical
- Pulmonary venous channel
  - Course
  - Drainage: Supracardiac/Cardiac/Infracardiac
  - Obstruction
- Interatrial communication and its adequacy
- Additional anomalies

8.1.3 Pulmonary Vein Stenosis

Pulmonary vein stenosis occurs either as diffuse hypoplasia or atresia of the pulmonary vein(s), or as a discrete stenosis at the point of entry of the veins into the LA [1].

Due to the increase in the flow velocity distal to the stenosis, turbulence is observed in the posterior LA at the site of the pulmonary venous inflow [3]. Apical or subcostal four-chamber views, or a posteriorly angulated suprasternal short-axis view ('crab view'), are useful in demonstrating this turbulent flow.

Spectral Doppler imaging aids in assessing the inflow pattern and flow velocity at the pulmonary vein orifice. Pulmonary venous hypertension results in elevated PASP and dilated RA-RV.

8.2 Anomalies of Left Atrium

Anomalous fibrous membranes in the left atrium can cause obstruction to the LV. This can result in: **(a) cor triatriatum sinister** (membrane located in the middle of the LA), or **(b) supravalvular stenosing ring** (membrane at or near the level of the mitral annulus).

8.2.1 Cor Triatriatum (Sinister)

Cor triatriatum in the left atrium is called **cor triatriatum sinister**, to distinguish it from **cor triatriatum dexter,** the right atrial equivalent of the same anomaly (rarer). In cor triatriatum, a thick fibro-muscular septum or membrane divides the atrium into two distinct chambers [1, 2]:

The proximal or superior chamber **(pulmonary venous chamber)** receives the pulmonary venous blood.

The distal or inferior chamber **(true atrium)** has three defining morphological features:

- The atrioventricular valve is connected to the distal chamber.
- The atrial appendage communicates with this chamber.
- The true atrial septum with the fossa ovalis forms a part of the distal chamber.

Passage of blood from the proximal chamber to the distal chamber is facilitated by the opening(s) in the dividing membrane. The membrane often has the shape of a windsock, which is directed towards the AV valve [7].

Loeffler's classification is based on the number and size of fenestrations in the fibromuscular membrane [8]:

- **Group 1**: There is no connection between the two chambers. The pulmonary venous chamber may connect directly to the RA. Otherwise, some of the pulmonary veins may drain anomalously.
- **Group 2**: There are a few small fenestrations in the fibromuscular membrane.
- **Group 3**: There is a large single opening in the fibromuscular membrane.

As is obvious, groups 1 and 2 present with features of pulmonary venous obstruction in infancy/childhood, while group 3 is mostly diagnosed in adults.

On the basis of the nature of pulmonary venous drainage into the proximal chamber and the communication between the proximal and the distal chambers, the anatomic variants of cor triatriatum can be categorized as follows [2, 3]:

1. **Proximal chamber receives all pulmonary veins**
   (a) Proximal chamber communicates with the distal chamber
      No anomalous connections present **(classic cor triatriatum)**
      Anomalous connections present: Either a direct connection between the proximal chamber and the RA, or anomalous pulmonary venous connection
   (b) Proximal chamber does not communicate with the distal chamber
      Proximal chamber connects directly to the RA or
      Total anomalous pulmonary venous connection
2. **Proximal chamber receives some pulmonary veins (Subtotal cor triatriatum)**
   (a) Proximal chamber communicates with the distal chamber.
      The remaining pulmonary veins connect either directly to the distal chamber or anomalously to a systemic vein.

(b) Proximal chamber connects directly to the RA, while the remaining pulmonary veins connect to the distal chamber.

When blood from the pulmonary venous chamber drains into the RA, the physiology is similar to that of TAPVC. When the fenestration in the dividing membrane is restrictive with no alternative pathway for the pulmonary venous blood, the physiology resembles that of pulmonary venous obstruction.

8.2.1.1 Echocardiographic Diagnosis

Fibromuscular Membrane

The perpendicular orientation of the ultrasound beam with the dividing membrane in the A4C view aids in clear visualization of this structure. The membrane is seen as a curvilinear, echogenic structure extending from the anterosuperior wall of the LA to the posterolateral wall (Fig. 8.4). The membrane may exhibit diastolic excursion towards the mitral valve [3, 7].

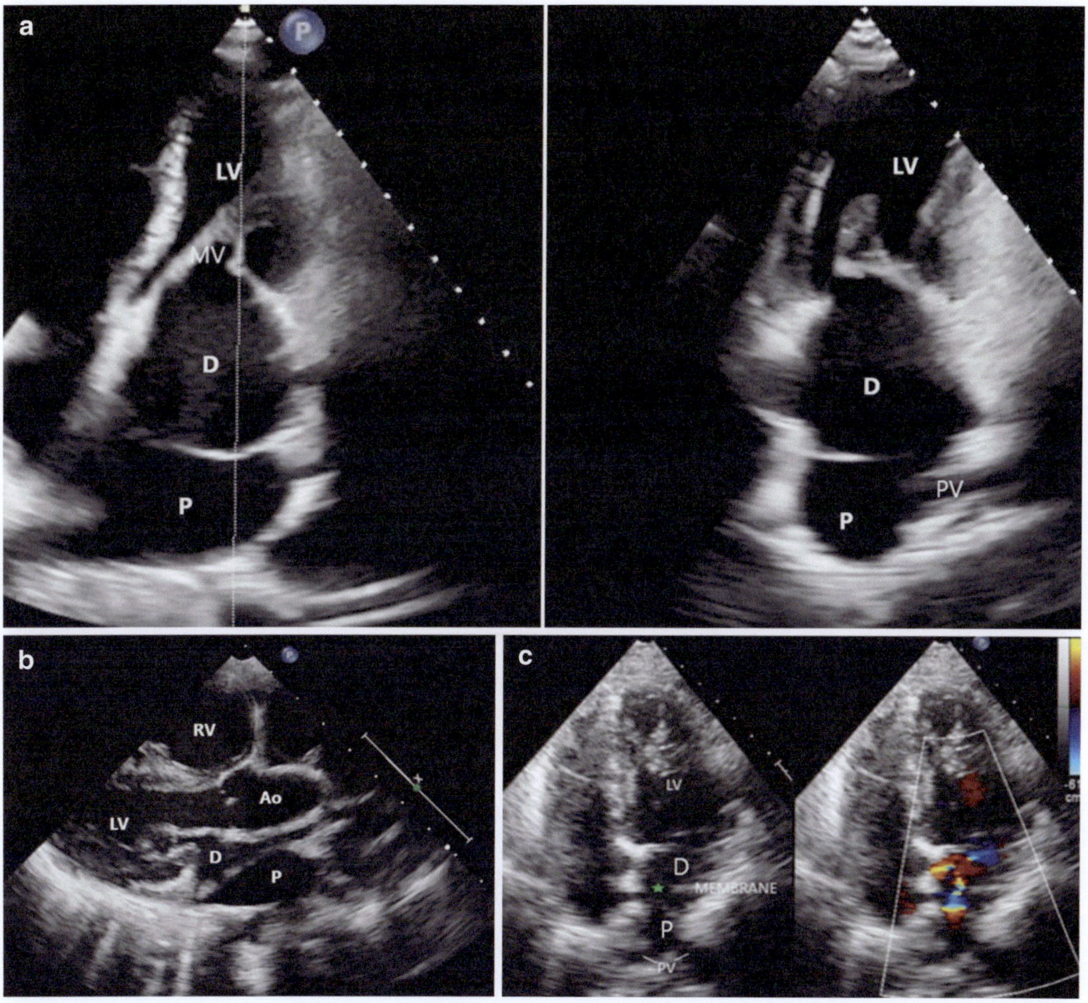

Fig. 8.4 Cor triatriatum: (**a**) Apical views depicting a fibromuscular membrane dividing the left atrium into proximal (P) and distal (D) chambers. Note the pulmonary vein (PV) draining into the proximal chamber. (**b**) PLAX view depicting the posterior alignment of the proximal chamber such that the distal chamber is in proximity to the interatrial septum. (**c**) Apical view showing the turbulent flow from the proximal chamber to the distal chamber across the membrane orifice (denoted by a green star)

Pulmonary Venous Chamber

- The attachments of all the pulmonary veins to the superiorly located proximal chamber need to be ascertained.
- Anomalous pulmonary venous connections need to be carefully searched for.
- Anomalous communication between the proximal chamber and the right atrium is particularly important in cases where the communication between the proximal and the distal chambers is inadequate, and a right-to-left shunt is obligatory.

True Atrium

The inferiorly located distal chamber encompasses the left atrial appendage and the foramen ovale. Anomalous connection of one or more pulmonary veins to the distal chamber needs to be searched for.

Communicating Orifice in the Membrane

Visualization of the orifice by 2D imaging is often difficult owing to the posterior location of the orifice. Colour Doppler imaging of the flow across the membrane aids in localizing the opening in the membrane [3].

Spectral Doppler interrogation across a restrictive orifice reveals a diastolic flow profile similar to that of mitral stenosis. The mean and peak diastolic gradients across the membrane need to be documented. The documented gradients may be falsely low if there is a non-restrictive communication between the proximal chamber and the RA, which decompresses the proximal chamber.

8.2.1.2 Pulmonary Hypertension and Ventricular Function

PA pressures need to be documented, as pulmonary venous hypertension may ensue as a result of restrictive outflow from the proximal chamber.

The RV is hypertrophic and dilated in most cases, while the RA is dilated in about 25% of cases [7]. RV function should be documented. LV function can be adversely affected due to chronic underfilling and/or pulmonary hypertension.

8.2.1.3 Associated Anomalies

- Anomalous pulmonary venous drainage
- Atrial septal defect
- Persistent LSVC

Points to Report

- Fibromuscular membrane: Location; Attachments; Presence of fenestration(s)
- Proximal pulmonary venous chamber:
  - Pulmonary venous connections
  - Abnormal communication with the RA
- Distal chamber (True atrium):
  - Anomalous pulmonary venous connections to the distal chamber
  - Pressure gradient across the mitral valve
- Communicating orifice:
  - Location of the orifice within the membrane
  - Size of the orifice
  - Pressure gradient across the orifice
- PA pressure
- RV size and function
- LV size and function
- Associated anomalies

8.2.2 Supravalvular Ring Mitral Stenosis (Supramitral Stenosing Ring)

Supramitral stenosing ring, which is a component of Shone's complex, seldom occurs as an isolated defect (10%) [1, 9]. In supramitral ring, the fibromuscular membrane in the left atrium is located in close proximity to the mitral valve leaflets. The attachment of the membrane can be at the MV annulus as well.

8.2.2.1 Echocardiographic Diagnosis (Fig. 8.5)

The echocardiographic features differentiating supramitral ring from cor triatriatum are:

- The anomalous membrane in supramitral ring is distal to the LA appendage and closer to the mitral valve.
- In supramitral ring, the membrane moves away from the mitral valve in diastole.

Supravalvular stenosing ring must be differentiated from thickened and calcified mitral annulus.

8.2.2.2 Functional Abnormality of Mitral Valve

Supramitral ring can cause obstruction to the LV inflow through various mechanisms [2, 9]:

- The membrane can extend into the region of MV orifice, causing obstruction to LV inflow.
- The membrane may adhere to the MV leaflets and restrict their mobility.
- The MV apparatus may be abnormal (fused leaflets, abnormal papillary muscle) resulting in valvular or subvalvular mitral stenosis.

The proximity of the membrane to the valve can result in mitral regurgitation as well.

The peak and mean gradients across the LV inflow need to be documented carefully. The level of the obstruction should be delineated using PWD and colour flow mapping.

Key point: In supravalvular ring mitral stenosis, the flow acceleration and turbulence occur at the level of the MV annulus, in contrast to valvular mitral stenosis, where obstruction is at the level of the leaflet tips.

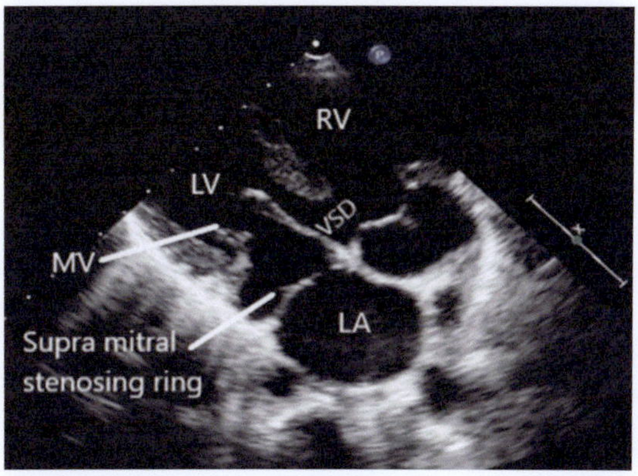

Fig. 8.5 Supramitral stenosing ring in a child with DORV

8.2.2.3 Associated Anomalies

- Atrial septal defect
- Ventricular septal defect
- Persistent left superior vena cava
- TOF, DORV

8.3 Anomalies of Mitral Valve

8.3.1 Congenital Mitral Stenosis

Congenital mitral stenosis, as an aetiologic entity, is much less common than rheumatic mitral stenosis. The different anatomic variants under this category are [1, 10]:

8.3.1.1 Mitral Valve Dysplasia and Hypoplasia

Dysplastic mitral valve is the most common cause of isolated congenital MS. Dysplastic mitral valve apparatus is characterized by:

- Thickened leaflets.
- Narrow or absent interchordal spaces.
- Deformed papillary muscles.
- Concomitant mitral regurgitation is common.
- Isolated dysplasia of PML often results in mitral regurgitation alone.

8.3.1.2 Parachute Mitral Valve

In a parachute mitral valve, all the chordae insert into a single dominant large papillary muscle [10, 11]. One or more small rudimentary papillary muscle(s) may be present. The associated leaflet dysplasia and chordal tethering can result in stenosis and/or regurgitation.

- The number, size, and location of the papillary muscles are assessed from PSAX view.
- PLAX view reveals mitral valve thickening, restricted mobility of the deformed leaflets, and chordal thickening and fusion.

8.3.1.3 Mitral Arcade (Hammock Mitral Valve)

Mitral valve arcade (Fig. 8.6) is characterized by [12, 13]:

- Thick, muscularized, poorly differentiated, and shortened chordae with direct attachment of the papillary muscles to the AML.
- Thickened papillary muscles are joined together by a fibromuscular band.
- Obliterated interchordal spaces and poor differentiation of the subvalvular apparatus.
- The chordae attached to the PML may be better differentiated.
- Poorly developed papillary muscle architecture with multiple small papillary muscle heads may be observed.

The most common functional abnormality in mitral arcade is mitral regurgitation, owing to the limited coaptation of the tethered valve. Valvular insufficiency can result from cleft leaflet, AML hypoplasia, shortened chordae, and/or annular dilatation as well [10]. Restricted leaflet excursion results in progressive mitral stenosis as well.

Key point: Apical views demonstrate the 'arcade-like' appearance from the LV aspect, while PSAX view reveals the 'hammock-like' appearance of the mitral valve from the LA aspect.

8.3.1.4 Double-Orifice Mitral Valve

In double orifice mitral valve, there is duplication of the mitral orifice with or without fusion of subvalvular structures [14]. All the chordae associated with one orifice insert into the same papillary muscle.

- Commonly seen in association with atrioventricular septal defect.
- Regurgitation is more common than stenosis.
- Visualization of two separate mitral valve orifices in the PSAX view clinches the diagnosis.
- The subvalvular apparatus of each orifice can be appreciated from PLAX and apical views.
- When one of the orifices is imperforate, the presence of the imperforate orifice is recognized by the subvalvular apparatus corresponding to the orifice in the LV.

8.3.2 Congenital Mitral Regurgitation

Congenital anomalies of the mitral valve usually occur in association with other cardiac anoma-

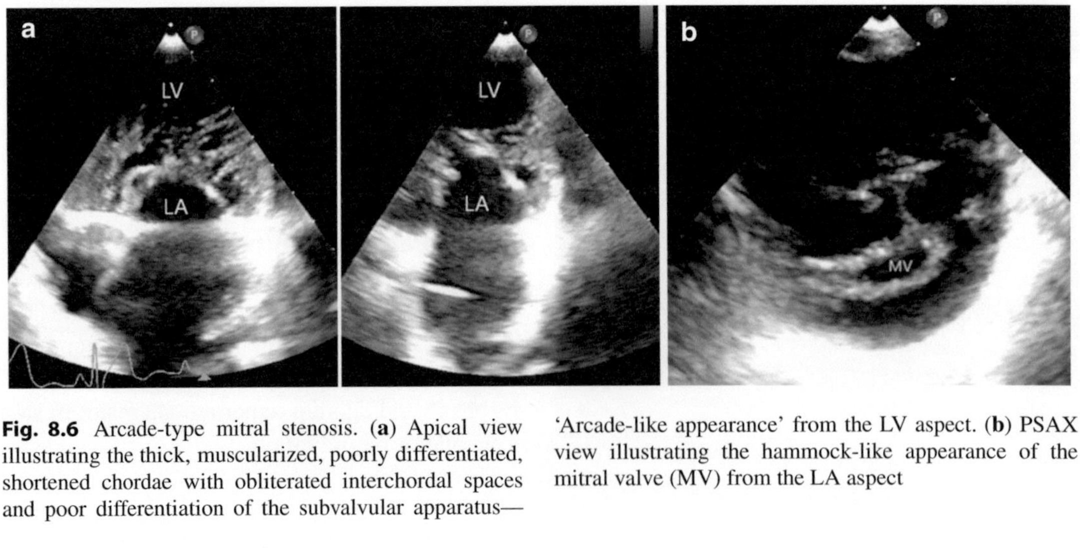

Fig. 8.6 Arcade-type mitral stenosis. (**a**) Apical view illustrating the thick, muscularized, poorly differentiated, shortened chordae with obliterated interchordal spaces and poor differentiation of the subvalvular apparatus—'Arcade-like appearance' from the LV aspect. (**b**) PSAX view illustrating the hammock-like appearance of the mitral valve (MV) from the LA aspect

lies. Most of the congenital anomalies involving the mitral valve apparatus discussed in the previous section cause mitral regurgitation as the predominant functional abnormality.

Anatomic classification [15]:

- **Type 1 (Supravalvar)**: Abnormality above the level of mitral leaflets and annulus, such as supravalvular mitral ring
- **Type 2 (Valvar)**: Abnormality at the level of mitral leaflets and annulus, such as dysplastic or myxomatous MV, cleft MV, and double-orifice MV
- **Type 3 (Subvalvar)**: Abnormality of chordae tendineae and/or papillary muscles, such as seen in mitral valve arcade
- **Type 4 (Mixed)**: Combination of two or more of the above types

Congenital causes of mitral regurgitation such as mitral valve dysplasia, mitral valve arcade, double orifice mitral valve, supramitral ring, and parachute mitral valve have been discussed in the previous sections.

8.3.2.1 Cleft Mitral Valve

Cleft mitral valve can occur in association with AV septal defect or as an isolated defect. The term 'zone of apposition' has been used to denote the cleft in AVSD, differentiating it from an isolated cleft of AML. The extent of the cleft in the AML can be variable, ranging from being a full-length defect to one that involves only the tip of the leaflet. When the edges of the cleft are not supported by chordae tendineae, mitral regurgitation results. Chordae tendineae may either insert into the crest of the interventricular septum or straddle an anterior VSD.

Key point: When the cleft extends along the entire length of the AML, the resultant MR tends to be more severe.

The features differentiating an isolated mitral valve cleft from the cleft of AV septal defect are [16]:

- Isolated cleft points towards the LVOT or aortic valve annulus (11 o'clock position in PSAX view), whereas the cleft of AVSD is directed towards the inlet septum (9 o'clock position in PSAX view).
- The papillary muscles in an isolated cleft are normally positioned, whereas the posterior papillary muscle is rotated laterally in AVSD.

Cardiac anomalies most commonly associated with an isolated cleft mitral valve are [17]:

- VSD (50%)
- LVOT obstruction (40%): valvular or subvalvular aortic stenosis
- Ostium secundum ASD (20%)
- Syndromic associations (60%): trisomy 21, CHARGE association, and heterotaxia

8.3.3 Mitral Valve Prolapse (Barlow's Syndrome)

Mitral valve prolapse (MVP) is characterized by abnormal protrusion of the whole or part of the mitral valve leaflet(s) into the left atrium during systole. The involvement of the leaflets can range from a single scallop of one leaflet to both leaflets in their entirety. The PML is involved more frequently than the AML.

MVP is classified into non-classic and classic variants based on the thickness of the MV leaflets:

- **Non-classic MVP**: Mitral valve leaflets are morphologically normal with a thickness of 0–5 mm.
- **Classic MVP**: Mitral valve is thickened (myxomatous) with a leaflet thickness of more than 5 mm.

Classic MVP is further classified based on the coaptation of the leaflet tips:

- **Symmetric**: The coaptation point of the prolapsed leaflet tips is above the plane of the mitral annulus, with both leaflets prolapsing almost symmetrically.
- **Asymmetric**: The involvement of the leaflets or the extent of the prolapse is not symmetric; one leaflet prolapses more than the other. Asymmetric MVP can be due to a flail leaflet.

- **Flail**: A flail leaflet results from an anatomic disruption of a portion of the mitral apparatus. There is unrestricted motion of a part of the leaflet, usually the tip. A flail leaflet tip usually points towards the LA. An eccentric jet of MR results, with the jet directed away from the flail leaflet.
- **Non-flail**: Asymmetric MVP without flail leaflets; the resultant MR is less severe.

Secondary mitral valve prolapse:
MVP can result as a consequence of a variety of aetiologic conditions such as:

- Connective tissue diseases such as Marfan syndrome and Ehlers-Danlos syndrome
- Hypertrophic cardiomyopathy
- Rheumatic heart disease
- Ischaemic heart disease

8.3.3.1 Echocardiographic Diagnosis

MVP is defined as displacement of mitral valve leaflet(s) ***more than 2 mm*** superior to the plane of the mitral annulus in the PLAX view. The plane of the mitral annulus is demarcated by the line joining the insertion points of the mitral leaflets in the long-axis view (Fig. 8.7). The echocardiographic view in which MVP is best appreciated depends on the portion of the MV that is involved [18].

The diagnosis of MVP is made when one or both leaflets transgress the plane of the mitral annulus, creating a 'buckled' appearance. The buckling can be either holosystolic or confined to late systole [18].

- Mitral valve bowing demonstrable in the A4C view alone has been considered less specific for the diagnosis of MVP, as the A4C view images the mitral annulus at the two low points. The A2C view is considered more specific than A4C for the diagnosis of MVP [19].
- Buckling of mitral leaflet(s) during systole can be demonstrated with M-mode echocardiography [18]:
  - The M-mode beam should be aligned to include the area just behind the mitral annulus.
  - The normal movement of the base of the heart can mimic or mask MVP in M-mode.
- Leaflet thickening and redundancy in severe myxomatous disease may result in:
  - Mass-like appearance, which could be confused with vegetation or tumour.
  - Appearance of a cystic structure.

  Flail leaflet is characterized by [10, 18]:

- Failure of leaflet coaptation
- Rapid systolic movement of the affected leaflet tip within the left atrium
- Severe eccentric jet MR

In children with clinical features of MVP, echocardiography may only show:

- Systolic straightening or doming of the mitral leaflets
- Posterosuperior displacement of the coaptation point of the MV

Key point: The PLAX view is the most reliable view for the diagnosis of MVP because this view images the mitral annulus at its two high points.

8.3.3.2 Grading of Severity of Prolapse

The following **echocardiographic grading for the severity of MVP** was proposed in 1987 by Cohen [20], based on the findings in either PLAX or A4C view. However, this is rarely used nowadays.

- **Grade I or 'mild prolapse**: The maximally prolapsed segment or the point of coaptation extends ≤5 mm above the plane of the mitral ring.
- **Grade II or 'moderate' prolapse**: The maximally prolapsed segment or the point of coaptation extends >5 mm above the plane of the mitral ring.
- **Grade III or 'severe' prolapse**: Grade II with at least one of the following:

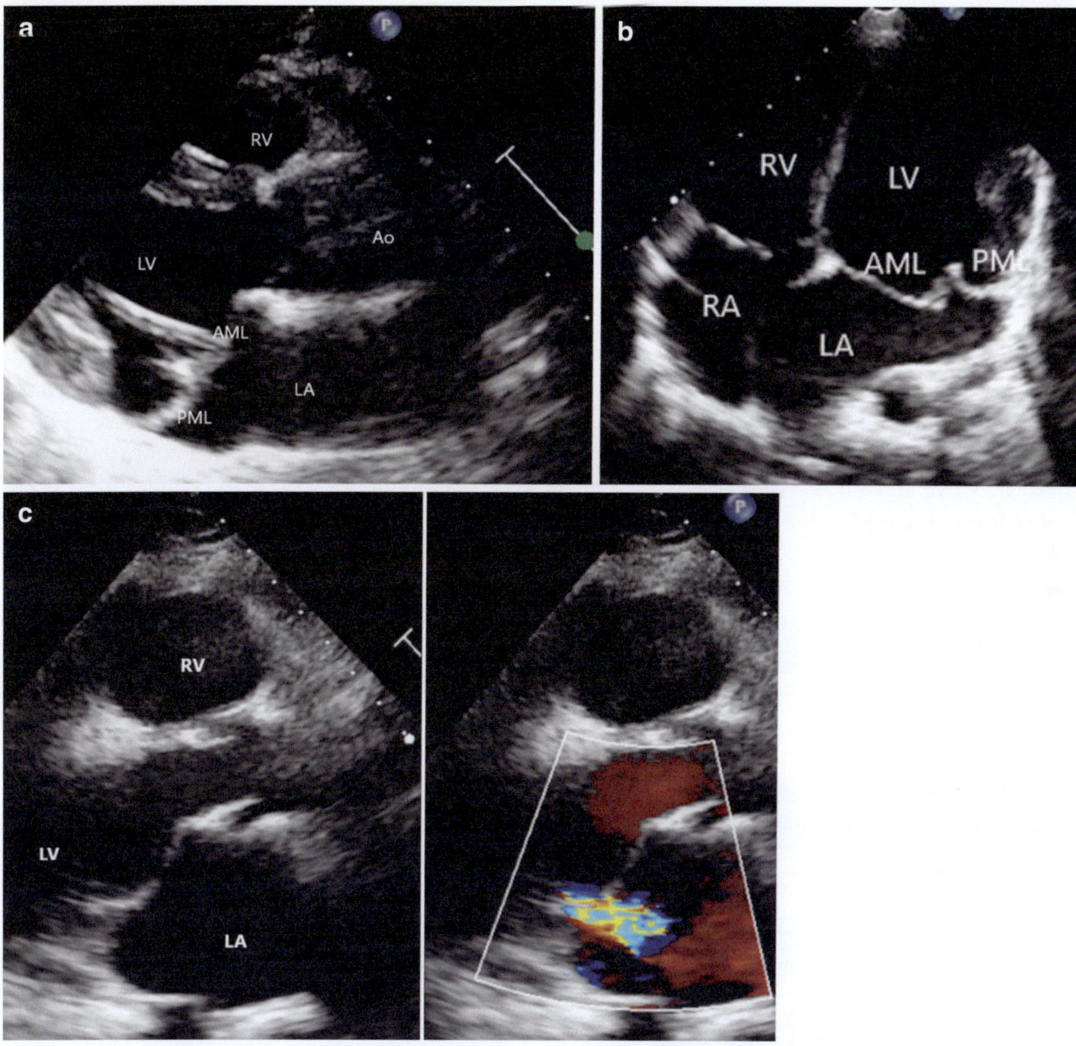

Fig. 8.7 Mitral valve prolapse. (**a**) Bileaflet prolapse in PLAX view. (**b**) AML prolapse as seen in the apical view. Note the dilated left atrium, suggestive of significant MR. (**c**) PLAX view depicting a case of MVP with the MR jet visualized on colour flow Doppler

- Evidence of redundancy (defined as infolding of tissue on itself in any view).
- Significant homogeneous thickening of the body of a leaflet (as opposed to the thickening of the surface alone, as in vegetation) in more than one view.

8.3.3.3 Severity of Mitral Regurgitation

The severity of mitral regurgitation is the most important functional aspect of MVP. The symmetric form of classic MVP may result in a central jet MR. However, the extent of prolapse of different scallops of the mitral leaflets is often asymmetric, resulting in an eccentric jet MR, with the jet directed away from the affected leaflet. The MR in a flail leaflet is almost always severe.

The echocardiographic features suggestive of severe MR are [21]:

- Dilated LA and LV
- Colour flow area of central jet MR >50% of LA area, or an eccentric wall-impinging jet of variable size

- Holosystolic, dense, and triangular CWD jet
- Vena contracta width ≥0.7 cm (>0.8 cm, if an average of measurements from A2C and A4C views are taken)
- E-wave dominant mitral inflow (>1.2 m/s)
- Systolic flow reversal in pulmonary veins, demonstrated by spectral Doppler
- Effective regurgitation orifice area (EROA) by PISA ≥0.40 cm^2
- Regurgitation volume ≥ 60 mL
- Regurgitation fraction ≥50%

8.3.3.4 Associated Abnormalities

- Aortic valve and aorta:
  - Bicuspid aortic valve may be associated with MVP.
  - Marfan's syndrome can be associated with aortic regurgitation and aortic aneurysm, in addition to MVP.
  - Rheumatic valvular heart disease can affect aortic valve as well.
- Hypertrophic cardiomyopathy and ischaemic heart disease.
- Infective endocarditis can cause a flail mitral valve.

Points to Report

- Leaflet(s) or parts of leaflets involved: AML/PML/Bileaflet/Scallops
- Morphology:
  - Leaflet thickness: Classic/ Non-classic
  - Coaptation of leaflet tip: Symmetric/ Asymmetric
  - Flail/Non-flail
- Severity of prolapse: Grade I, II, III
- Severity of MR
- LA and LV size; LV function; Pulmonary hypertension
- Associated cardiac abnormalities: Hypertrophic cardiomyopathy; Bicuspid aortic valve; Aortic aneurysm; Infective endocarditis; Ischaemic heart disease

8.3.4 Miscellaneous Anomalies

8.3.4.1 Ebsteins-Like Anomaly of the Mitral Valve

- Posterior mitral leaflet is apically displaced.
- Atrialized portion of LV is not thinned [10].
- Rare as compared to Ebstein's anomaly of the tricuspid valve.

8.3.4.2 Straddling of the Mitral Valve

Straddling of the mitral valve refers to the anomalous insertion of MV chordae tendineae into the contralateral ventricle through a conoventricular-type malalignment VSD [10]. Straddling is commonly associated with conotruncal anomalies such as DORV and TGA.

Straddling usually involves the AML. The chordal apparatus has variable attachments in the RV, the crest of the IVS, or a papillary muscle on the proximal or distal IVS. There is usually an associated cleft in the MV with an eccentric orifice, pointing more towards the VSD.

Echocardiographic Diagnosis

- A straddling mitral valve is diagnosed by identifying the MV subvalvular apparatus in both ventricles.
- A4C view is useful in determining the extent of straddling.
- PSAX view aids in identifying the presence and location of accessory papillary muscles, if any.
- Off-axis images are often necessary to delineate the attachments of the straddled valve:
  - Counterclockwise rotation of the transducer from the standard subcostal sagittal plane produces an oblique coronal image (subcostal inflow–outflow view) that displays both ventricles and the infundibulum.
  - Straddling MV leaflets and chordae can be visualized from this view.

References

1. Perloff JK, Marelli AJ. Congenital obstruction to left atrial flow: mitral stenosis, cor triatriatum, pulmonary vein stenosis. In: Clinical recognition of congenital heart disease. 6th ed. Philadephia: Saunders; 2012. p. 129–40.
2. Brown DW, Geva T. Anomalies of the pulmonary veins. In: Allen HD, Driscoll DJ, Shaddy RE, Feltes TF, editors. Moss and Adams' heart disease in infants, children, and adolescents. 8th ed. Philadelphia: Lippincott Williams & Wilkins; 2013. p. 809–39.
3. Brown DW. Pulmonary venous anomalies. In: Lai WW, Mertens LL, Cohen MS, et al., editors. Echocardiography in pediatric and congenital heart disease: from fetus to adult. 1st ed. Hoboken: Wiley; 2009. p. 119–42.
4. Neill CA, Ferencz C, Sabiston DC. The familial occurrence of hypoplastic right lung with systemic arterial supply and venous drainage: "scimitar syndrome". Johns Hopkins Med J. 1960;107:1–21.
5. Craig JM, Darling RC, Rothney WB. Total pulmonary venous drainage into the right side of the heart; report of 17 autopsied cases not associated with other major cardiovascular anomalies. Lab Investig. 1957;6:44–64.
6. Delisle G, Ando M, Calder AL, et al. Total anomalous pulmonary venous connection: report of 93 autopsied cases with emphasis on diagnostic and surgical considerations. Am Heart J. 1976;91:99–122.
7. Niwayama G. Cor triatriatum. Am Heart J. 1960;59:291–317.
8. Loeffler E. Unusual malformation of the left atrium: pulmonary sinus. Arch Pathol. 1949;48(5):371–6.
9. Sullivan ID, Robinson PJ, De Leval M, et al. Membranous supravalvular mitral stenosis: a treatable form of congenital heart disease. J Am Coll Cardiol. 1986;8:159–64.
10. Mackie AS, Smallhorn JF. Anatomical and functional mitral valve abnormalities in the pediatric population. In: Allen HD, Driscoll DJ, Shaddy RE, Feltes TF, editors. Moss and Adams' heart disease in infants, children, and adolescents. 8th ed. Philadelphia: Lippincott Williams & Wilkins; 2013. p. 1003–22.
11. Shone JD, Sellers RD, Anderson RC, et al. The developmental complex of "parachute mitral valve," supravalvular ring of left atrium, subaortic stenosis, and coarctation of aorta. Am J Cardiol. 1963;11:714–25.
12. Pacileo G, Russo MG, Calabro R. Anomalous mitral arcade: echocardiographic and color flow findings. Echocardiography. 1991;8:657–9.
13. Layman TE, Edwards JE. Anomalous mitral arcade. A type of congenital mitral insufficiency. Circulation. 1967;35:389–95.
14. Trowitzsch E, Bano-Rodrigo A, Burger BM. Two-dimensional echocardiographic findings in double orifice mitral valve. J Am Coll Cardiol. 1985;6:383–7.
15. Mitruka SN, Lamberti JJ. Congenital heart surgery nomenclature and database project: mitral valve disease. Ann Thorac Surg. 2000;69:S132–46.
16. Kohl T, Silverman NH. Comparison of cleft and papillary muscle position in cleft mitral valve and atrioventricular septal defect. Am J Cardiol. 1996;77:164–9.
17. Timóteo A, Galrinho A, Fiarresga A, et al. Isolated cleft of the anteriormitral valve leaflet. Eur J Echocardiogr. 2007;8:59–62.
18. Feigenbaum H. Echocardiography in the management of mitral valve prolapse. Aust NZ J Med. 1992;22:550–5.
19. Levine RA, Stathogiannis E, Newell JB, et al. Reconsideration of echocardiographic standards for mitral valve prolapse: lack of association between leaflet displacement isolated to the apical four chamber view and independent echocardiographic evidence of abnormality. J Am Coll Cardiol. 1988;11:1010–9.
20. Cohen IS. Two-dimensional echocardiographic mitral valve prolapse: evidence for a relationship of echocardiographic morphology to clinical findings and to mitral annular size. Am Heart J. 1987;113(4):859–68.
21. Bonow R, O'Gara P, Adams D, et al. 2020 Focused update of the 2017 ACC expert consensus decision pathway on the management of mitral regurgitation: a report of the American College of Cardiology Solution Set Oversight Committee. JACC. 2020;75(17):2236–70.

9 Left Ventricular Outflow Anomalies

9.1 Left Ventricular Outflow Tract Obstruction

LVOTO can occur at multiple levels—subvalvular (proximal to the aortic valve), valvular, and/or supravalvular (distal to the aortic valve). The morphologic variants of LVOTO at different anatomic levels can be categorized as follows [1, 2]:

- **Subvalvular:**
  - Discrete: Membrane or fibromuscular ridge
  - Diffuse: Fibromuscular tunnel
  - Dynamic: Hypertrophic obstructive cardiomyopathy
- **Valvular:**
  - Unicuspid/Bicuspid
  - Dysplastic
- **Supravalvular:**
  - Discrete (membranous or 'hourglass')
  - Aortic hypoplasia or atresia
  - Interrupted aortic arch
  - Coarctation of the aorta

9.1.1 Subvalvular Stenosis

Subvalvular obstruction to left ventricular outflow occurs proximal to the aortic valve (subaortic stenosis). Subaortic stenosis can be dynamic or fixed. Dynamic subaortic stenosis, which results from asymmetric septal hypertrophy and systolic anterior motion (SAM) of the AML, is discussed in the section on hypertrophic cardiomyopathy (Chap. 15).

Fixed subvalvular aortic stenosis can be discrete or diffuse.

Discrete subaortic stenosis is caused by an anomalous fibrous membrane or a fibromuscular ridge, which forms a barrier within the LVOT. The fibromuscular ridge is thicker and less discrete, as compared to a membrane [1]. Discrete subaortic stenosis may also occur as a consequence of VSD closure or PA banding.

The obstructing membrane, which usually extends from the anterior septum to the AML, may be tethered to the AML or the aortic valve [2]. Aortic valve thickening and regurgitation occur in half of the cases. The commonly observed mitral valve anomalies include [3]:

- Papillary muscle insertion into the septum or AML
- Accessory mitral valve tissue
- Muscularization of the subaortic portion of the AML

Diffuse tunnel-type subaortic stenosis is characterized by diffuse thickening and narrowing of the LVOT, along with associated concentric LVH.

Supplementary InformationThe online version contains supplementary material available at https://doi.org/10.1007/978-981-96-0243-8_9.

A. Paul, *Manual of Echocardiography for Congenital Heart Diseases*,
https://doi.org/10.1007/978-981-96-0243-8_9

9.1.1.1 Echocardiographic Diagnosis (Fig. 9.1)

- The fibromuscular membrane is visualized as a discrete linear echogenic structure in the LVOT perpendicular to the IVS (Fig. 9.1).
- A5C view facilitates better visualization of the thin membrane, as the ultrasound beam is perpendicular to the membrane in this view, although multiple transducer positions may be required (Fig. 9.1).
- Features that need to be assessed and documented include [4]:
  - **Length** of the membrane
  - **Distance** of the membrane from the aortic valve (<6 mm is a risk factor for recurrence)
  - **Extension** of the membrane onto the aortic valve cusp or AML

The characteristic echocardiographic features of fixed subvalvular aortic stenosis are:

- M-mode echocardiography depicts [5]:
  - Partial closure and fluttering of the aortic leaflets in mid-systole and
  - Reopening of the leaflets in late systole

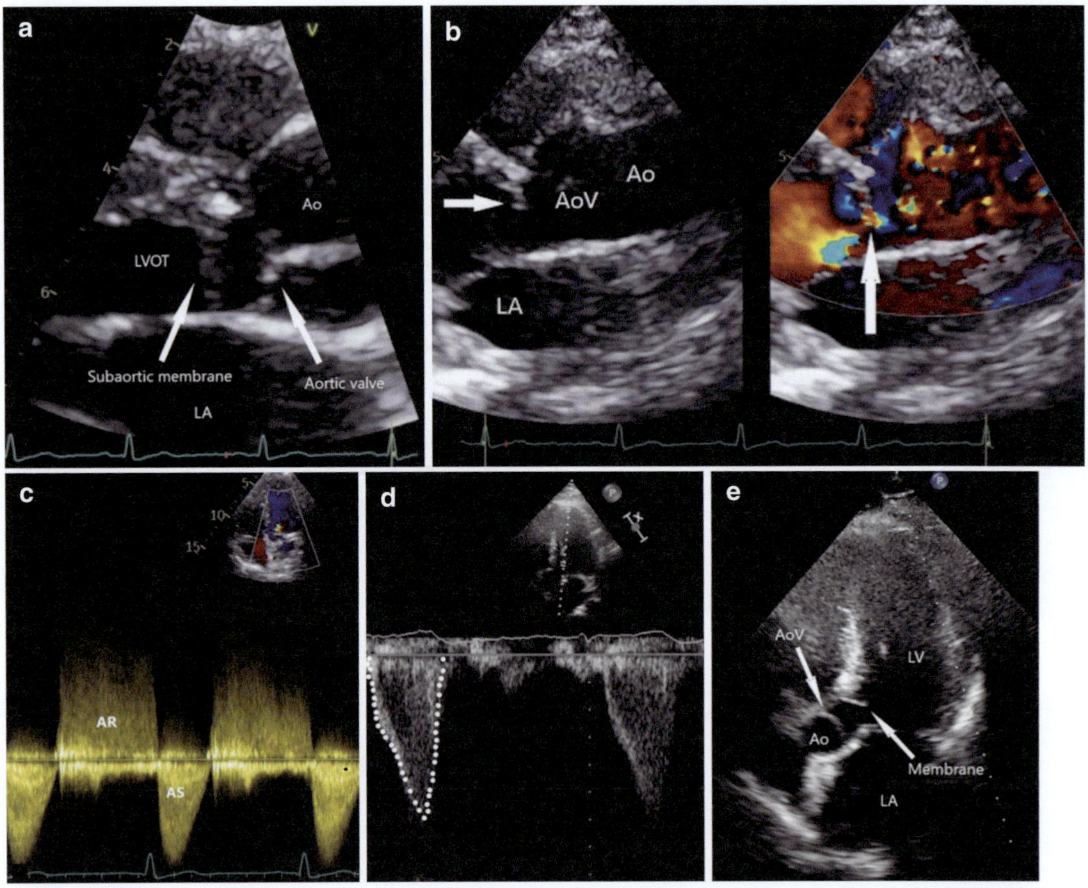

Fig. 9.1 Subvalvular aortic stenosis. (**a**) PLAX view (magnified image) depicting a thin membrane in the LVOT proximal to the aortic valve and perpendicular to the IVS. (**b**) PLAX view with colour flow Doppler depicting the turbulent flow originating at the level of the membrane (horizontal arrow), proximal (vertical arrow) to the aortic valve. (**c**) CWD tracing of the flow across the LVOT showing fixed subvalvular stenosis (characterized by rounded contour and mid-systolic peaking) and aortic regurgitation. (**d**) CWD tracing from a case of HCM with dynamic subaortic stenosis, depicting the late systolic peaking of flow velocity. (**e**) A5C view depicting the subaortic membrane in the LVOT proximal to the aortic valve

Table 9.1 Grading of severity of aortic stenosis at all levels (subvalvular/valvular/supravalvular) based on CWD-derived flow velocity and pressure gradients

Severity of aortic stenosis	Mild	Moderate	Severe
Peak flow velocity across LVOT	2–3 m/s	3–4 m/s	> 4 m/s
Peak pressure gradient across LVOT	< 36 mmHg	36–64 mmHg	> 64 mmHg
Mean pressure gradient across LVOT	<20 mmHg	20–40 mmHg	> 40 mmHg

- Colour Doppler imaging reveals a turbulent jet in the LVOT originating proximal to the aortic valve.
- Spectral Doppler demonstrates velocity acceleration and aliasing in the LVOT, proximal to the valve. The mid-systolic peaking of fixed subvalvular stenosis is in contrast to the late systolic peaking of dynamic subvalvular stenosis (Fig. 9.1).

9.1.1.2 Severity of Stenosis

CWD-derived peak pressure gradient across the LVOT determines the severity of LVOTO at any level. Apical, right parasternal, and suprasternal windows aid in obtaining the maximal velocity (by aligning the jet parallel to the ultrasound beam) for accurate estimation of the Doppler gradient.

For patients with normal LV function without other defects, such as a VSD, the severity of aortic stenosis (subvalvular/valvular/supravalvular) is graded on the basis of CWD-derived parameters as shown in Table 9.1 [6].

Co-existent valvular stenosis is evidenced by an additional velocity increase at the level of the aortic valve by PWD.

The estimated pressure gradients may be:

- **Overestimated** in the presence of stenoses at multiple levels
- **Underestimated** in the settings of [2]:
  - Depressed LV systolic function
  - Non-restrictive VSD with left-to-right shunting, which acts as a 'pop-off' for LV ejection
  - RV providing a significant portion of the systemic cardiac output via a leftward-shunting PDA

Indications for intervention in subvalvular AS include [4]:

- Peak gradient ≥50 mmHg
- Peak gradient <50 mmHg associated with either of:
  - Moderate or severe AR
  - LV dysfunction attributable to obstruction

Key point: A leftward-shunting non-restrictive VSD, a rightward-shunting large PDA with retrograde flow in the distal aortic arch, and/or significant LV dysfunction need to be ruled out before confirming the peak pressure gradients across the LVOT as truly low.

9.1.1.3 Presence and Severity of Concomitant Aortic Regurgitation

Peak gradient >50 mmHg and increased age at diagnosis are risk factors for developing significant AR.

The echocardiographic features suggestive of severe AR are:

- Dilated LV
- Dense AR jet by CWD
- Jet width/LVOT width ≥ 65%
- Jet CSA/LVOT CSA ≥ 60%
- Vena contracta width ≥ 0.6 cm
- Pressure half-time < 200 ms
- Holodiastolic flow reversal in descending aorta, as seen by PWD
- Effective regurgitation orifice area (EROA) by PISA ≥0.30 cm^2
- Regurgitation volume ≥ 60 mL
- Regurgitant fraction ≥50%

9.1.1.4 Associated Anomalies

- Discrete subaortic stenosis is often associated with:
  - VSD, PDA, or aortic coarctation
- Tunnel-like subaortic stenosis may be associated with:

- – Shone complex (supramitral ring, parachute mitral valve, subaortic stenosis, COA)

Key point: A peak instantaneous gradient > 50 mmHg, proximity of the membrane to the aortic valve (<6 mm), and extension of the membrane to the aortic or mitral valves are risk factors for recurrence of LVOTO after surgical excision of the subaortic membrane.

9.1.2 Valvular Aortic Stenosis

In congenital forms of valvular aortic stenosis, the stenosis may be present at birth (a congenitally stenotic aortic valve) or may develop over time in a congenitally abnormal non-stenotic valve [2]. The various morphologic types of congenitally abnormal aortic valve are [1]:

- **Unicuspid valve**: Congenitally stenotic; may result in symptomatic neonatal aortic valve stenosis. The aortic valve ring, ascending aorta, MV, and the LV cavity are usually hypoplastic. A unicuspid valve can be of two types:
  - Acommissural (pinhole opening; volcano-like small central orifice).
  - Unicommissural (slit-like orifice extending to the annulus, resembling an exclamation mark).
- **Bicuspid valve**: Rarely stenotic at birth. It becomes stenotic over time because of progressive fibrosis and calcification or may develop predominant regurgitation due to degeneration (Fig. 9.2a).
- **Tricuspid valve**: Stenosis in a tricuspid aortic valve can result from a hypoplastic annulus with normal cusps, myxoid dysplasia, and/or cuspal inequality (Fig. 9.2b).
- **Quadri** (four) or **hexa** (six) **cuspid valve** [rare] (Fig. 9.2c, d).

9.1.2.1 Echocardiography: Valve Morphology

Anatomy of the aortic valve and mechanisms of dysfunction should be analysed by imaging in different planes, with a focus on:

- Number of aortic cusps visible (should be verified in diastole as well)
- Maximum aortic cusp separation (MACS): Maximum distance between inner margins of aortic valve cusps at the tips during systole
- Mobility of the cusps and their excursion during systole

In the PLAX view, a normal aortic valve appears as a thin echogenic structure opening completely in systole and aligning parallel along the aortic walls. Congenital aortic stenosis is characterized by:

- Thickened cusps
- Decreased valve mobility
- Systolic doming (due to restricted motion of the tips relative to the more mobile bodies of the cusps)

In the PSAX view, the commissures of a normal tricuspid valve appear as a 'Y' during diastole (referred to as the 'inverted Mercedes-Benz sign'); while a unicuspid valve has a single slit-like commissure, with a restricted and eccentric opening.

Key point: Aortic valve annulus dimension is best measured from the PLAX view at mid-systole, from the hinge point of the right coronary cusp to that of the non-coronary cusp.

The bicuspid aortic valve (BcAV) is characterized by two cusps, two sinuses, and a single linear commissure. This results from congenital fusion of any two of the three aortic valve cusps, as follows [1, 2]:

- Fusion of right and left coronary cusps (70–80%) results in a right–left systolic opening in the PSAX view.
- Fusion of right and non-coronary cusps (20–30%) results in an anterior–posterior opening in the PSAX view.
- Fusion of left and non-coronary cusps (rare).

A prominent raphe between the conjoined cusps may render the closed bicuspid valve appear as a trileaflet valve in diastole. Imaging of the valve during systole confirms the diagnosis, as the raphe does

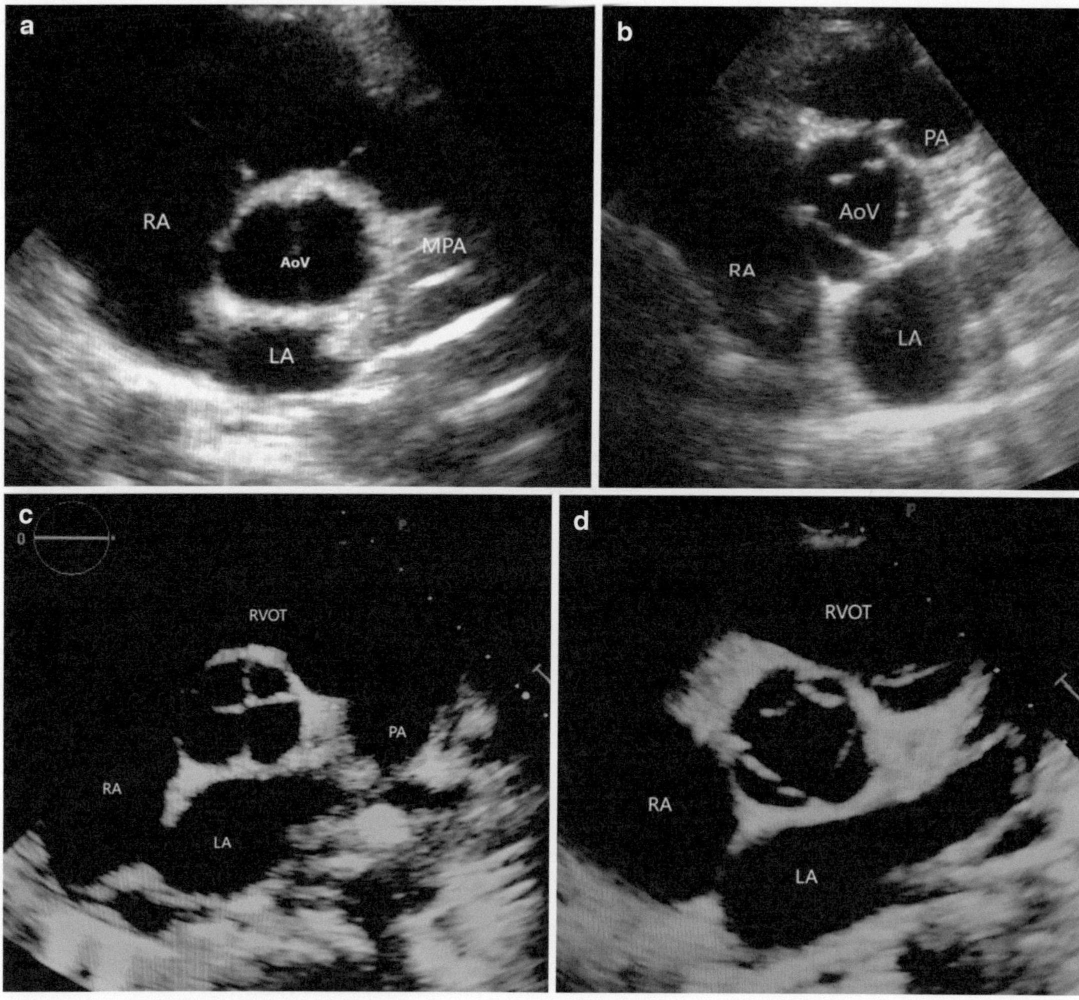

Fig. 9.2 Aortic valve in PSAX view: (**a**) Bicuspid aortic valve during diastole depicting two cusps and a single linear commissure. (**b**) Tricuspid aortic valve during systole depicting the triangular opening. (**c**, **d**) Quadricuspid aortic valve during diastole and systole, respectively

not open up during systole. In the PSAX view (Fig. 9.3a), BcAV is characterized by an elliptical/oval 'fish mouth appearance' during systole.

The characteristic features of BcAV in the PLAX view are:

- Eccentric opening of the valve
- Systolic doming of the cusps
- Diastolic prolapse of the cusps into the LVOT (Fig. 9.3b)

Sclerosis of the valve begins as early as the second decade of life, and calcification is noted by the fourth decade [2]. An anomalous origin of the left circumflex artery from the right sinus is a common anomaly of coronary artery origin in cases of bicuspid aortic valve.

M-mode:

Although of limited utility, the following M-mode features may aid the recognition of an abnormal aortic valve [7]:

- **Eccentricity index** (EI):
  - EI = ½ A/a

    (A = aortic root diameter at the onset of diastole, and a = distance from the aortic cusp closure line to the nearest aortic wall at the onset of diastole).

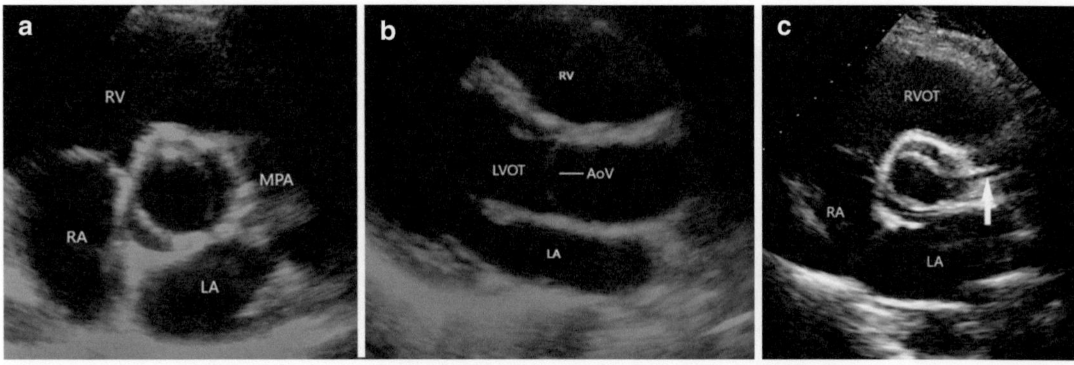

Fig. 9.3 Bicuspid aortic valve. (**a**) PSAX view showing oval (fish mouth) opening of bicuspid aortic valve during systole. (**b**) PLAX view showing prolapse of aortic valve cusps into the LVOT during diastole. (**c**) PSAX view depicting a bicuspid valve with a non-bifurcating left coronary artery (arrow). Anomalous origin of the left circumflex artery from the right sinus is a common anomaly in bicuspid aortic valve.

  - The EI of a normal tricuspid aortic valve is 1–1.25.
  - EI ≥ 1.3 is diagnostic of BcAV, although EI ≥ 1.5 is more specific.
  - EI < 1.3 does not rule out the possibility of BcAV.
  - In the presence of perimembranous VSD with aortic cusp prolapse, even a tricuspid aortic valve may have EI ≥ 1.3.
- Increased leaflet thickness.
- Multiple diastolic cusp lines within the aortic root may indicate valve cusp redundancy.
- Progression of stenosis is faster in eccentric cusps (width ratio of major to minor cusps >1.2) [8].

9.1.2.2 Haemodynamic Severity of Aortic Stenosis

(a) ***2D echocardiographic features*** suggestive of severe aortic stenosis are (Fig. 9.4):

- Increased thickness and severe immobility of the cusps
- Poor leaflet tip separation in systole
- Severe degree of left ventricular hypertrophy
- Presence of post-stenotic aortic root dilation

Key point: The PSAX view should never be used to exclude severe congenital AS, as movement of the cusps out of the imaging plane during systole may lead to an underestimation of severity [9].

(b) ***CWD-derived flow velocities and pressure gradients*** are used to classify the severity of aortic stenosis as mild, moderate, or severe [Table 1]. Please refer to the section on subaortic stenosis for further details.

(c) ***Aortic valve area (AVA):*** The normal aortic valve area is approximately 2 cm^2/m^2 in children and 3–4 cm^2 in fully grown adolescents and adults.

AVA is calculated using the continuity equation as follows:

$$AVA\ \left(cm^2\right) = \text{LVOT cross sectional area}\ \left(cm^2\right) \times \text{LVOT}\,VTI\left(cm/\text{s}\right) / \ AoV\ VTI\ \left(cm/\text{s}\right)$$

[LVOT VTI is measured from the A5C view (and/or A3C view) by placing the PWD sample volume in the LVOT just proximal to the aortic valve (within 5 mm). LVOT diameter is measured from the PLAX view in early to mid-systole at about 5 mm proximal to the aortic annulus.]

Based on AVA, the severity of AS may be graded as:

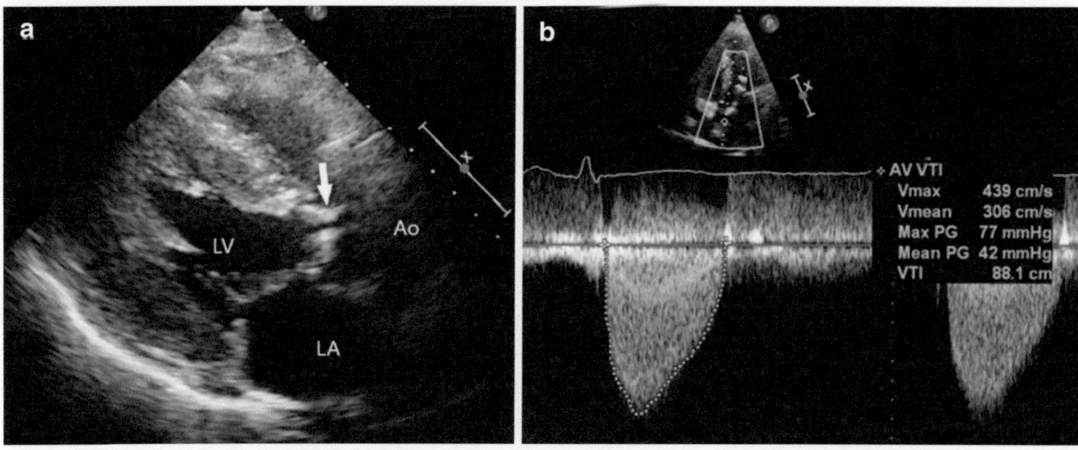

Fig. 9.4 Severe aortic stenosis. (**a**) PLAX view depicting thickened and stenotic aortic valve [arrow] with poor leaflet tip separation in systole. Concentric left ventricular hypertrophy can be noted. (**b**) CWD tracing of the flow across the stenotic aortic valve with a relatively rounded contour and mid-systolic peaking

- **Mild**: AVA > 1.5 cm^2
- **Moderate**: AVA 1.0–1.5 cm^2
- **Severe**: AVA <1.0 cm^2 (<0.6 cm^2/m^2 in children)

9.1.2.3 Aortic Dimensions [10]

- **Subvalvular LVOT and aortic annulus:** Best measured in the PLAX view during early to mid-systole. LVOT dimension is measured 0.5–1 cm proximal to the aortic annulus. Aortic annulus diameter is measured between the hinge points of the right- and non-coronary cusps of the aortic valve.
- **Proximal aortic diameters:** Best measured in the PLAX, high left parasternal, or high right parasternal views during mid-systole at the levels of the aortic root, sinotubular junction, and ascending aorta.
- **Aortic arch and aortic isthmus diameters:** Best measured in the suprasternal long-axis view during mid-systole.
- **Descending aorta diameter:** Best measured in the subxiphoid short-axis view at the level of the diaphragm during mid-systole.

9.1.2.4 Feasibility of Biventricular Repair in Critical Neonatal Aortic Stenosis

The adequacy of the left heart to support a two-ventricle circulation needs to be determined in neonates with critical aortic stenosis and a small left ventricle.

The parameters associated with a low likelihood of success with two-ventricle repair are [11]:

- Non-apex-forming left ventricle in infants
- Small aortic annulus (<5 mm)
- Small mitral valve annulus (<9 mm)
- Significant retrograde flow in the distal aortic arch via PDA

The other parameters that have been found to be predictive of failure of biventricular circulation include [12]:

- Ratio of the long axis of the left ventricle to the long axis of the heart: ≤ 0.8
  (Long axis of the heart is measured from the A4C view as the distance from the crux of the heart to the apical endocardium of the apex-forming ventricle: RV or LV)
- Aortic root diameter indexed to BSA: ≤ 3.5 cm/m^2
- Mitral valve area indexed to BSA: ≤ 4.75 cm^2/m^2
- LV mass index: ≤ 35 g/m^2

Predictors of greater survival benefit for the Norwood procedure (vs biventricular repair) include [13, 14]:

- Lower z-score of the aortic valve at the level of the sinuses of Valsalva
- Lower z-score of the left ventricular length
- Higher grade of endocardial fibroelastosis
- Larger diameter of the ascending aorta
- Absence of significant tricuspid regurgitation

Key point: Balloon size for balloon aortic valvuloplasty is determined based on the diameter of the aortic annulus (usually in a 1:1 ratio).

9.1.2.5 Presence and Severity of Aortic Regurgitation

Please refer to the section on subaortic stenosis.

9.1.2.6 Associated Anomalies

- Shunt lesions (VSD, PDA) and direction of flow
- Features of HLHS: Mitral valve abnormalities; Endocardial fibroelastosis

9.1.3 Supravalvular Aortic Stenosis

Supravalvular aortic stenosis (SVAS) refers to the obstruction of LV outflow due to narrowing of the aorta just distal to the aortic valve. SVAS accounts for about 10% of congenital AS. The most common site of stenosis is the sinotubular junction.

The three morphologic types of supravalvular aortic stenosis are [1]:

- ***Fibromuscular thickening*** produces an ***hourglass-shaped narrowing*** above the sinuses, characterized by a segment of gradual tapering followed by widening of the lumen. This is the most common form of supravalvular AS.
- ***Discrete fibrous membrane*** in a normal-sized aorta, usually located near the sinotubular junction.
- ***Diffuse hypoplasia*** of the ascending aorta, often involving the origins of the brachiocephalic arteries, and in some cases the transverse aorta.

SVAS is commonly associated with William's syndrome (William-Beuren syndrome). Non-syndromic SVAS has been reported in neurofibromatosis type 1, Noonan syndrome, Alagille syndrome, aortic hypoplasia, and atherosclerotic plaques of homozygous familial hypercholesterolemia [1, 2].

9.1.3.1 Echocardiographic Diagnosis

- Colour Doppler imaging reveals turbulence originating distal to the aortic valve.
- PLAX view, high right parasternal view, and suprasternal view aid in the diagnosis and estimation of severity by CWD-derived peak gradient.

Adherence of the cusps to the narrowed sinotubular junction may result in [1]:

- Aortic regurgitation
- Ostial stenosis and/or dilation of the coronary arteries
- Impaired coronary perfusion, involving the left coronary artery

In supravalvular AS, the aortic wall and the aortic valve cusps usually appear thickened, which may result in concomitant valvular stenosis.

9.1.3.2 Associated Anomalies

Supravalvular PS (MPA, branch PAs), and stenosis of the branches of the aortic arch, may be associated with syndromic as well as non-syndromic SVAS.

Points to Report

- **Level of obstruction:** Subvalvular/Valvular/Supravalvular/Multiple levels
- **Anatomy of the relevant structure:**
  - Subvalvular stenosis: Discrete subaortic membrane/Diffuse tunnel-like stenosis
  - Valvular stenosis: Uni/Bi/tricuspid aortic valve
  - Supravalvular stenosis: Hourglass narrowing/Discrete membrane
- **Severity of obstruction:**
  - CWD: Peak velocity; Peak and mean gradients
  - Aortic valve area
- **Measurements:**
  - LVOT, annulus, aortic root at the level of the sinus, sinotubular junction, ascending aorta, arch of aorta, aortic isthmus
  - Subaortic membrane: Length; Distance from aortic annulus
  - Mitral valve annulus: Lateral diameter in A4C view and anteroposterior diameter in PLAX view
- **Left ventricular size and function**
- **Concomitant aortic regurgitation and its severity**
- **Associated anomalies:**
  - Coarctation of the aorta
  - Supravalvular/peripheral pulmonary stenosis
  - VSD; PDA
  - Mitral valve abnormality
- **Various indices determining the surgical options in neonatal critical AS** (see above)

9.1.4 Coarctation of the Aorta

Coarctation of the aorta (CoA) is a discrete stenosis in the aorta.

- The commonest site is at or near the point of insertion of the ductus (or ligamentum) arteriosus, just distal to the origin of the left subclavian artery, referred to as 'juxtaductal' coarctation [15].
- Coarctation may be called 'preductal' or 'postductal' based on its position relative to the ductus.
- Coarctation involving the ascending aorta or abdominal aorta is rare.

9.1.4.1 Echocardiographic Diagnosis (Fig. 9.5)

- Thoracic aorta is visualized from the suprasternal long-axis view.
- Coarctation is seen as localized narrowing of the aorta just beyond the origin of the left subclavian artery.
- Fibrous tissue is seen protruding from the posterior aspect of the aorta towards the site of ductal insertion ('posterior shelf')[15, 16].
- Peak instantaneous pressure gradient >20 mmHg across the narrowing is diagnostic of CoA.

Associated confirmatory findings of coarctation [16]:

- Hypoplasia of the aortic isthmus
- Dilation and exaggerated pulsation of the proximal aortic arch
- Post-stenotic dilation of the descending aorta
- Delayed rate of systolic upstroke and persistent diastolic flow in the abdominal aorta
- Flow acceleration and turbulence within the narrowed segment (by colour Doppler)

Diagnosis of neonatal CoA [4]:

- Aortic isthmus diameter < 3 mm in the absence of PDA
- Aortic isthmus diameter < 4 mm in the presence of PDA
- The ratio of the aortic isthmus diameter to the descending aorta diameter at the level of the diaphragm <0.64

Pseudo-coarctation: Artefactual appearance of aortic narrowing may result from a tangential imaging plane. This possibility can be ruled out by the following findings:

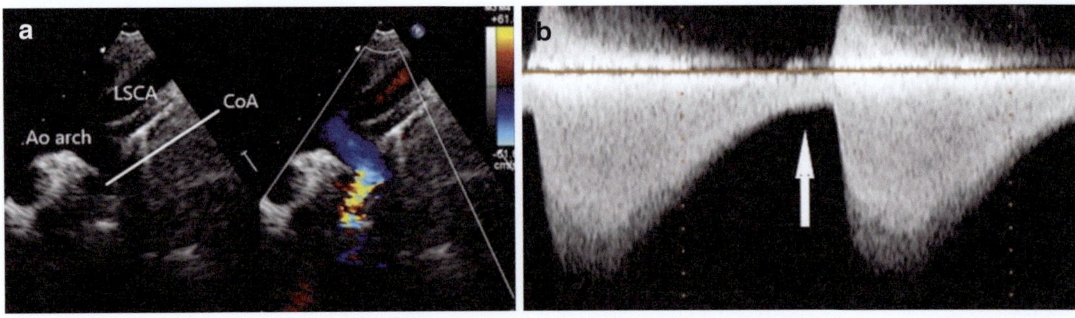

Fig. 9.5 Coarctation of aorta. (**a**) Suprasternal view showing narrowing of the DTA with turbulence on colour flow Doppler. (**b**) CWD tracing across a severe coarctation, showing a diastolic gradient (arrow). Note the shark-tooth pattern of the tracing.

- Visualizing the aortic lumen beyond the suspected region of narrowing opens up the segment.
- Doppler imaging demonstrates no evidence of flow acceleration or turbulence.

Key point: A mild increase (1.5–2 m/s) in descending aortic flow velocity occurs due to normal flow acceleration around the arch, in the absence of aortic coarctation.

9.1.4.2 Severity

- A significant obstruction is suggested by [17]:
  - Persistence of a high-velocity flow signal into diastole (diastolic gradient).
  - Diastolic velocity (DV): velocity measured at the end of the 'T' wave >193 cm/s
  - Diastolic/systolic velocity ratio > 0.53.
- The CWD tracing across the coarctation demonstrates diastolic runoff (continuous slowly tapering antegrade flow in diastole, known as a ***diastolic 'tail'***) (Fig. 9.5b).
- The presence of a diastolic tail is suggestive of:
  - Severe coarctation or
  - Extensive collateral circulation.
- The CWD profile distal to the coarctation is composed of two superimposed signals representing the proximal and distal flows.
- The flow velocities proximal and distal to the coarctation should be considered while estimating pressure gradients using the Bernoulli equation because the proximal flow velocity is often greater than 1.5 m/s and hence not negligible [16].

CWD-derived pressure gradients may ***underestimate the actual severity*** of obstruction in the presence of:

- Patent ductus arteriosus: Left-to-right runoff of blood through the ductus reduces the jet velocity through the coarctation and leads to an underestimation of the pressure gradient.
- Well-developed collaterals.

Leg raising test:

- Since coarctation gradients are flow-dependent, the leg raising test can be performed in borderline cases for further clarification.
- Although there may not be a significant increase in the peak systolic gradient, an increase in the diastolic gradient may point towards the possibility of significant coarctation.

Flow-independent indices: Since the CWD-derived pressure gradients are highly flow-dependent, which could vary with heart rate or associated abnormalities such as dysfunctional bicuspid aortic valve, the following flow-independent indices have been suggested to detect significant CoA [18, 19]:

- CoA Doppler velocity index (DVI): The ratio of peak PW velocity pre-CoA to peak CW velocity through CoA <0.6
- CoA effective orifice area (EOA) using the continuity equation <3.0 cm^2
- Diastolic velocity half-time index (DVHTi): Time taken for DV to fall to half its value, indexed to heart rate > 200 ms
- Diastolic pressure half-time index (DPHTi): Time taken for the diastolic gradient (4[DV]^2) to fall to half its value, indexed to heart rate > 75 ms

9.1.4.3 Associated Cardiac Anomalies

- Perimembranous, muscular, or malalignment VSDs may be associated with coarctation.
- Posterior malalignment VSDs may cause significant LVOTO.
- Bicuspid aortic valve occurs in up to 85% of patients with a coarctation, associated with valvular stenosis or annular hypoplasia [15, 16].
- Diffuse hypoplasia of the transverse aortic arch and isthmus, proximal to a discrete coarctation.
- **Shone complex**: Multiple left-sided obstructive lesions, including supra-valvar mitral ring, parachute mitral valve, subaortic membrane, and aortic coarctation.
  - Mitral stenosis may result from thickening and dysplasia of the mitral leaflets or short, dysplastic chordae tendineae.
  - **Shone's variant**, or incomplete Shone's syndrome, consists of two or three left-sided obstructive lesions in the setting of left heart hypoplasia.
- Atrioventricular septal defects, d-TGA, l-TGA, tricuspid atresia, Taussig-Bing type of DORV.

Key point: Coarctoplasty may be considered in the presence of significant collateral vessels, left ventricular hypertrophy, and/or LV dysfunction due to CoA, even when the systolic gradient is < 20 mmHg.

9.1.4.4 Coarctoplasty and Post-repair Evaluation [20]

- The maximum permissible size of the balloon for coarctoplasty is equal to the diameter of the abdominal aorta at the level of the diaphragm, which is usually 2–4 times the diameter of the narrowed segment.
- A covered stent is preferred for cases with a high risk of rupture as suggested by:
  - Bicuspid aortic valve with aortopathy.
  - Atretic isthmus (<3 mm in diameter).

  Post-repair evaluation focuses on evidence of:

- Residual narrowing or restenosis at the site of coarctation
- Residual pressure gradient across the coarctation site

In patients without residual narrowing of the coarctation segment, lack of compliance of the vessel wall may result in a resting or inducible Doppler gradient at the repair site. Other Doppler parameters suggested for post-coarctoplasty follow-up include [20, 21]:

- Diastolic velocity (DV) of aortic flow
- Diastolic velocity half-time index (DVHTi)
- Diastolic pressure half-time index (DPHTi)

9.1.5 Interrupted Aortic Arch

Interrupted aortic arch is an extreme form of aortic coarctation, present in 1% of critically ill infants with congenital heart disease.

Types [15]**:**

- Type A: The interruption is distal to the left subclavian artery (30%).
- Type B: The interruption is between the left carotid and left subclavian arteries (43%). DiGeorge syndrome is present in about 50% of patients with type B interruption.

- Type C: The interruption is between the innominate and left carotid arteries (17%).

9.1.5.1 Echocardiographic Evaluation

Similar to that of aortic coarctation.

9.1.5.2 Associated Defects

- PDA and VSD are almost always associated
- Bicuspid aortic valve (60%)
- Subaortic stenosis (20%)
- Mitral valve deformity (10%)
- Persistent truncus arteriosus (10%)

9.2 Truncus Arteriosus

Truncus arteriosus is an anomalous single arterial outlet that arises from the base of the heart. This anomalous vessel gives origin to the coronary, the pulmonary, and the systemic arteries.

Collett and Edwards classification [22]:

Based on the anatomic origin of the pulmonary arteries, four types of truncus are described:

- **Type I** (50–70%): A short main pulmonary trunk originating from the truncus arteriosus gives rise to both pulmonary arteries.
- **Type II** (30–50%): No MPA. The RPA and LPA arise separately from the posterior part of the truncus, close to one another.
- **Type III** (6–10%): No MPA. The RPA and LPA arise separately from the lateral part of the truncus, at a distance from one another.
- **Type IV**: No pulmonary arterial branch arising from the common trunk. The lungs are supplied by collaterals (pseudotruncus). This is now considered to represent ***a form of pulmonary atresia with ventricular septal defect***.

Van Praagh and Van Praagh have proposed an expanded classification system [23]:

- **Type A1**: Corresponds to type I of Collett and Edwards classification.
- **Type A2**: Encompasses types II and III of the Collett and Edwards classification.
- **Type A3**: A single pulmonary artery arises from the truncus, while the other lung derives its blood supply from the ductus arteriosus or from an aortopulmonary collateral artery.
- **Type A4**: Associated with underdevelopment of the aortic arch, including tubular hypoplasia, discrete coarctation, or complete interruption.

Each type may be subclassified into 'A' (presence of VSD) or 'B' (intact ventricular septum).

Hemitruncus: One pulmonary artery arises from the ascending aorta, and the other arises from the right ventricle with a well-developed pulmonary valve at its origin. **This is not a type of truncus arteriosus**.

9.2.1 Echocardiographic Diagnosis (Figs. 9.6 and 9.7)

9.2.1.1 Truncal Artery

The cardinal echocardiographic features of truncus arteriosus are:

- A single great arterial trunk (truncal artery) arises from the heart
- Absence of a pulmonary outflow tract from either ventricle
- Origin of pulmonary arteries from the truncal artery

The truncal artery can be visualized using the subcostal coronal view, the PLAX view (Fig. 9.6) and the A5C view. The ventricular commitment of the truncal artery is variable [24]:

- Biventricular (70–80%) origin is the commonest (Fig. 9.6)
- Right ventricular (10–30%)
- Left ventricular (4–6%)

Echocardiographically, truncus may be confused with pulmonary atresia because no separate origin of pulmonary arteries is seen from either of the ventricles. The key differentiating feature of truncus arteriosus is the origin of the pulmonary arteries from the ascending portion of the single great artery. In pulmonary atresia, PAs do not arise directly from the visualized single great artery (i.e. aorta).

High PSAX and high PLAX views visualize the pulmonary artery origin, commonly from the

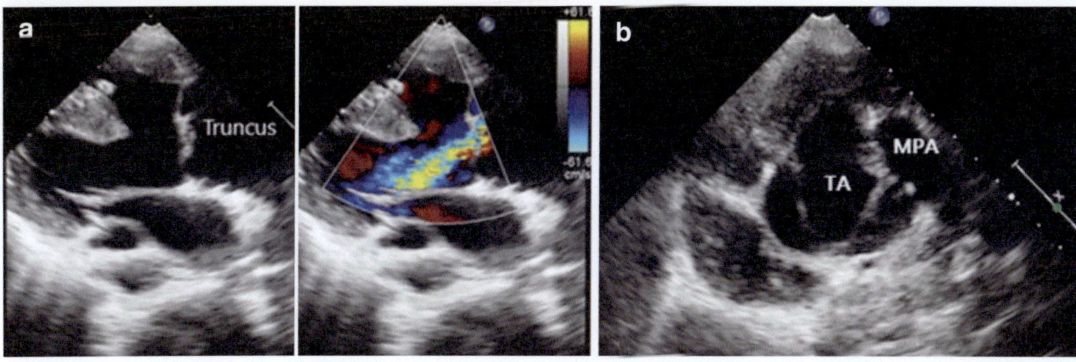

Fig. 9.6 Truncus arteriosus. (**a**) PLAX view showing truncus arteriosus with biventricular origin. Note the thickened and nodular truncal valve cusps, with the colour Doppler demonstrating valvular insufficiency. (**b**) PSAX view showing the valve of truncus arteriosus [TA] in cross-section. The main pulmonary artery [MPA] is also seen.

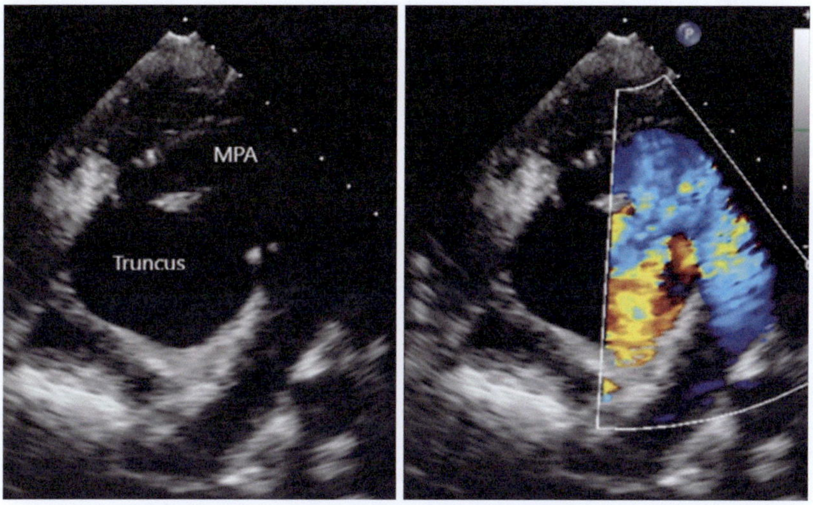

Fig. 9.7 Truncus arteriosus with the main pulmonary artery [MPA] taking origin from the truncus

left posterolateral aspect of the truncus, just above the truncal valve. The truncal valve tissue can obstruct the pulmonary artery ostia during ventricular systole.

When one PA is absent (type A3), it is usually absent on the side of the aortic arch in truncus arteriosus (in contrast to tetralogy of Fallot, where the PA is usually absent on the side opposite the aortic arch). In such cases, the source of pulmonary blood supply to the side of the absent PA (ductus or aortopulmonary collateral) should be confirmed.

Key point: The origin of the PA from the truncal artery differentiates truncus arteriosus from pulmonary atresia.

9.2.1.2 Truncal Valve

Morphology and number of truncal valve leaflets can be ascertained from subcostal sagittal and PSAX views. The valve morphology is variable [24]:

- Tricuspid (70%)
- Quadricuspid (20%)
- Bicuspid, pentacuspid, or unicommissural (rare)

The fibrous continuity of the truncal valve with the mitral valve can be seen from the PLAX view in all patients. Fibrous continuity with the tricuspid valve occurs if there is a membranous VSD.

Functional abnormality: The truncal valve is more likely to be incompetent than stenotic. Subcostal views permit visualization of the truncal root and assessment of truncal valve function.

- **Truncal valve regurgitation** can result from dysplasia of valve cusps, prolapse of unsupported or conjoined cusps, unequal cusp size, and annular dilation [25].
  - Regurgitation of the truncal valve results in:
    Doppler flow reversal in the abdominal aorta (could also be due to pulmonary artery runoff)
    Pulmonary venous hypertension
    Dilation of ventricular chambers
- **Truncal valve stenosis** is associated with nodular and dysplastic cusps
  - Peak and mean gradients should be estimated using CWD
  - When associated with significant truncal valve insufficiency, systolic Doppler gradient may overestimate the degree of valve stenosis, owing to the greater volume of flow across the valve. Hence, it is important to correlate the 2D morphology with the Doppler findings.

9.2.2 Associated Defects

(a) ***Ventricular septal defect*** in truncus arteriosus is usually a large defect involving the infundibular septum. Less commonly, the membranous septum may be defective. Small, restrictive defects are rare.
PLAX view demonstrates the VSD with the overriding great artery (Fig. 9.6). The location and extension of the VSD can be confirmed from the subcostal sagittal and PSAX views.

(b) ***Aortic arch anomalies*** seen in truncus are [26]:
  - Right aortic arch with mirror-image brachiocephalic branching (25%)
  - Double aortic arch
  - Aortic arch hypoplasia with or without coarctation
  - Interrupted aortic arch with ductal continuation of the descending thoracic aorta

(c) ***Ductus arteriosus:***
  - Patent in one-third of the cases.
  - There is an inverse relationship between the relative sizes of the aorta and ductus. The ductus arteriosus is larger in patients with underdeveloped aortic arch (type A4 truncus) [24].
  - The pulmonary blood supply on the side of the absent PA in type A3 truncus may be from the PDA.

(d) ***Anomalies of coronary arterial origin*** [27]:
Visualization of coronary artery origin may be accomplished from the PSAX view.
  - The left coronary artery tends to arise from the left posterolateral truncal surface, and right coronary artery from the right anterolateral surface.
  - A single coronary ostium, two coronary ostia arising from the same truncal sinus, or a high ostial origin above the truncal sinotubular junction may be seen.
  - The left coronary artery may originate from the pulmonary trunk.
  - Dysplastic valvular tissue may obstruct the coronary artery ostia.

(e) ***Other associated anomalies:***
  - Shunt lesions: Ostium secundum ASD
  - Vascular anomalies: Aberrant subclavian artery; Persistent LSVC draining into the coronary sinus; TAPVC
  - Rarer associations: Tricuspid stenosis/atresia, mitral atresia, ventricular inversion, complete AV septal defect, asplenia

9.2.3 Secondary Complications

- Massive biventricular hypertrophy may result in chronic subendocardial myocardial ischaemia.
- Pulmonary arterial hypertension: Chronic exposure of the pulmonary vasculature to systemic arterial pressure may result in plexogenic pulmonary arteriopathy [24].

- Significant truncal valve stenosis/regurgitation can result in dilation and dysfunction of the ventricles.

Key point: AP window can be differentiated from truncus in that (1) AP window is usually not associated with a VSD; (2) RVOT and pulmonary valve are in the normal positions in AP window.

Key views	Key features assessed
Subcostal coronal/sagittal	Detection of a single great vessel arising from the heart Truncal valve function, truncal root Pulmonary artery branch anatomy, and atrial septum
Parasternal	
– Long axis	VSD and the overriding great artery; Continuity between the truncal valve and the mitral valve Higher transducer position visualizes the origin of the pulmonary trunk and/or the branches
– Short axis	Number and morphology of the truncal valve leaflets Coronary arteries origin Location and extension of the VSD High PSAXview: Origin and bifurcation of PA
Suprasternal notch	Aortic arch anatomy Evaluation of PA branches for significant stenoses

References

1. Perloff JK, Marelli AJ. Congenital aortic stenosis, congenital aortic regurgitation. In: Clinical recognition of congenital heart disease. 6th ed. Philadephia, PA: Saunders; 2012. p. 74–100.
2. Schneider DJ, Moore JW. Aortic stenosis. In: Allen HD, Driscoll DJ, Shaddy RE, Feltes TF, editors. Moss and Adams' heart Diseasein infants, children, and adolescents. 8th ed. Philadelphia, PA: Lippincott Williams & Wilkins; 2013. p. 1023–43.
3. Cohen L, Bennani R, Hulin S, et al. Mitral valvar anomalies and discrete subaortic stenosis. Cardiol Young. 2002;12:138–46.
4. Park MK. Obstructive lesions. In: Park's pediatric cardiology for practitioners. 5th ed. Philadephia, PA: Saunders; 2016. p. 116–32.
5. Sabbah HN, Stein PD. Mechanism of early systolic closure of the aorticvalve in discrete membranous subaortic stenosis. Circulation. 1982;65:399–402.
6. Bonow RO, Carabello B, de Leon ACJR, et al. ACC/AHA guidelines for themanagement of patients with valvular heart disease: a report of the AmericanCollege of cardiology/American Heart Association task force on PracticeGuidelines (writing committee to revise the 1998 guidelines for the Management of Patients with Valvular Heart Disease). J Am Coll Cardiol. 2006;48:598–675.
7. Radford DJ, Bloom KR, Izukawa T, Moes CA, Rowe RD. Echocardiographic assessment of bicuspid aortic valves. Angiographic and pathological correlates. Circulation. 1976;53(1):80–5.
8. Beppu S, Suzuki S, Matsuda H, Ohmori F, Nagata S, Miyatake K. Rapidity of progression of aortic stenosis in patients with congenital bicuspid aortic valves. Am J Cardiol. 1993;71(4):322–7.
9. Armstrong WF, Ryan T. Feigenbaum's echocardiography. 8th ed. Philadelphia: Wolters Kluwer; 2019.
10. Lopez L, Colan SD, Frommelt PC, Ensing GJ, Kendall K, Younoszai AK, et al. Recommendations for quantification methods during the performance of a pediatric echocardiogram: a report from the pediatric measurements writing Group of the American Society of echocardiography pediatric and congenital heart disease council. J Am Soc Echocardiogr. 2010;23:465–95.
11. Leung MP, Mckay R, Smith A, et al. Critical aortic stenosis in early infancy: anatomic and echocardiographic substrates of successful openvalvotomy. J Thorac Cardiovasc Surg. 1991;101:526–35.
12. Rhodes LA, Colan SD, Perry SB, et al. Predictors of survival in neonateswith critical aortic stenosis. Circulation. 1991;84:2325–35.
13. Lofl GK, McCrindle BW, Williams WG, et al. Critical aortic stenosis in the neonate: a multi-institutional study of management, outcomes, andrisk factors. J Thorac Cardiovasc Surg. 2001;121:10–27.
14. Kovalchin JP, Brook MM, Rosenthal GL, et al. Echocardiographic hemodynamic and morphometric predictors of survival after two-ventriclerepair in infants with critical aortic stenosis. J Am Coll Cardiol. 1998;32:237–44.
15. Perloff JK, Marelli AJ. Coarctation of the aorta and interrupted aortic arch. In: Clinical recognition of congenital heart disease. 6th ed. Philadephia, PA: Saunders; 2003. p. 101–28.
16. Beekman HR. Coarctation of the aorta. In: Allen HD, Driscoll DJ, Shaddy RE, Feltes TF, editors. Moss and Adams' heart Diseasein infants, children, and adolescents. 8th ed. Philadelphia, PA: Lippincott Williams & Wilkins; 2013. p. 1044–60.
17. Tan JL, Babu-Narayan SV, Henein MY, Mullen M, Li W. Doppler echocardiographic profile and indexes in the evaluation of aortic coarctation in patients before and after stenting. J Am Coll Cardiol. 2005;46(6):1045–53.
18. Keshavarz-Motamed Z, Garcia J, Maftoon N, Bedard E, Chetaille P, Kadem L. A new approach for the evaluation of the severity of coarctation of the aorta using Doppler velocity index and effective orifice area: in vitro validation and clinical implications. J Biomech. 2012;45(7):1239–45.

19. Lim DS, Ralston MA. Echocardiographic indices of Doppler flow patterns compared with MRI or angiographic measurements to detect significant coarctation of the aorta. Echocardiography. 2002;19(1):55–60.
20. Stout KK, Daniels CJ, Aboulhosn JA, Bozkurt B, Broberg CS, Colman JM, et al. 2018 AHA/ACC guideline for the Management of Adults with Congenital Heart Disease: a report of the American College of Cardiology/American Heart Association task force on clinical practice guidelines. J Am Coll Cardiol. 2019;73:e81–e192.
21. Aboulhosn J, Child JS. Echocardiographic evaluation of congenital left ventricular outflow obstruction. Echocardiography. 2015;32(Suppl 2):S140–7.
22. Collett RW, Edwards JE. Persistent truncus arteriosus; a classification according to anatomictypes. Surg Clin North Am. 1949;29(4):1245–70.
23. Van Praagh R, Van Praagh S. The anatomy of common aorticopulmonary trunk (truncus arteriosus communis) and its embryologic implications. A study of 57 necropsy cases. Am J Cardiol. 1965;16(3):406–25.
24. Cabalka AK, Edwards WD, Dearani JA. Truncus arteriosus. In: Allen HD, Driscoll DJ, Shaddy RE, Feltes TF, editors. Moss and Adams' heart Diseasein infants, children, and adolescents. 8th ed. Philadelphia, PA: Lippincott Williams & Wilkins; 2013. p. 990–1002.
25. Fuglestad SJ, Puga FJ, Danielson GK. Surgical pathology of the truncal valve: a study of 12 case. Am J Cardiovasc Pathol. 1988;2:39–47.
26. Butto F, Lucas RV, Edwards JE. Persistent truncus arteriosus: pathologic anatomy in 54 cases. Pediatr Cardiol. 1986;7:95–101.
27. Shrivastava S, Edwards JE. Coronary arterial origin in persistent truncusarteriosus. Circulation. 1977;55:551–4.

10 Tetralogy of Fallot

10.1 Tetralogy of Fallot

Tetralogy of Fallot (TOF) is the most commonly encountered form of cyanotic CHD. The cardinal morphological abnormality in TOF is the anterior and cephalad deviation of the outlet (infundibular) septum relative to the septomarginal trabeculation. This results in the characteristic features of TOF, namely:

- Large non-restrictive outlet VSD.
- Aorta overriding the VSD.
- Right ventricular outflow tract obstruction (RVOTO).
- Hypertrophied right ventricle.

Physiologically, the magnitude of the right-to-left shunt through the large VSD in TOF depends on the severity of RVOTO. The more severe the obstruction of the RVOT, the less the pulmonary blood flow will be, resulting in greater right-to-left shunt. This produces severe cyanosis, whereas minor degrees of RVOTO result in minimal right-to-left shunting and less severe cyanosis (so-called pink TOF).

Supplementary Information The online version contains supplementary material available at https://doi.org/10.1007/978-981-96-0243-8_10.

10.1.1 Echocardiography: Ventricular Septal Defect (Fig. 10.1)

The VSD in TOF is essentially a defect of the outlet (infundibular) septum, bordered by the aortic valve (***outlet*** VSD) [1]. However, the extent and margins of the defect are variable.

In about 80% of the cases, the defect involves the membranous septum as well (hence the term '***perimembranous***' VSD). The outlet VSD can also extend to the inlet septum, with or without a common AV valve (***inlet*** VSD).

When there is near-total absence of the infundibular septum, the semilunar valves, which are in fibrous continuity with each other, form the superior border of the VSD (termed as '***doubly committed, subarterial or juxta-arterial***' VSD). These are prone to aortic regurgitation and post-operative residual RVOTO.

By definition, the VSD of TOF is large and non-restrictive. However, these defects may be made smaller (restrictive) by the septal tricuspid leaflet or accessory tricuspid valve tissue.

The size, the extent, and the margins of the VSDs, along with the gradient across the defect and the direction of flow, should be assessed and documented from the PLAX, PSAX, A4C, A5C, and subcostal views.

Key point: Since equalization of RV and LV pressures minimizes the flow across concomitant muscular VSDs, colour Doppler imaging of the ventricular septum must be done with a reduced colour scale to rule out these defects.

A. Paul, *Manual of Echocardiography for Congenital Heart Diseases*,
https://doi.org/10.1007/978-981-96-0243-8_10

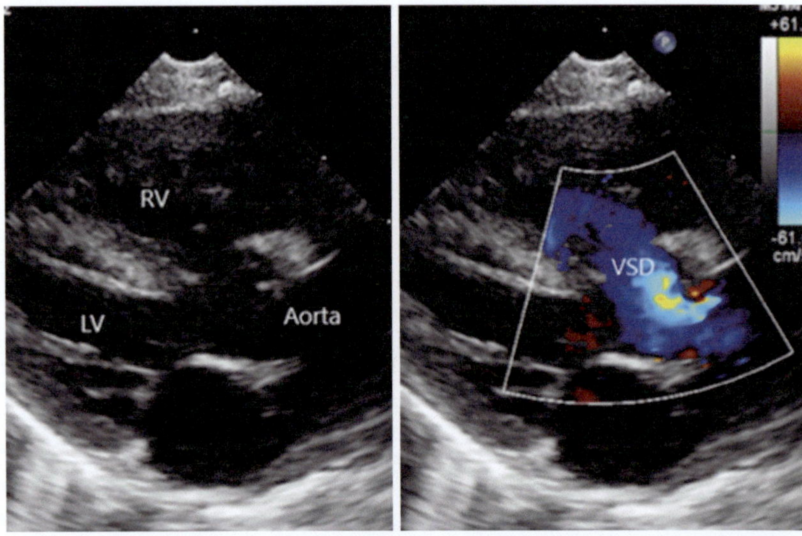

Fig. 10.1 Tetralogy of Fallot: PLAX view showing a large non-restrictive perimembranous VSD with a right-to-left shunt, overriding aorta, and right ventricular hypertrophy

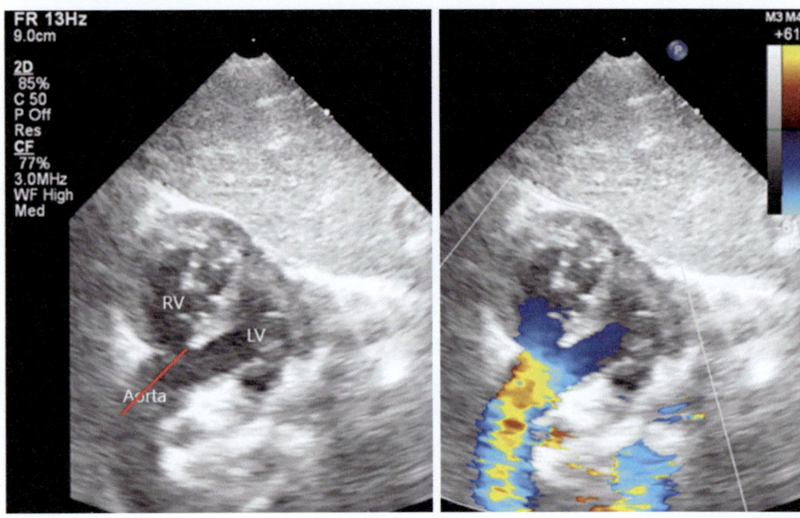

Fig. 10.2 Subcostal short-axis view depicting an overriding aorta. An imaginary line (red), drawn from the crest of the ventricular septum along the septal plane, transects the aortic valve cusps and extends into the aortic root

10.1.2 Overriding of Aorta

Overriding aorta is the morphological feature in which the mouth of the ascending aorta overlies the outlet VSD such that the aorta receives blood from both ventricles (Fig. 10.2).

Although the degree of aortic override varies between 15% and 95%, the degree of override in TOF is conventionally accepted to be ≤50%. Aortic override >50% in the presence of bilateral conal tissue favours the diagnosis of DORV [1].

Aortic override, which is believed to be an effect of rightward malposition and clockwise rotation of the aortic root, is best appreciated from the PLAX view.

10.1.3 RV Outflow Tract Obstruction (RVOTO)

RVOTO can occur at one or more of the following levels in TOF:

- Right ventricular infundibulum (most common).
- Pulmonary valve (usually thickened, can be bicuspid).
- Main pulmonary artery.
- Branch pulmonary arteries (commonly discrete stenosis at the LPA origin).

Valvular and subvalvular obstructions can be distinguished by the typical appearances of their spectral Doppler patterns, which may be seen as overlapping in cases of obstruction at both levels (Fig. 10.3).

Each level of potential obstruction should be evaluated from PSAX and subcostal coronal views. Colour Doppler imaging aids in localizing the origin of turbulent flow in the RVOT. Spectral Doppler aids in determining the peak gradient across the RVOT (CWD), as well as the pressure gradients across various levels of obstruction (PWD).

The severity of RVOTO is graded based on the pressure gradient (discussed in Chap. 7). It is important to document the pressure gradient carefully, as progression of the severity of RVOTO is well known in TOF.

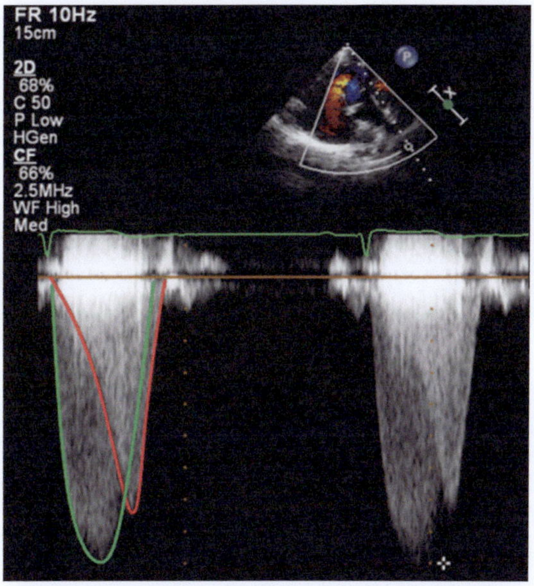

Fig. 10.3 CWD tracing of RVOT flow in TOF demonstrating overlapping patterns of valvular [traced in green] and infundibular [traced in red] stenosis. The mid-systolic peaking of valvular stenosis is seen distinctly from the late peaking Doppler signal of infundibular stenosis ('lobster claw' shape)

10.1.4 Pulmonary Artery Anatomy

The size and anatomy of the pulmonary valve, MPA, and branch PAs are of critical importance in deciding the strategy for surgical repair. Valve-sparing (or annulus-sparing) repair [VSR] can be performed in cases without severe hypoplasia of the pulmonary annulus. In cases of severe hypoplasia, the RVOT may require a transannular patch [TAP] augmentation. Various indices have been proposed to assess the adequacy of the RVOT for total repair, namely [2, 3]:

(a) ***Pulmonary Valve Annulus (PVA) or Pulmonary Annulus (PA) Z-score*:** PVA Z-score represents the number of standard deviations by which the measured PVA diameter differs from the mean value for the reference population. This is derived from established nomograms based on the height, body weight, and body surface area of the patient [2].

    The PV annulus diameter is measured from the PSAX view. A PVA Z-score below −2 (i.e. a PVA diameter, which is more than two standard deviations below the mean) generally indicates that the RVOT may need transannular patch augmentation, although lower cut-offs have also been suggested.

(b) ***Pulmonary Annulus Index (PAI):*** The ratio between measured and expected pulmonary annulus diameters. The proposed cut-off values for adequacy for VSR vary between 0.53 and 0.78. [3, 4]

(c) ***Main Pulmonary Artery (MPA) Z-score*:** Z-score of the MPA diameter measured from the PSAX view. A cut-off of −1 has been found to have a good correlation with better surgical outcomes [3].

(d) ***Great Arterial (GA) ratio (Great artery annulus ratio):*** The ratio of the pulmonary annulus diameter to the aortic annulus diameter. The cut-off values varying between 0.55

and 0.64 have been shown to suggest good outcomes with VSR [5].

(e) ***Pulmonary valve Annulus Area Index (PAAI):*** The ratio between the pulmonary valve annulus cross-section and the aortic valve annulus cross-section. A cut-off value of 0.4 has been suggested for VSR [6].

(f) ***Non-echocardiographic indices*** such as the Nakata index and McGoon's ratio are traditionally used to evaluate the adequacy of pulmonary arteries.

- **Nakata index** is calculated from the diameter of PAs measured immediately proximal to the origin of their upper lobe branches. The sum of the cross-sectional area (CSA) of the right and left PAs is divided by the body surface area of the patient. Although the normal value is >200 mm^2/m^2, a Nakata index >150 mm^2/m^2 is deemed adequate for complete TOF repair [7].
- **McGoon's ratio** is calculated by dividing the sum of the diameters of the RPA (at the level of crossing the lateral margin of the vertebral column on angiogram) and the LPA (just proximal to its upper lobe branch) by the diameter of the aorta at the level above the diaphragm. A value >1.2 is associated with acceptable RV systolic pressure after VSD closure in TOF, whereas a value <0.8 is deemed inadequate [7].

As is evident, the Nakata index and McGoon's ratio are derived from cross-sectional imaging and angiographic estimates, respectively.

10.1.5 Coronary Artery Anomalies (5–7%) and Aortic Arch

Coronary anatomy needs to be carefully delineated because of the vulnerability of the anomalous vessels to injury during surgical repair, particularly the branches crossing the RVOT (discussed in Chap. 14). The PSAX view is the most useful view for this purpose.

Right aortic arch with mirror-image branching is seen in up to 25% of patients with TOF. Suprasternal notch views aid in the diagnosis (please refer to the chapter on segmental anatomy).

10.1.6 Anatomic Variants of TOF

(a) ***TOF with pulmonary atresia*** (10%) with or without major aortopulmonary collaterals [discussed in Chap. 7 under 'Pulmonary atresia with VSD (PA-VSD)'].

(b) ***TOF with absent PV*** (3–6%): TOF with absent pulmonary valve (also known as Hoffman's variant) is associated with aneurysmal MPA and branch PAs. RVOTO is usually caused by a ring of tissue at the level of the pulmonary valve leaflet [8].

Right-sided aortic arch is observed in 50% of these cases. Often, one of the branch PAs is either absent or takes origin from the aorta. The ductus arteriosus is always absent.

(c) ***TOF with Atrioventricular Septal Defect*** (2%) is common in children with Down syndrome. The only clue to the presence of AVSD in these cases may be the presence of a cleft left AV valve, as the primum defect is often absent, and the AV junction is usually divided to form two orifices [1].

(d) ***TOF with origin of one pulmonary artery from the right ventricle and one from the ascending aorta (usually LPA)***—sometimes called 'Hemitruncus'.

10.1.7 Associated Anomalies

- Persistent left SVC draining into the coronary sinus, or rarely directly into the left atrium (10%).
- Abnormalities of subaortic and aortic valve morphology and function.

Key point: The source of pulmonary blood flow (PDA/AP collaterals) needs to be delineated, especially in cases of pulmonary atresia.

Key views	Key features assessed
Subcostal	
– Coronal	Outlet VSD, ASD, aortic override
– Sagittal	VSD, great arterial relationship, aortic override, AV valve cleft, outflow tract obstruction, anomalous systemic venous drainage
Parasternal	
– Long axis	Conal septal malalignment, outlet VSD, aortic override, dimensions of aortic valve annulus/ascending aorta
– Short axis	VSDs, RV outflow tract obstruction, dimensions of PVA/MPA/branch PAs, ventricular function, coronary artery origin/course across the RVOT
Apical	
– Four chamber	Ventricular size/function, VSD—Inlet extension/muscular
– Five chamber	Outlet VSD, aortic override, outflow tract obstruction, relationship of the septal tricuspid leaflet with the VSD
Suprasternal	
– Long axis	Aortic arch anomalies, aberrant subclavian artery, PDA, AP collaterals
– Short axis	Aortic arch sidedness, branch PA size, anomalous systemic/pulmonary venous drainage

Points to Report

- VSD: Location, size, degree of overriding of the aorta
- RVOTO
  - Level(s) of obstruction
  - Severity
- PA anatomy and dimensions (Z-score): PVA, MPA, branch PAs
- Aortic dimensions: Annulus, ascending aorta
- PDA, AP collaterals: Especially in pulmonary atresia
- Coronary artery anomalies
- Anatomic variants: Pulmonary atresia/Absent PV/AVSD
- Associated abnormalities: LSVC, Right aortic arch

10.2 Post-Surgical Evaluation

Surgical repair is indicated in all patients with TOF. The surgical options are determined by the age of the patient, severity of symptoms/cyanosis, and associated anomalies. Generally, early corrective repair is the recommended strategy, although the optimal age preferences may vary according to institutional policies.

Surgery in asymptomatic infants is usually undertaken around 6 months of age. If pulmonary valve preservation is likely, surgical repair may be performed as early as 3 months of age. Symptomatic infants with favourable anatomy are treated with primary neonatal or early repair [1, 9].

Repair (transatrial/transpulmonary) consists of VSD closure and relief of RVOTO by a combination of RV muscle resection and infundibulotomy, with or without patch enlargement of the RVOT and PAs [1, 9].

- **Transannular RVOT patch** repair results in complete relief of RVOTO. However, the resultant free pulmonary regurgitation leads to progressive RV dilation and dysfunction, with or without LV dysfunction. Long-standing chronic RV volume overload causes dilation of the tricuspid annulus, resulting in functional TR.
- **Pulmonary valve sparing/preserving repair** results in mild to moderate residual pulmonary stenosis and associated insufficiency, leading to progressive RV hypertrophy.

Important determinants of long-term prognosis after repair [1, 9]**:**

- Pulmonary regurgitation and its severity.
- Residual lesions (VSD, RVOTO, PA branch stenosis).
- RVOT aneurysms.
- Aortic root dilatation.
- Ventricular dysfunction.
- Conduction abnormalities and arrhythmias.

Symptomatic neonates, who are at high risk for an early repair, may undergo an early palliative procedure—surgical or interventional [1, 9]:

- **Surgical palliation** involves the creation of a shunt between a systemic artery and a pulmonary artery, in order to increase the pulmonary blood flow in a controlled manner. This improves oxygenation and exercise tolerance.
  - **Classic Blalock-Taussig shunt (CBTS)**: End-to-side anastomosis between the subclavian artery (contralateral to the side of the aortic arch) and the ipsilateral PA.
  - **Modified Blalock-Taussig shunt (MBTS)**: Anastomosis between the subclavian artery (contralateral to the side of the aortic arch) and the ipsilateral PA using a PTFE (or Gore-Tex) graft.
  - **Mee's (central) shunt**: Anastomosis between the ascending aorta and the MPA using a short PTFE conduit.
  - **Waterson shunt**: Anastomosis between the ascending aorta and the RPA.
  - **Potts shunt**: Anastomosis between the descending aorta and the LPA.
  - **Cooley shunt**: Intrapericardial anastomosis from the ascending aorta to the RPA.
- **Interventional palliative procedures:**
  - RVOT stenting.
  - Pulmonary valve balloon valvuloplasty.
  - Stenting of the ductus arteriosus.

In TOF with an absent pulmonary valve, repair of the dilated/aneurysmal PA branch is also done, apart from TOF repair, around 1 year of age. In cases with an anomalous LAD crossing the RVOT, an RV–PA conduit is used.

Before performing an echocardiogram in a post-surgical case of TOF, the following details should be sought:

- Pre-operative imaging reports: Anatomical and physiological attributes of the anomaly; associated defects.
- Operation notes: The type of surgery performed; residual defects after surgery.
- Imaging reports from the immediate post-operative period.

10.2.1 Echocardiographic Evaluation of Repaired TOF

10.2.1.1 Evaluation of RVOT and PAs

Evaluation of the RVOT is vital in the echocardiographic evaluation of patients who have undergone intracardiac repair (ICR) for TOF. This can be performed from PLAX, PSAX, subcostal coronal, and subcostal sagittal views. The subcostal inflow–outflow view, where the transducer notch points towards the 1–2 o' clock position, is similar to the angiographic right anterior oblique view [9].

- MPA can be visualized from PSAX view, PLAX RV outflow view, and subcostal views.
- RPA can be visualized from subcostal coronal view, high left PSAX view, right PSAX view, and suprasternal long-axis and short-axis views.
- LPA can be visualized from high left PSAX view and suprasternal long-axis view with a tilt to the left.

(a) ***Structural abnormalities***, such as infundibular aneurysm and/or residual PV tissue, may be seen at the RVOT. Measurement of RV diastolic wall thickness in the infundibular free wall may generate an erroneous impression of RV hypertrophy if a patch has been placed in that segment during surgery.

(b) ***Residual RVOTO:*** Variable degrees of residual RVOTO may be present after TOF repair, especially in those who underwent valve-sparing repair. A high TR jet velocity should alert the echocardiographer towards the possibility of residual RVOTO at some level, including the RV-to-PA conduit or the PAs. Hence, the peak and mean gradients across the RVOT, including the RV-to-PA conduits, should be documented.

The level of the obstruction needs to be ascertained using PWD, spectral Doppler, and colour flow Doppler (Fig. 10.4). Obstruction may be present at more than one level [discussed in Chap. 7].

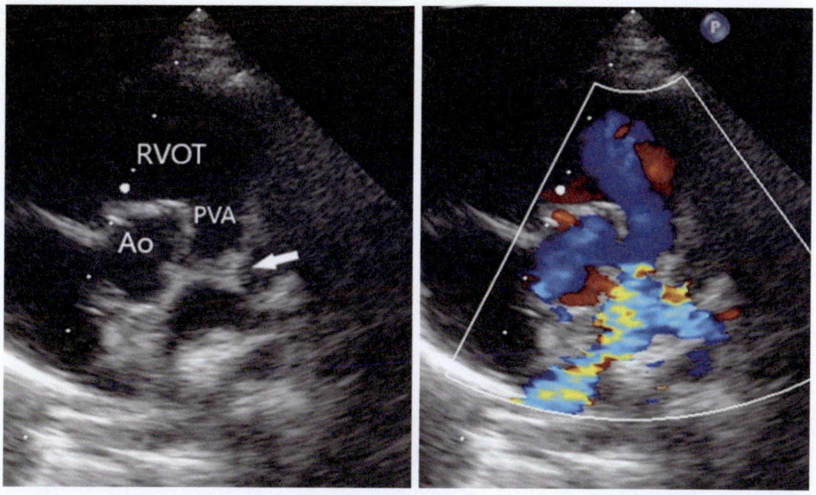

Fig. 10.4 PSAX view depicting supravalvular pulmonary stenosis (arrow) in a case of TOF after intracardiac repair. It may be noted that there is no residual VSD or subvalvular stenosis. *RVOT* Right Ventricular Outflow Tract, *Ao* Aorta, *PVA* Pulmonary valve annulus

(c) ***Dimensions:*** RVOT dimension at the site of the native PV should be documented in those who have undergone transannular patch repair, especially when transcatheter valve implantation is being considered.
   - The dimensions of the MPA at the mid-point and branch PAs at the origins should be documented in all cases.
   - The diameter of the narrowest segment needs to be reported in cases of supravalvular stenosis.

10.2.1.2 Degree of Pulmonary Regurgitation

Pulmonary regurgitation (PR) may occur as a result of transannular patching or pulmonary valvotomy/valvectomy. The severity of PR is an important determinant of long-term outcomes after TOF repair. It should be remembered that in the context of ***'wide open/free PR', the jet(s) may be overlooked by colour Doppler*** because the flow is laminar with low velocity (Fig. 10.5a).

Various echocardiographic indices suggestive of severe PR include:

- **Diastolic flow reversal** in the MPA or branch PAs (ascertained by colour or spectral Doppler) (Fig. 10.5b).
- **PR jet width** on colour Doppler ≥50% of the pulmonary annulus [**PR jet width/pulmonary annulus ratio cut-off of > 0.7** corresponds to CMR PR fraction ≥40%] [9, 10].
- **Short deceleration time** in the PR spectral Doppler tracing, reflective of rapid equilibration of PA and RV diastolic pressures, suggests severe PR. This may result in a period of no flow or even antegrade flow across the RVOT in the latter part of diastole.
- **Pressure half-time < 100 msec** is reflective of severe PR. This measurement may not be reliable in the presence of high RVEDP, which results in early equilibration between the MPA and RV pressures.
- **PR index**: The ratio of the duration of the PR signal in the spectral Doppler tracing to the total duration of diastole (Fig. 10.5c). A value <0.77 has been shown to correlate with CMR-derived PR fractions >25% [9, 10].
- Pulmonary Regurgitation Index by M-mode echocardiography (**PRIME**) or **RPA pulsatility** is the systolic-to-diastolic variation in RPA diameter. M-mode echocardiography of the proximal RPA is performed from the suprasternal window, after visualizing the RPA in its long axis. The systolic and diastolic diameters of the RPA are measured edge-to-edge. The ratio between the systolic and diastolic diameters is defined as PRIME (or RPA pulsatility). ***PRIME value > 1.2 suggests significant PR*** [11].
- **DSTVI** (the ratio of the diastolic to systolic time velocity integral of the main pulmonary artery) has been found to have moderate correlation with PR fraction by CMR, with values ≥0.8 corresponding to severe PR [12].

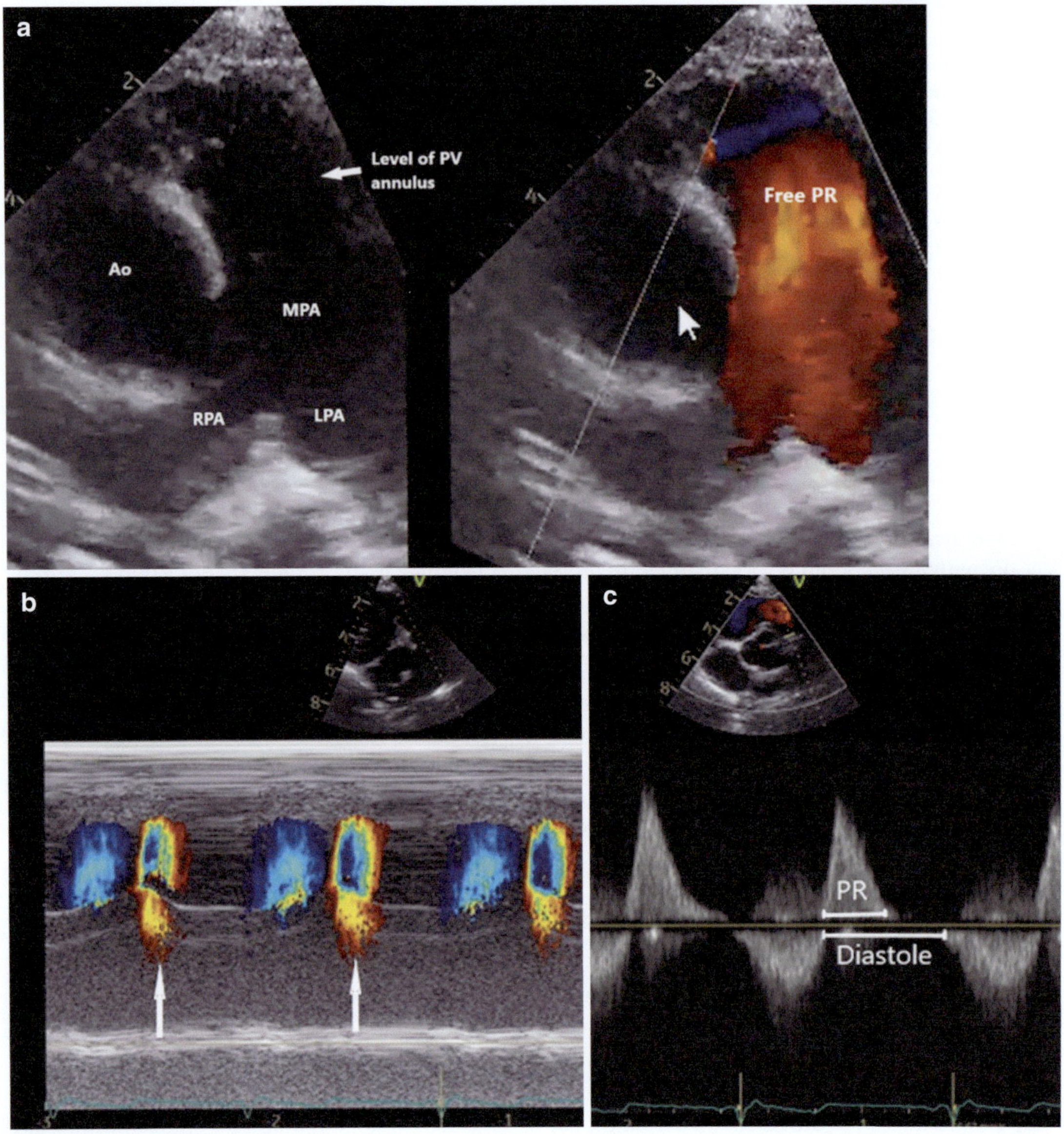

Fig. 10.5 Post-transannular patch repair of TOF. (**a**) Colour Doppler image depicting free PR. (**b**) Colour M-mode imaging depicting diastolic flow reversal in the branch PA [arrow]. (**c**) Spectral Doppler tracing of the flow across the pulmonary valve annulus. The early diastolic PR jet decelerates rapidly, with the duration of the PR signal lasting less than 50% of the total duration of diastole

10.2.1.3 Tricuspid Valve Morphology and Function

(a) ***Tricuspid regurgitation*** may result from direct damage to the valve, basal-lateral displacement of the free-wall papillary muscles, or annular dilatation secondary to RV enlargement [9]. TR can also result from impingement of the VSD patch on the septal-anterior commissure of the TV, in which case the TR jet originates at the junction between the VSD patch and the septal attachment of the TV, and extends along the atrial septum [9]. A vena contracta width > 0.7 cm indicates severe TR.

(b) ***b) TV annular diameter:*** The septolateral tricuspid annular dimension should be measured in end-diastole from the A4C view (Fig. 10.6a).

(c) ***TV tethering [tenting] height and area:*** Measurements for the tricuspid tenting height and tenting area are performed in end-systole from the A4C view (Fig. 10.6b).

- Tethering [tenting] height > 0.76 cm and tethering [tenting] area > 1.63 cm^2 are predictive of residual TR after TV annuloplasty [13].

10.2.1.4 Right Ventricle

(a) ***Dimensions:***

RV enlargement is suggested by:

- RV to LV basal diameter ratio > 1, in A4C view.
- Anteroposterior (AP) diameter of the RV is greater than that of the LV in PSAX view at the level of the papillary muscles. RV-focused A4C view, with both the crux and the apex visible, is ideal for measurements:
- RV diameter > 42 mm at the base and > 35 mm at the midventricular level indicate RV dilation [14].
- RV end-diastolic cross-sectional area < 20 cm^2/m^2 body surface area (BSA) corresponds with CMR-measured RV end-diastolic volume index <170 mL/m^2 [15].

(b) ***RV function:***

- Percentage **RV fractional area change** is defined as:

  ([End-diastolic area minus End-systolic area]/End-diastolic area) /100.

  RV endocardium is traced in end-systole and end-diastole from the RV-focused A4C view.

  RV fractional area change <35% is suggestive of RV systolic dysfunction [16].
- Indices of RV systolic dysfunction, such as TAPSE [< 16 mm] and tricuspid annulus systolic excursion velocity [< 10 cm/s], have not been studied particularly in this subset of patients.
- The clinical utility of non-geometric indices of RV function (such as dP/dt and myocardial performance or Tei index) is unclear in patients with repaired TOF [9].
- RV volume and pressure overload are associated with flattening of the interventricular septum in diastole and systole, respectively, resulting in a D-shaped left ventricle.

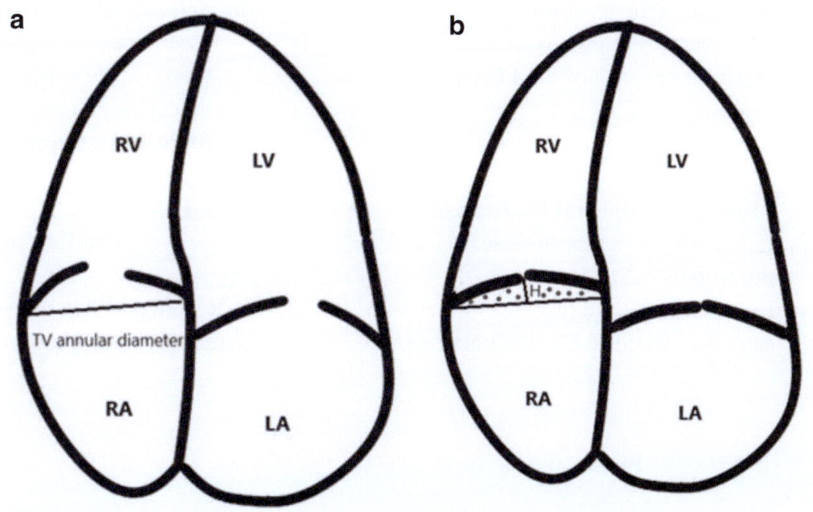

Fig. 10.6 Schematic representation of the measurement of (**a**) TV annular diameter in end-diastole [left]; and (**b**) TV tenting height [H] and area in end-systole [dotted area] [right]

10.2.1.5 Right Atrium

(a) ***Dimensions:***

RA dimensions are measured from the A4C view at end-systole. The major axis extends from the middle of the bisected TV annulus to the posterior wall of the RA, and the minor axis is perpendicular to the major axis. In adults, an RA area > 18 cm^2 indicates RA enlargement [9].

(b) ***RA pressure:***

RA pressure is estimated, on the basis of RA size and respiratory variation in IVC dimension, as follows [14]:

- 3 mmHg: IVC diameter < 2.1 cm; >50% collapse (sniff) or > 20% (normal inspiration)
- 15 mmHg: IVC > 2.1 cm; <50% collapse (sniff) or < 20% (normal inspiration)
- 8 mmHg: When IVC size and respiratory change do not fit the above scheme.

10.2.1.6 Pulmonary Artery Pressure

PA pressures may be estimated using the modified Bernoulli equation:

- PA systolic pressure (in the absence of residual RVOT obstruction) = 4 V^2 + RA pressure (where V is the peak tricuspid regurgitation velocity).
- PA diastolic pressure: Estimated from the end-diastolic velocity of the PR jet.
- PA mean pressure: Estimated from the early diastolic velocity of the PR jet.

Elevated PASP may be seen in:

- Branch or peripheral PA stenosis.
- Pulmonary vascular disease.
- Pulmonary vein stenosis.
- Left heart disease (e.g. cor triatriatum, mitral stenosis, diastolic LV dysfunction).

10.2.1.7 LV Size and Function

LV systolic dysfunction is a marker of premature death in patients with repaired TOF. The biplane area-length method is the preferred method for the estimation of LV systolic function [9]. In patients with marked RV dilation, where the apex is formed by the RV, the transducer position needs to be adjusted accordingly to avoid foreshortening of the LV in the apical windows.

Decreased mitral annular plane systolic excursion (MAPSE), measured by M-mode imaging of the lateral mitral annulus from the A4C view, is another index of LV systolic dysfunction, which has been found to be associated with adverse outcomes.

10.2.1.8 Aorta and Aortic Valve

(a) ***Aortic dimensions:*** Dilated aortic root is common in adults with repaired TOF, particularly in those with prior shunts and late repair.
   - The PLAX view is ideal for measuring the aortic root dimension.
   - The right parasternal window is preferred for the proximal ascending aorta and mid ascending aorta (measured at the level of the RPA).
   - The internal diameter (inner edge to inner edge) is measured during maximal expansion (mid to late systole) [17].

(b) ***Severity of aortic regurgitation*** needs to be ascertained by standard measures [discussed in Chap. 9]. Spectral Doppler evaluation of the descending thoracic aorta and abdominal aorta for diastolic run-off is useful in this regard.

10.2.1.9 Residual Intracardiac and Extracardiac Shunts

- Residual VSDs may be located at any part of the interventricular septum. The superior

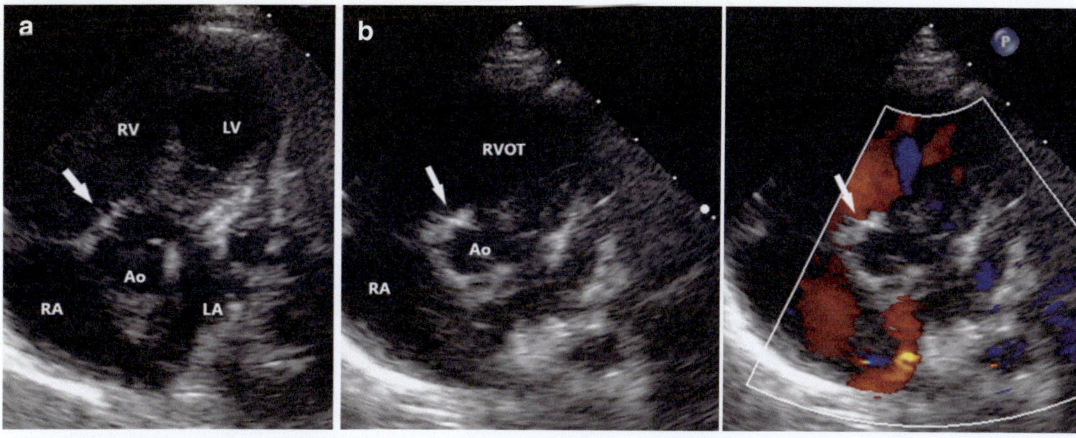

Fig. 10.7 Post-intracardiac repair for TOF. An intact VSD patch (denoted by arrow) with no residual shunt is seen in the A5C view (**a**) and PSAX view (**b**)

portion of the VSD patch is a common site of residual VSD. The VSD patch is seen in the PLAX view as a linear structure passing obliquely from the septum to the anterior aortic root (Fig. 10.7).

- Other shunt lesions: PFO, secundum ASD, unrepaired sinus venosus defect, coronary sinus ASD, and PAPVC.
- Systemic-to-pulmonary collateral vessels may be assessed by colour Doppler interrogation.

Points to Report

- RVOT
  - Location and mechanism of residual RVOTO
  - Peak and mean gradients
  - Presence of infundibular aneurysm and residual PV tissue
- MPA and Branch PAs
  - Standard dimensions, as well as the diameter of the narrowest segments
  - Location and severity of obstruction
- RV-to-PA conduit: Peak and mean gradients
- Severity of PR
- Tricuspid valve
  - Annulus diameter
  - Severity and mechanism of TR
  - TV tethering height and area
- RV pressure, RV dimensions, RV function
- Pulmonary artery pressure
- LV size and function
- VSD patch integrity; residual VSDs and other shunts
- Aortic dimensions and aortic valve regurgitation
- Systemic-to-pulmonary collateral vessels

10.2.2 Echocardiographic Evaluation of Palliative Procedures

10.2.2.1 Systemic-to-Pulmonary Artery Shunts

Visualizing the Shunt Vessel/Conduit

The shunts can be detected by 2D echocardiography and colour flow Doppler. The first step in visualizing the shunt flow is the identification of the systemic arterial orifice. This can be done by visualizing the aortic arch and/or the descending aorta from the suprasternal window. The ascending aorta can be visualized from the right or left parasternal window.

- **Left BT shunt**: After visualizing the aortic arch from the suprasternal or high parasternal window, the proximal opening of the shunt is noted in the LSCA, or the left brachiocephalic artery. The distal segment of the conduit can be imaged by clockwise rotation of the transducer by 30^0. Colour Doppler imaging depicts the shunt flow in a direction away from the transducer.

- **Right BT shunt**: After visualizing the aortic arch from the suprasternal window, the proximal opening of the shunt is noted in the RSCA. The distal segment of the conduit can be imaged by tilting the transducer by 15° to the right. Colour Doppler imaging depicts the shunt flow in a direction away from the transducer.
- **Central shunt or Waterston shunt**: The ascending aorta is visualized in its long axis from the left parasternal window. The shunt flow can be imaged till the RPA, with the aid of colour flow Doppler. The RPA can be imaged in its long axis from the suprasternal window.

The shunt vessel/conduit should be imaged along its entire length with the aid of colour flow Doppler. The internal diameter of the shunt conduit should be measured, and segmental stenosis should be thoroughly searched for (Fig. 10.8).

Key point: A narrow jet with turbulence on colour flow is suggestive of shunt stenosis.

Spectral Doppler Assessment of Gradients Across the Shunt

CWD imaging of normal systemic-PA shunt flow depicts a high-velocity (>4 m/s) flow with a characteristic saw-tooth pattern, which peaks during early systole [9]. Owing to the continuous pressure gradient between the aorta and pulmonary artery, the shunt flow is continuous, with a gradual decline in velocity prior to the next systole (Fig. 10.9). A reduced peak flow velocity in the shunt often suggests an increase in PA pressure.

The peak flow velocity should be measured at the proximal and distal parts of the shunt vessel/conduit, as peak flow velocity will be higher in the distal portion of segmentally stenosed shunt vessels [9]. It needs to be remembered that Doppler velocity underestimates the pressure gradient in a conduit due to the inherent limitations of the modified Bernoulli equation in assessing long segmental narrowing.

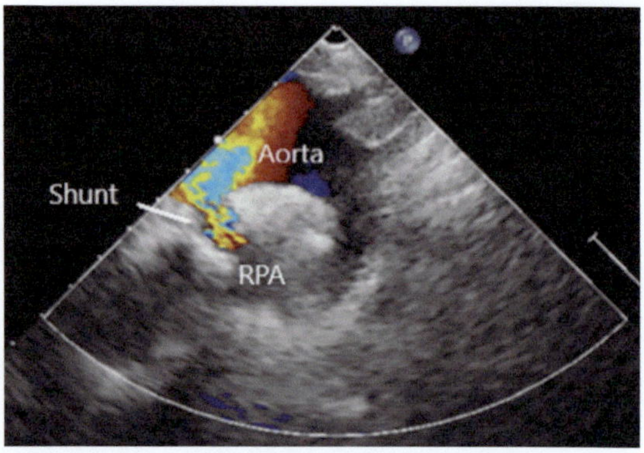

Fig. 10.8 Suprasternal aortic arch view depicting a narrow jet with turbulent flow in the aorta-to-RPA central shunt, suggesting stenosis of the shunt

Key points: The presence of continuous shunt flow with no significant gradient rules out the possibility of significant segmental stenosis in the shunt vessel/conduit. Non-segmental uniform narrowing may be detected by a significant reduction in the internal diameter of the shunt.

Functional Assessment

The presence of pulmonary hypertension, post-surgical PA anatomy, and the presence of ventricular dysfunction are important factors to consider prior to late repair after shunt surgery.

10.2.2.2 Interventional Palliative Procedures

Echocardiographic evaluation of an interventional palliative procedure depends on the type of procedure performed:

- RVOT stenting: Evaluate for residual RVOTO/stent re-stenosis; PA size and pressures; RV size and function.
- Pulmonary valve balloon valvuloplasty: Evaluate for residual pulmonary stenosis; PA size and pressures; RV size and function.
- Stenting of the ductus arteriosus: Evaluate for patency of the ductal stent; MPA and branch PA dimensions; PA pressure; RV size and function.

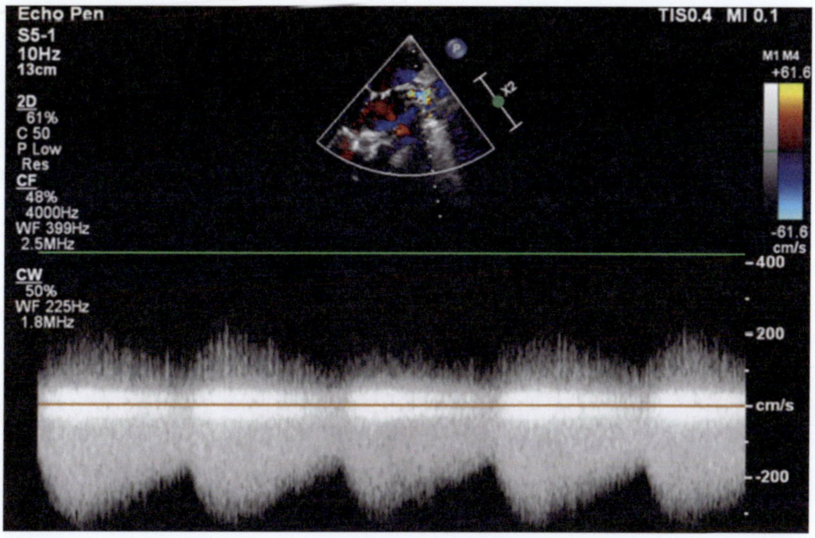

Fig. 10.9 CWD Doppler tracing depicting continuous flow with the characteristic saw-tooth pattern in an unobstructed left BT shunt

References

1. Roche SL, Greenway SC, Redington AN. Tetralogy of Fallot with pulmonary stenosis and tetralogy of Fallot with absent pulmonary valve. In: Allen HD, Driscoll DJ, Shaddy RE, Feltes TF, editors. Moss and Adams' heart disease in infants, children, and adolescents. 8th ed. Philadelphia, PA: Lippincott Williams & Wilkins; 2013. p. 969–89.
2. Awori MN, Leong W, Artrip JH, O'Donnell C. Tetralogy of Fallot repair: optimal z-score use for transannular patch insertion. Eur J Cardiothorac Surg. 2013;43(3):483–6.
3. Wang Z, Li Z, Ding N, Zhu Y, Li X, Yi H. When do patients with tetralogy of Fallot need a transannular patch. J Card Surg. 2022;37(12):5041–51.
4. Lyu Z, Jin M, Yang Y. Value of pulmonary annulus index in predicting transannular patch in tetralogy of Fallot repair. J Card Surg. 2021;36(7):2197–203.
5. Choi KH, Sung SC, Kim H, Lee HD, Ban GH, Kim G, Kim HY. A novel predictive value for the Transannular patch enlargement in repair of tetralogy of Fallot. Ann Thorac Surg. 2016;101(2):703–7.
6. Zhao J, Cai X, Teng Y, Nie Z, Ou Y, Zhuang J, et al. Value of pulmonary annulus area index in predicting transannular patch placement in tetralogy of Fallot repair. J Card Surg. 2020;35(1):48–53.
7. Gao Y, Yang ZG, Shi K, Diao KY, Xu HY, Guo YK. Computed tomography for evaluating right ventricle and pulmonary artery in pediatric tetralogy of Fallot: correlation with post-operative pulmonary regurgitation. Sci Rep. 2018;8(1):7515.
8. Donofrio MT, Jacobs ML, Rychik J. Tetralogy of Fallot with absent pulmonary valve: echocardiographic morphometric features of the right-sided structures and their relationship to presentation and outcome. J Am Soc Echocardiogr. 1997;10:556–61.
9. Valente AM, Cook S, Festa P, Ko HH, Krishnamurthy R, Taylor AM, et al. Multimodality imaging guidelines for patients with repaired tetralogy of fallot: a report from the American society of echocardiography: developed in collaboration with the Society for Cardiovascular Magnetic Resonance and the Society for Pediatric Radiology. J Am Soc Echocardiogr. 2014;27(2):111–41.
10. Mercer-Rosa L, Yang W, Kutty S, Rychik J, Fogel M, Goldmuntz E. Quantifying pulmonary regurgitation and right ventricular function in surgically repaired tetralogy of Fallot: a comparative analysis of echocardiography and magnetic resonance imaging. Circ Cardiovasc Imaging. 2012;5(5):637–43.
11. Festa P, Ait-Ali L, Minichilli F, Kristo I, Deiana M, Picano E. A new simple method to estimate pulmonary regurgitation by echocardiography in operated fallot: comparison with magnetic resonance imaging and performance test evaluation. J Am Soc Echocardiogr. 2010;23(5):496–503.
12. Bhat M, Goldmuntz E, Fogel MA, Rychik J, Mercer-Rosa L. Longitudinal validation of the diastolic to systolic time-velocity integral ratio as a Doppler-derived measure of pulmonary regurgitation in patients with repaired tetralogy of Fallot. Pediatr Cardiol. 2017;38(2):240–6.
13. Fukuda S, Song JM, Gillinov AM, McCarthy PM, Daimon M, Kongsaerepong V, Thomas JD, Shiota T. Tricuspid valve tethering predicts residual tricuspid regurgitation after tricuspid annuloplasty. Circulation. 2005;111(8):975–9.
14. Rudski LG, Lai WW, Afilalo J, Hua L, Handschumacher MD, Chandrasekaran K, et al. Guidelines for the echocardiographic assessmentof the right heart in adults: a report from the American Societyof Echocardiography endorsed by the European Association of Echocardiography, a registered branch of the European Society of Cardiology,

and the Canadian Society of Echocardiography. J Am Soc Echocardiogr. 2010;23:685–713.

15. Alghamdi MH, Grosse-Wortmann L, Ahmad N, Mertens L, Friedberg MK. Can simple echocardiographic measures reduce the numberof cardiac magnetic resonance imaging studies to diagnose right ventricularenlargement in congenital heart disease? J Am Soc Echocardiogr. 2012;25:518–23.
16. Wang J, Prakasa K, Bomma C, Tandri H, Dalal D, James C, et al. Comparison of novel echocardiographic parameters of right ventricular functionwith ejection fraction by cardiac magnetic resonance. J Am Soc Echocardiogr. 2007;20:1058–64.
17. Lopez L, Colan SD, Frommelt PC, Ensing GJ, Kendall K, Younoszai AK, et al. Recommendations for quantification methods during the performance of a pediatric echocardiogram: a report from the pediatric measurements writing Group of the American Society of echocardiography pediatric and congenital heart disease council. J Am Soc Echocardiogr. 2010;23:465–95.

11 Transposition of Great Arteries

11.1 Transposition of Great Arteries [d-TGA]

The morphological characteristics of d-TGA are:

- Atrioventricular concordance (d loop) with ventriculo-arterial discordance.
- An anteriorly positioned right-sided aorta arises from the RV with a subaortic conus [leading to aorto-mitral discontinuity].
- A posteriorly positioned left-sided pulmonary artery arises from the LV with mitro-pulmonary fibrous continuity.

Physiologically, d-TGA is characterized by parallel circulation, where systemic and pulmonary circulations are completely separated from each other in the absence of intercirculatory shunting.

Classification [1, 2]

- **I**: TGA with intact ventricular septum (or small VSD)—increased pulmonary blood flow and small intercirculatory shunting.
- **II**: TGA with large VSD—increased pulmonary blood flow and large intercirculatory shunting.
- **III**: TGA with VSD and LVOTO—restricted pulmonary blood flow.
- **IV**: TGA with VSD and Pulmonary Vascular Obstructive Disease—restricted pulmonary blood flow.

Nomenclature and Position of Aortic Sinuses in TGA [1, 3] (Fig. 11.1)

- The two aortic sinuses of Valsalva that 'face' the pulmonary artery are called septal or **facing sinuses**.
- The facing sinuses contain the coronary artery ostia in almost all cases.
- When the great arteries are positioned side-by-side, the facing sinuses are termed as 'anterior' and 'posterior' facing sinuses (Fig. 11.1a).
- When the aorta is anterior and to the right of the PA (**the typical relationship in d-TGA**), the facing sinuses are termed as 'leftward anterior' and 'rightward posterior' facing sinuses (Fig. 11.1b).
- When the great arteries are oriented antero-posteriorly, the facing sinuses are termed as 'left' and 'right' facing sinuses (Fig. 11.1c).

A. Paul, *Manual of Echocardiography for Congenital Heart Diseases*,
https://doi.org/10.1007/978-981-96-0243-8_11

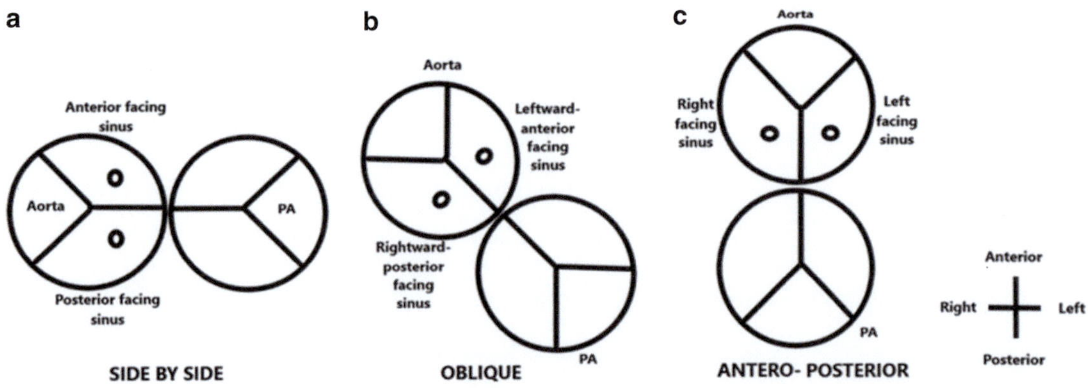

Fig. 11.1 Schematic representation of the various types of great arterial relationships in d-TGA and the nomenclature of facing sinuses in each type

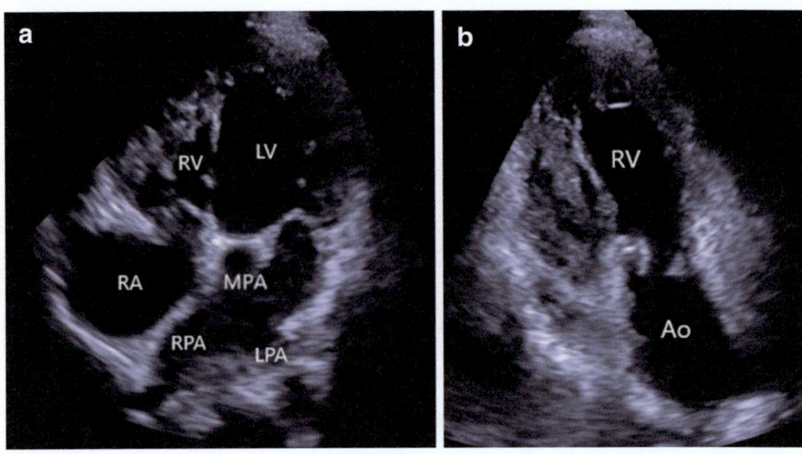

Fig. 11.2 Anterior sweep from A4C view showing d-TGA with posterior PA (identified by its early bifurcation) arising from the smooth-walled LV (**a**), and the aorta seen originating from the morphologic RV on sweeping further anteriorly (**b**)

11.1.1 Echocardiographic Diagnosis

(a) **Great arterial origin:** The subcostal coronal or apical four-chamber view, on anterior sweep, reveals that the first visible great artery originates from the LV and bifurcates into two branches, identifying it as the PA (Fig. 11.2). Further anterior sweep reveals the aorta originating from the anteriorly located RV. In the PLAX view, the origin and proximal course of both the great arteries can be seen in TGA (in contradistinction to the normal situation, where only the great artery arising from the posteriorly placed LV is seen in the PLAX view).

(b) **Great arterial relation:**
- In contrast to the normal spiralling pattern, the proximal parts of the great arteries **are parallel to each other** in TGA.
- The PLAX view depicts the parallel course of the proximal great arteries as they arise from the two ventricles.
- The PSAX (or subcostal sagittal) view reveals a **'double-barrel' appearance** instead of the normal 'circle and sausage appearance' of the great arteries.

(c) **Conal morphology** in d-TGA is different from normal and takes one of the following forms [1, 4]:

- Persistent subaortic conus with regression of subpulmonary conus is the commonest (90%).
- Bilateral conus (5%).
- Bilaterally absent conus, or subpulmonary conus with absent subaortic conus (rare).

In the majority of cases, the fibrous continuity between the mitral and pulmonary valves can be visualized from the PLAX view, indicative of regression of the subpulmonary conus.

Key point: Visualization of parallelly aligned great arteries in the PLAX and PSAX views suggests the diagnosis of TGA. The classical 'circle and sausage' appearance of the great arteries in the PSAX view rules out the possibility of TGA.

11.1.2 Adequacy of Intercirculatory Mixing

Intercirculatory mixing of blood in d-TGA is facilitated by communications at the level of the atria, ventricles, and/or great arteries. The adequacy of this communication needs to be ascertained in the early neonatal period. If the communication is deemed inadequate, balloon atrial septostomy (BAS) may be necessary to facilitate intercirculatory mixing.

(a) **Interatrial communication:** A patent foramen ovale is almost always present, whereas a true ostium secundum ASD is present in 5% of cases [1, 2]. The size of this communication, as well as the mean pressure gradient across the interatrial septum, should be documented.

(b) **Ventricular Septal Defect:** A VSD is present in 40% of cases, enhancing the intercirculatory mixing of blood [1, 2]. Identification of multiple VSDs by colour Doppler is often difficult when there is no significant pressure difference between the ventricles. The location of the VSD may vary as follows [1, 5]:
   - Outlet (Doubly committed subarterial) VSD: These are conal septal defects, which result in fibrous continuity of the aortic and pulmonary valves. Closure of these defects poses the risk of semilunar valve distortion and regurgitation.
   - Perimembranous defects with malalignment are associated with outflow tract obstruction. Posterior malalignment of the conal septum results in LVOTO (subpulmonary), while anterior malalignment results in RVOTO (subaortic) and distal arch obstruction.
   - Muscular defects.
   - Inlet (AV canal type) VSD: Relatively uncommon in TGA, except in cases of heterotaxy syndrome. In cases of inlet VSDs, straddling of the TV needs to be searched for, which may preclude a two-ventricle repair.

   The location, number, size, and margins of VSDs, along with the associated anomalies, should be ascertained from subcostal, apical, and parasternal views.

(c) **Patent ductus arteriosus:** In the absence of adequate intercirculatory mixing at the atrial or ventricular levels, a large PDA may facilitate adequate mixing of blood (ductus-dependent circulation). This is best evaluated in the high left parasternal view or the suprasternal sagittal view due to the parallel orientation of the aorta and PA. When there is no significant pressure difference between the ventricles, it may be challenging to visualize the ductal flow (discussed in Chap. 4).

Key point: High pulmonary vascular resistance and pulmonary vein stenosis can cause marked cyanosis, despite adequate communication at the atrial, ventricular, and/or great arterial level [1].

11.1.3 LV Outflow Tract (Subpulmonic) Obstruction

Occurs in 10% of TGA-VSD cases. LVOTO is evaluated from the subcostal and apical views. Careful assessment reveals the mechanism of obstruction:

- **Fixed obstruction**: Can result from basal septal hypertrophy, protrusion of an aneurysmal mem-

branous septum into the LVOT, posterior malalignment of the outlet septum, a subvalvular fibromuscular ridge, pulmonary valve stenosis, and/or abnormal attachment of the mitral subvalvular apparatus to the ventricular septum [6, 7].
- **Dynamic obstruction**: Can result from bowing of the ventricular septum into the LVOT. This occurs in TGA with an intact ventricular septum owing to RV hypertension [1, 6].

11.1.4 Associated Lesions [1, 2, 6, 7]

- Persistent left SVC (4%), partial or total anomalous pulmonary venous return.
- Leftward juxtaposition of the atrial appendages (2%) (discussed in Chap. 15).
- Complete AV canal defect: common AV valve, cleft mitral valve, straddling AV valve.
- Straddling of the tricuspid valve with abnormal attachments to the crest of the VSD or to the conal septum.
- Anterior malalignment of the conal septum into the RVOT (less common) can be associated with aortic outflow tract obstruction at multiple levels.
- **Aortopulmonary collateral vessels** arising from the descending aorta can be visualized from subcostal and suprasternal views. These are seen as multiple flow signals around the DTA with colour Doppler.

11.1.5 Coronary Arteries (CAs)

Coronary arteries are always in an abnormal position in TGA due to the anomalous position of the aorta. CA ostia are almost always located in one or both of the 'facing sinuses' of the aorta [1, 3, 6]. The various CA patterns in TGA are [3, 6] (Fig. 11.3):

- The **'usual' CA pattern** (65%): The LCA originates from the leftward/anterior facing aortic sinus, and the RCA from the rightward/posterior facing aortic sinus. The LCA gives rise to the LAD and circumflex CA.
- **Common CA variants** include [1, 3, 6]:
  - Circumflex CA from RCA (13%), where the circumflex CA courses posterior to the PA.
  - Single CA originating from the rightward/posterior facing sinus [designated as ***'single RCA'***], which gives rise to the LCA that courses behind the PA (7%).
  - ***Inverted RCA and circumflex***: RCA and anterior descending CA from the leftward/anterior facing sinus, while the circumflex CA from the rightward/posterior facing sinus (6%).
- Rare variants include [1, 3, 6]:
  - ***Inverted CAs***: The LCA originates from the rightward/posterior facing sinus, and the RCA from the leftward/anterior facing sinus (3%).
  - ***Single*** CA originating from the leftward/anterior facing sinus [designated as ***'single LCA'***] gives rise to the RCA, which courses anterior to the aorta (2%).
  - **Intramural CA**: The coronary artery anomaly in which a segment of the proximal CA runs within the wall of the aorta. CAs with an intramural course usually originate from the opposite sinus of Valsalva [i.e. LCA/LAD from the rightward sinus; RCA from the leftward sinus] and pass between the semilunar valves.

    Intramural LCA: Both the RCA and LCA originate from the rightward/posterior facing sinus, with the LCA pursuing an intramural course between the great arteries (3%).

    Intramural RCA: Both the RCA and LCA originate from the leftward/anterior facing sinus, with the RCA taking an intramural course between the great arteries (1%).

Intramural CA and single CA ostium carry the highest risk for early post-operative mortality [6].

Key point: Identification of a CA branch coursing posterior to the PA and/or anterior to the aorta suggests that the coronary artery anatomy is not in the usual pattern.

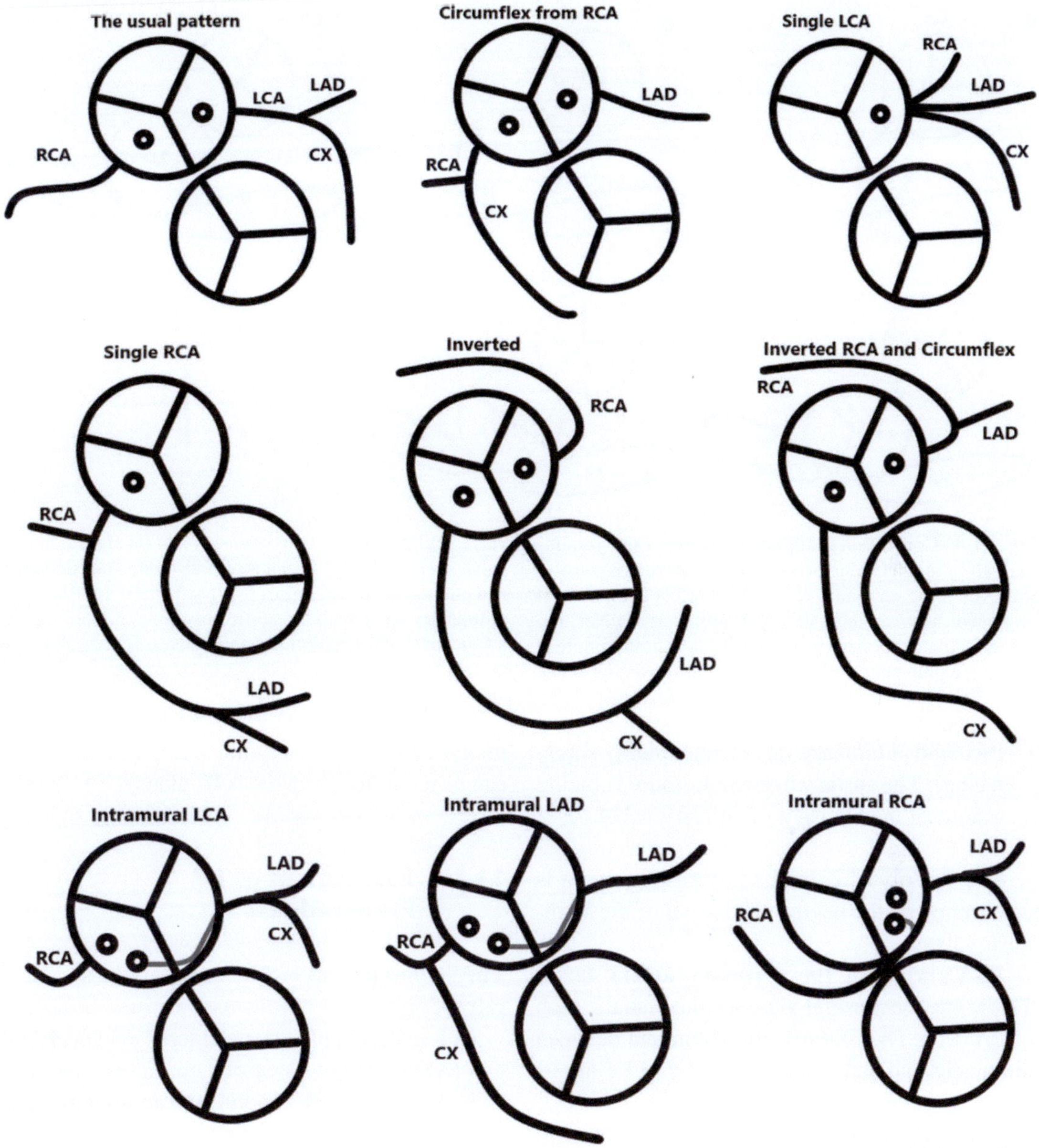

Fig. 11.3 Schematic representation of the short axis view of the great arteries, depicting various anomalies of coronary artery origin/course seen in d-TGA. *LCA* left coronary artery, *RCA* right coronary artery, *LAD* left anterior descending artery, *CX* circumflex artery. (Source: Reference [6])

Commissural Malalignment:

Alignment between the aortic and pulmonary valve commissures is a key consideration in the pre-operative assessment of TGA. Malalignment between the commissures of the semilunar valves poses a significant challenge to the translocation of CA. On the basis of the rotation of the semilunar valves, the various patterns of commissural malalignment are classified as [8] (Fig. 11.4):

- Sinus-facing pulmonary valve: The aortic valve commissure is facing the pulmonary valve sinus.
- Sinus-facing aortic valve: The pulmonary valve commissure is facing the aortic valve sinus.
- Sinus-facing of both valves: The aortic valve sinus faces the pulmonary valve sinus.

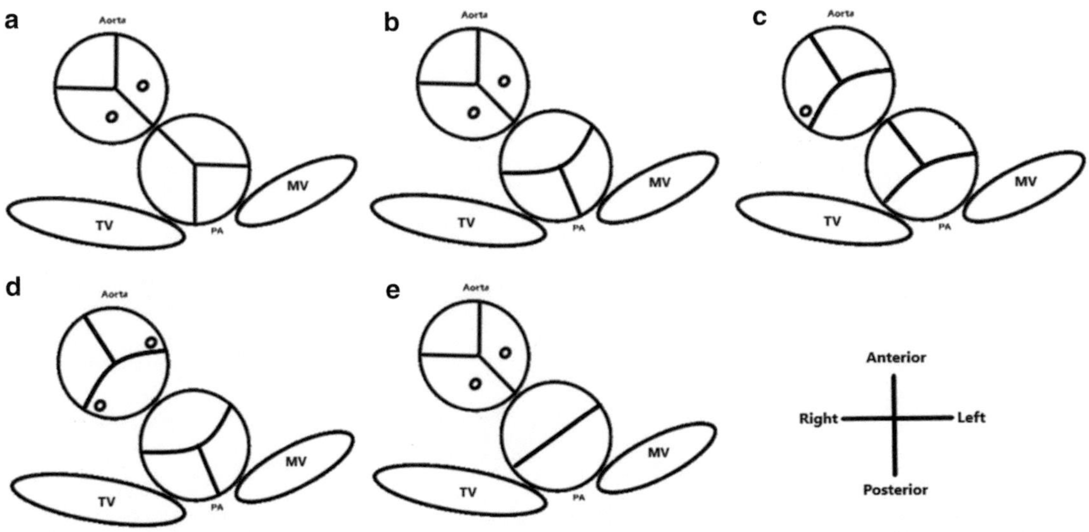

Fig. 11.4 Commissural alignment between the semilunar valves in d-TGA, as seen from the short axis view at the level of the semilunar valves. (**a**) Perfect commissural alignment between the aortic and pulmonary valves. (**b**) Sinus-facing pulmonary valve (aortic valve commissure is facing the pulmonary valve sinus). (**c**) Sinus-facing aortic valve (pulmonary valve commissure is facing the aortic valve sinus). (**d**) Sinus-facing of both valves (aortic valve and pulmonary valve sinuses face each other). (**e**) Bicuspid pulmonary valve with the aortic valve commissure facing a sinus of the bicuspid pulmonary valve

- Bicuspid pulmonary valve (functionally sinus-facing): The aortic valve commissure is facing a sinus of the bicuspid pulmonary valve.

The severity of commissural malalignment is determined by the degree of rotation of the semilunar valves.

PSAX view at the coronary artery level: This is the most useful view for the visualization of CA ostia. The commissural alignment between the aortic and pulmonary valves can be ascertained from this view (Fig. 11.4).

Intramural CAs coursing between the two great arteries can be identified in this view. A juxtacommissural origin at an acute angle from the aortic root, or a high take-off from the sinotubular junction, are also features associated with intramural CAs [9].

PLAX view: Useful in the visualization of the bifurcation of the LCA.

Subcostal coronal and apical views: Posterior to anterior sweeps in these views are useful in visualizing the CA branches coursing posterior to the PA [9–11]. The circumflex CA from the RCA can be identified from the A4C view.

11.1.6 Assessment of 'LV Preparedness'

The concept of 'LV preparedness and retraining':

Before the era of neonatal arterial switch operation (ASO), ASO was carried out in two stages—an initial stage of PA banding and aortopulmonary shunt, followed by a delayed ASO. This delay in ASO posed a risk of regression of LV mass. Hence, it was important to determine the 'preparedness' of the LV to function as the systemic ventricle, prior to proceeding with ASO. When there is significant regression of LV mass, LV retraining is done in preparation for the late arterial switch. In the current era of neonatal ASO, the strategy of LV retraining is relevant in cases of ccTGA [12].

The specificity of the criteria for the determination of 'LV preparedness' is controversial. The ability of the LV to support the systemic circulation after the early neonatal period depends on factors such as relative pulmonary and systemic vascular resistances, flow across intracardiac communications such as the PDA, and the presence of LVOTO. In patients who present beyond the early neonatal period, the following parameters may be used to assess LV preparedness [12–14]:

- **LV muscle mass** (by MRI) of 50–65 g/m^2 is generally considered adequate. LV mass < 35 g/m^2 is an indication for LV retraining. This may be estimated by M-mode echo as well.
- **Geometry of the IVS** is an indicator of relative pulmonary and systemic vascular resistances. In the PSAX view, a circular configuration of the LV with convexity of the IVS towards the RV is reassuring, while IVS flattening, banana-shaped configuration of the LV, and/or bulging of the IVS to the left are considered as indications for LV retraining.
- Normal left ventricular systolic function: LVEF >55% is a good indicator of LV preparedness.
- Left AV (mitral) valve function: Moderate or greater degrees of AV valve insufficiency may serve as a harbinger of post-operative complications.

Key views	Key features assessed
Subcostal	
— Coronal	ASD, VSD, great arterial relationship, juxtaposition of atrial appendages, abnormal CA coursing posterior to the PA
— Sagittal	AV valve abnormalities, outflow tract obstruction, VSD, great arterial relationship, anomalous systemic or pulmonary venous drainage
Parasternal	
— Long axis	Parallel origin of great arteries, outflow tract obstruction, left AV valve abnormality, conal septal malalignment, left coronary artery bifurcation
— Short axis	Great arterial relationship, ventricular function, VSD, coronary artery origin/intramural course, commissural alignment, semilunar valve morphology and location, outlet VSD
Apical	
— Four chamber	Ventricular function/hypoplasia, CA anatomy [origin of circumflex CA from RCA], VSD, AV valve function/straddling
— Five chamber	Great arterial origin, outflow tract obstruction, semilunar valves, outlet VSD, abnormal AV valve attachment to the outlet septum
Suprasternal	
— Long axis	Aortic arch obstruction, PDA flow
— Short axis	Aortic arch sidedness, great arterial relation, branch PA size, anomalous systemic/pulmonary venous drainage

Points to Report [in Segmental Order]

- Venous connections: Systemic/pulmonary.
- Interatrial communication and its adequacy.
- Juxtaposition of the atrial appendages, if present.
- AV valve anatomy and function.
- RV and LV size and function.
- VSD: Type, size, and direction of flow.
- Outflow tract obstruction: Type, severity.
- Semilunar valve anatomy and function.
- Aortic arch sidedness, branching, and obstruction.
- Patency of ductus arteriosus: Size, direction of flow.
- Branch pulmonary artery obstruction, if present.
- Coronary artery origin and anomalies, including intramural course and commissural malalignment.
- LV preparedness: In cases of late presentation beyond the early neonatal period.

11.2 Post-Operative Evaluation of d-TGA

The surgical approaches in d-TGA have evolved over time. While evaluating a post-operative case of d-TGA, it is important to gather all available data pertaining to the surgical procedure.

11.2.1 Arterial Switch Operation (ASO)

Neonatal ASO is the most widely prescribed surgical procedure for d-TGA in the current era. The procedure involves transection of the great arterial roots above the sinuses, corrective re-anastomosis of the great arteries to establish ventriculo-arterial concordance, and translocation of the CAs to the neoaortic root. A relatively normal pulmonary valve is a necessary prerequisite for successful ASO.

LeCompte manoeuvre, which involves relocating the aorta posterior to the PA, is often performed simultaneously to minimize distortion of the branch PAs during the arterial switch [6]. The medical and surgical records should be carefully reviewed to determine if a VSD was closed and if LeCompte manoeuvre was performed.

11.2.1.1 Echocardiography: Coronary Artery Stenosis

Occlusion or kinking of the coronary artery during translocation can result in early or late CA stenosis [15, 16]. This can manifest as low cardiac output syndrome in the immediate post-operative period, as well as myocardial ischaemia due to late CA stenosis. The implanted CA ostia in the anterior left and right sinuses of the neoaortic root should be imaged from the PSAX view.

(a) Myocardial ischaemia may be evidenced by LV systolic and/or diastolic dysfunction, and RWMA.
(b) Colour flow mapping of coronary arteries [17]:
  - Lowering the scale and narrowing the colour sector enhances imaging of CA blood flow.
  - LMCA VTI > 0.14 and LMCA peak systolic velocity > 0.6 cm/s are indicative of significant stenosis.

11.2.1.2 Pulmonary Hypertension

Accelerated development of PH in infancy occurs in the setting of a large left-to-right shunt through [18, 19]:

(a) Large VSD.
(b) Abnormal aortopulmonary collateral circulation.

The classic signs of PH, such as flattening of the interventricular septum and elevated TR gradient, are usually present. Aortopulmonary collaterals may be identified by colour Doppler imaging from subcostal, parasternal, or suprasternal windows, as multiple flow signals around the DTA.

11.2.1.3 Supravalvular Pulmonary Stenosis and Neopulmonary Valve Regurgitation

Supravalvular pulmonary stenosis can occur as an early or late sequelae (15%) [20, 21]. This usually occurs as:

(a) Discrete stenosis of the anastomotic suture line.
(b) Diffuse hypoplasia of MPA.
(c) Branch PA stenosis.

The A5C, PSAX, and subcostal coronal views facilitate imaging of the neopulmonary valve and supravalvular region. The neopulmonary artery appears mildly narrowed in comparison with the neoaorta [6]. Branch PAs are seen in the subcostal coronal view, with an anterior sweep. LPA stenosis is more common than RPA stenosis, as the branch PAs straddle across the aorta after LeCompte manoeuvre. This is best visualized from a high PSAX view, where the branch PAs, located anterior to the aorta, are seen descending down on either side.

Significant neopulmonary regurgitation is rare, except in cases with pre-operative structural abnormalities of the neopulmonary (native aortic) valve.

Key point: Peak Doppler velocities < 2 m/s across the branch PAs are normal after ASO [6].

11.2.1.4 Supravalvular Aortic Stenosis and Neoaortic Valve Regurgitation

The most common cause of supravalvular AS after ASO is anastomotic suture line narrowing [22]. This can occur as an early or late sequelae. Evaluation of the supravalvular aortic region is best performed from the subcostal LAO view and the A5C view. Sluggish flow and dampened signals in the descending aorta may be demonstrated from the subcostal sagittal view in cases of aortic coarctation.

Congenitally abnormal neoaortic (native pulmonary) valve and prior PA banding predispose to significant neoaortic valve regurgitation in the early post-operative period [22, 23]. Neoaortic root dilation and progressive neoaortic valve regurgitation can occur as late sequelae.

Neoaortic valve regurgitation is best demonstrated from the A3C, A5C, PLAX, and PSAX views. Severe neoaortic valve regurgitation is characterized by retrograde flow in the descending aorta, which can be visualized from suprasternal and subcostal sagittal views.

The severity of neoaortic regurgitation is to be assessed using standard parameters for AR (discussed in Chap. 9). The diameter of the neoaortic root is measured from the PLAX view during mid-systole (inner edge to inner edge). Z-scores are used to follow up children over time [6]. LV dimensions should be documented.

11.2.1.5 Assessment of Ventricular Function After ASO

Assessment of LV function includes measurement of LV dimensions, fractional area shortening, and ejection fraction (EF).

Assessment of RV function includes RV dimensions, tricuspid annular plane systolic excursion (TAPSE), fractional area change (FAC), and tissue Doppler measurements (discussed in Chap. 15).

Key point: Post-operative septal dyskinesia can result in underestimation of LV function by M-mode.

Points to Report

- Supravalvular pulmonary stenosis; Branch PA patency (especially after LeCompte procedure).
- Neopulmonary valve regurgitation.
- Supravalvular aortic stenosis; Residual aortic arch obstruction (in cases with COA).
- Neoaortic root dilation (dimension with z-score) and regurgitation.
- LV and RV dimensions and function; Regional wall motion abnormalities.
- Coronary artery size and flow, evidence of CA stenosis.
- Residual shunt lesions (ASD/VSD): Size, flow, and gradients.
- Severity of tricuspid regurgitation and RVSP.
- Evidence of pulmonary hypertension.
- Presence of aortopulmonary collateral vessels.

11.2.2 Atrial Switch Operation (AtrSO)

Atrial switch operations (Mustard/Senning) are rarely performed in the current era. In this procedure, the great arteries are not transected, and ventriculo-arterial discordance is not corrected. Instead, a systemic venous baffle is created, which diverts systemic venous return to the left atrium, i.e. systemic venous return-LA-MV-LV-PA (the pulmonary circulation). Similarly, the pulmonary venous return is diverted to the right atrium through a pulmonary venous baffle, i.e. pulmonary venous return-RA-TV-RV-Ao (the systemic circulation) [24, 25]. Thus, a physiological correction is achieved whereby systemic venous return reaches the lungs through LV-PA, and oxygenated blood from the lungs is channelled towards the systemic circulation through RV-Aorta.

The Senning procedure uses autologous tissue to construct the atrial baffles, while the Mustard procedure uses prosthetic material [24, 25]. TEE is used more frequently after AtrSO because it is difficult to interrogate the venous baffles by TTE.

Key point: The Mustard procedure entails a greater predisposition for systemic venous baffle obstruction, whereas pulmonary venous baffle obstruction is commoner after the Senning procedure.

11.2.2.1 Assessment of the Atrial Baffle (Figs. 11.5 and 11.6)

The best echocardiographic views to visualize the venous baffles are the ***apical and the subcostal four-chamber views***. Most of the pulmonary venous atrium (anatomic RA) and the pulmonary venous baffle can be visualized from these views. An anterior sweep of the transducer from the A4C view may visualize the systemic venous baffle, wrapping around the pulmonary venous baffle. A posterior tilt of the transducer with slight anticlockwise rotation visualizes the IVC limb in the systemic venous atrium (anatomic LA).

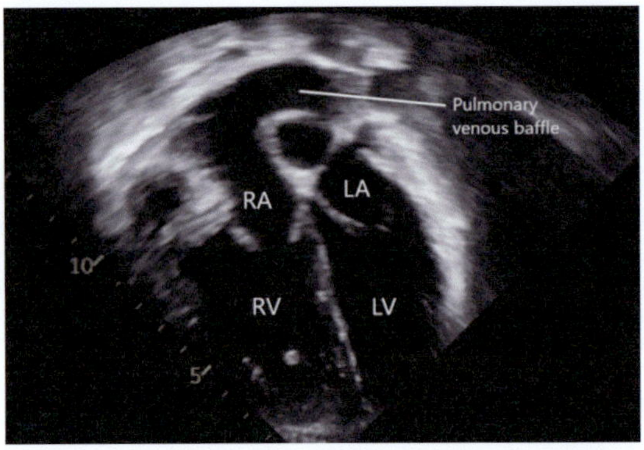

Fig. 11.5 Apical view showing an unobstructed pulmonary venous baffle draining into the RA

In the **PLAX view**, the systemic venous baffle may be visualized within the LA. The limb that crosses the plane of the atrial septum to the mitral valve is the SVC limb. A permanent pacing lead, when present, confirms the vessel to be the SVC. The **PSAX view** is ideal for the assessment of the pulmonary venous baffle.

A long axis sweep from the **suprasternal window** visualizes the innominate vein and SVC proximal to the baffle. Colour Doppler demonstrates low-velocity flow from the venae cavae to the systemic venous atrium. Colour Doppler assessment of pulmonary veins can be done from the suprasternal short axis view (crab view).

Key views:

- Pulmonary venous baffle to anatomic RA: A4C, PSAX, and subcostal four-chamber views.
- IVC limb of the systemic venous baffle to anatomic LA: A4C (with posterior tilt and slight

Fig. 11.6 Apical view showing a systemic venous baffle draining into the left atrium

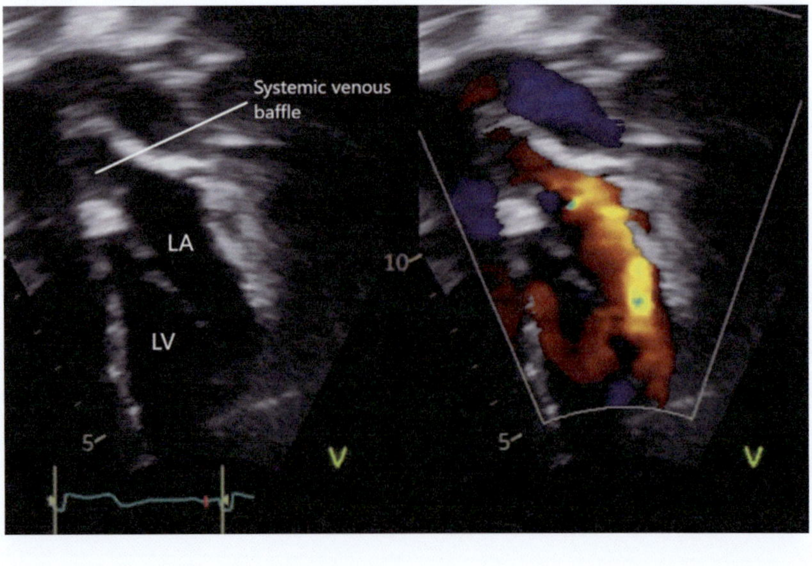

anticlockwise rotation), A2C, and subcostal views.

- SVC limb of the systemic venous baffle to anatomic LA: A4C, A3C, A2C, PLAX, and suprasternal short axis views.

Key point: Normal flow in a venous baffle is a phasic flow with a velocity < 1 m/s.

Baffle Obstruction

Pulmonary venous baffle obstruction (7%) commonly occurs at the junction of the pulmonary venous baffle with the RA. Obstruction of the pulmonary venous baffle is characterized by [26]:

- Turbulent flow in the obstructed baffle.
- Diastolic flow velocity greater than 2 m/s.
- Increasing MR jet velocity and left ventricular hypertrophy due to back-pressure from the obstructed pulmonary venous baffle.

Systemic venous baffle obstruction: The superior (SVC) limb of the systemic venous baffle, which is usually smaller than the inferior (IVC) limb, is more prone to obstruction (5–30%) than the inferior limb (1–2%). Obstruction of the systemic venous baffle is characterized by [27]:

- Turbulent non-phasic flow in the affected limb.
- Diastolic flow velocities >1.2 m/s.
- Mean gradient more than 2–3 mmHg.
- Dilated IVC may suggest the possibility of IVC baffle obstruction.

Saline contrast echocardiography is useful in cases of suspected systemic venous baffle obstruction [28]. Agitated saline contrast is injected into an upper-extremity vein, and the junction between IVC and systemic venous atrium (morphologic LA) is imaged from the subcostal long axis sweep.

- Unobstructed SVC limb: Contrast appears in the LA through the SVC, with no opacification of the IVC.
- Partially obstructed SVC limb: Contrast reaches the LA through the SVC; however, after a few cardiac cycles, contrast appears in the IVC as well (through IVC-azygous collaterals), and drains into the LA through the IVC.
- Obstructed SVC limb: Contrast reaches the LA through the IVC (via IVC-azygous collaterals), with no opacification of the SVC.

Baffle Leaks

Baffle leaks are commoner than baffle obstruction, seen in up to 25% of cases. The leak results in a physiological left-to-right shunt from the pulmonary venous RA to the systemic venous LA (physiologically similar to an ASD), causing progressive dilation of the LA and LV.

- The Nyquist limit needs to be lowered for colour Doppler imaging of these leaks.
- Saline contrast echocardiography from the four-chamber view aids in the detection of baffle leaks.

Key point: Progressive dilation of the systemic venous atrium (LA) and the corresponding ventricle (LV) should raise suspicion of a baffle leak.

11.2.2.2 Systemic AV (Tricuspid) Valve Regurgitation and Systemic Ventricular (RV) Dysfunction

Chronic pressure and volume overload of the RV results in severe functional TR, which in turn causes worsening RV dysfunction. The RV may appear hypokinetic and dilated.

The usual indices of RV function, such as TAPSE, FAC, and tissue Doppler velocities, may be used (discussed in Chap. 15). However, it should be remembered that progressive TR may increase tricuspid annular motion and FAC, masking RV dysfunction [6].

Points to Report

- Specify the pre-operative anatomy, as well as the type of procedure performed.
- Patency of baffles and leaks: Superior and inferior limbs of the systemic venous baffle, and pulmonary venous baffle—turbulence of flow, flow velocity, gradient.
- AV valve function: Tricuspid (systemic AV) valve and Mitral (pulmonary AV) valve.
- RV (systemic ventricle) size and function.
- LV (subpulmonary ventricle) size and function.
- Residual VSD, if present.
- Aortic and pulmonary valve function.
- Estimated pressure gradients:
  Mean PA pressure from PR gradient.
  LV systolic pressure (equivalent to PASP in the absence of pulmonary valve stenosis) from MR gradient.

11.2.3 Rastelli or Nikaidoh Procedure

Complex TGA with VSD and significant LVOTO (i.e. pulmonary stenosis at any level) requires alternative surgical strategies [29]:

(a) **Rastelli procedure**: LV output is rerouted through an intracardiac tunnel into the aorta; interventricular communication is closed; and the RV is connected to the PA through an extracardiac conduit.

(b) **REV procedure**: Similar to the Rastelli procedure, except that RVOT reconstruction is done without using an extracardiac prosthesis.

(c) **Nikaidoh procedure**: The aortic root is resected from the RV (usually along with the coronary arterial attachments), translocated posteriorly, and anastomosed to the pulmonary annulus. Pulmonary stenosis is resolved by dividing the conal septum and resecting the pulmonary valve. LVOT is reconstructed using the posteriorly translocated aortic root and a VSD patch. A conduit is placed from the RV to the PA. The risk of development of subaortic obstruction is less with the Nikaidoh procedure.

11.2.3.1 Echocardiography: Residual VSD

Residual VSD is common, although the small communications around the VSD patch are insignificant. Erroneous attachment of the VSD patch to the RV trabeculations instead of the RV free wall can result in an intramural defect [30]. The pressure gradient across the defect needs to be documented.

11.2.3.2 Subaortic [LVOT] Obstruction

LVOTO may result from hypertrophy of the subaortic conus, progressive narrowing of the VSD (which is now the opening of the LVOT), fibrous tissue formation, and/or abnormal AV chordal attachments [6].

LVOT may be visualized from subcostal left anterior oblique, A5C, and PLAX views. The peak LVOT gradient determines the severity of the obstruction. When direct measurement is not feasible, the peak LVOT gradient may be estimated as the difference between peak MR gradient and systolic blood pressure.

11.2.3.3 Conduit Dysfunction

The RV-PA conduits used in these procedures (pulmonary or aortic homografts, or prosthetic conduits) are prone to progressive degeneration of the conduit valve, which results in pulmonary regurgitation or stenosis [6]. Severe conduit regurgitation can lead to progressive RV dilation and dysfunction.

The subcostal sagittal view visualizes the RVOT and the proximal portion of the RV-PA conduit. Gradients estimated across the conduit may be inaccurate owing to the limitations of the modified Bernoulli equation in estimating the gradient across a long narrow tube. Hence, **RV pressures derived from TR or VSD jets are likely to provide a more reliable estimate of the severity of conduit stenosis** [6]. The A4C view aids in the assessment of TR and RV function.

11.2.3.4 CA Stenosis

CA stenosis can occur if CA reimplantation is performed with Nikaidoh procedure (discussed in post-ASO evaluation).

11.2.3.5 Branch PA Stenosis

Branch PA stenosis can occur if LeCompte manoeuvre is performed with Nikaidoh procedure.

Points to Report

- Specify the pre-operative anatomy, as well as the type of procedure performed.

- Residual VSD: Location, size, and pressure gradient.
- Subaortic [LVOT] obstruction.
- Patency and regurgitation of RV-PA conduit or alternative pathways.
- RV systolic pressure (estimated from TR jet).
- AV valve function.
- Aortic root dimension.
- LV and RV size and function.
- Branch pulmonary artery stenosis [in cases of Nikaidoh with LeCompte manoeuvre].
- Coronary artery stenosis [in cases of Nikaidoh with CA reimplantation].

11.3 Congenitally Corrected Transposition of Great Arteries [L-TGA]

The morphologic characteristics of l-TGA are [31, 32]:

- Ventriculo-arterial discordance.
- Atrioventricular discordance/ventricular inversion (l- loop)—morphologic RV is to the left of morphologic LV.
- The circulation is physiologically ‘corrected’ because of ventricular inversion (double discordance).
- Anteriorly positioned left-sided aorta with subaortic conus arises from the morphologic RV [which is located in the left side, connected to the LA] i.e. **LA → RV → Ao**.
- Posteriorly positioned right-sided PA with mitro-pulmonary fibrous continuity arises from the morphologic LV [which is located in the right side, connected to the RA] i.e. **RA → LV → PA**.

11.3.1 Echocardiographic Diagnosis

- Abnormal atrioventricular and ventriculo-arterial connections can be established by a systematic segmental assessment from the subcostal and apical views by posterior to anterior sweep (Fig. 11.7).
- The presence of apically placed tricuspid valve in the left side (A4C view) gives the first indication of ventricular inversion (i.e. an l-loop, where RV is to the left of LV).
- **Significant malalignment between the atrial and ventricular septa** is another clue towards the possibility of AV discordance in ccTGA [33] (Fig. 11.7a).
- The side-by-side positioning of the ventricles results in anomalous echocardiographic appearances such that [26]:
  - In the PLAX view, the **heart appears to be more vertically oriented than usual, and visualization of the LV may be difficult**.
  - In the short-axis view, the interventricular **septum appears more vertical** and perpendicular to the frontal plane, with side-by-side ventricles.
- Fibrous continuity between the pulmonary valve and the anterior mitral leaflet can be appreciated in the PLAX view.
- The parallel origin of the great arteries may be evident in the PLAX and apical views.
- The great arterial relationship, where the aorta is usually positioned leftward, anterior, and superior to the pulmonary valve, is appreciable in the PSAX view.
- Anterior tilt of the transducer from the A4C view reveals that the pulmonary outflow tract

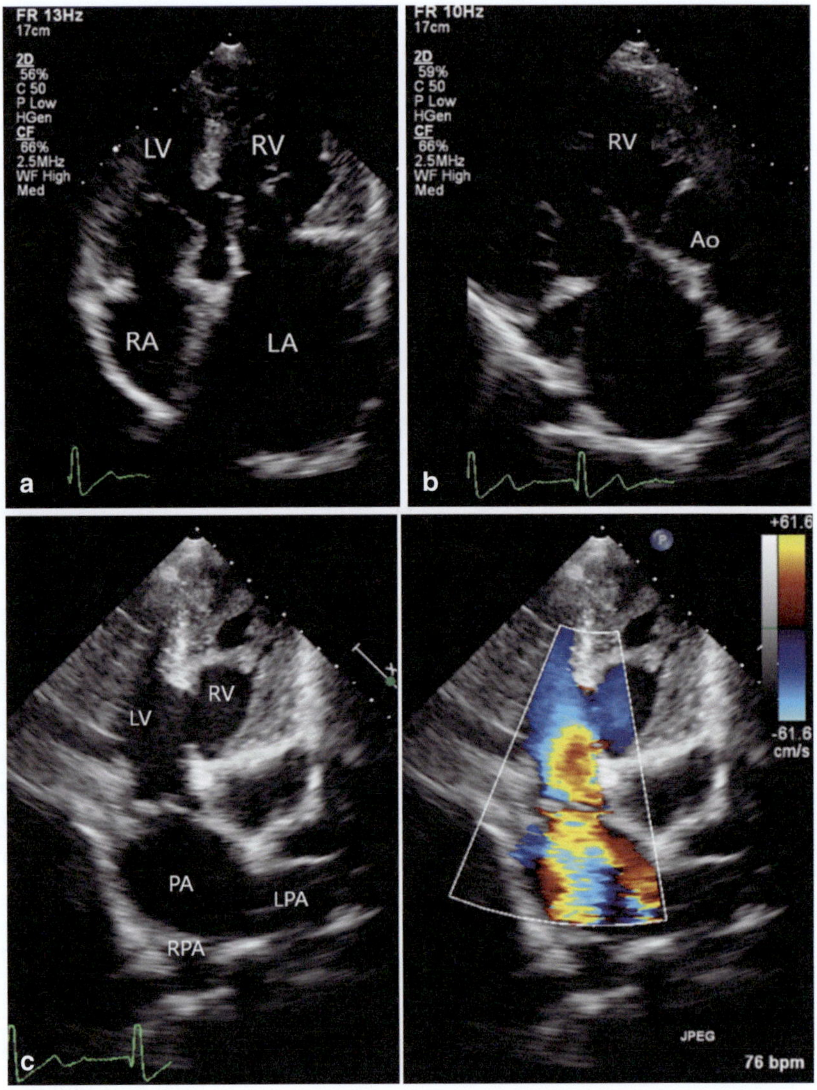

Fig. 11.7 (**a**) A4C view showing ventricular inversion (trabeculated RV on the left side and smooth-walled LV on the right side) and VSD in ccTGA. Note the malalignment between the interatrial and interventricular septa. (**b**) Anterior and leftward aorta is seen originating from the left-sided RV. (**c**) Rightward-positioned PA originates from the right-sided LV and bifurcates immediately. There is turbulence of flow across the thickened semilunar valve in the LVOT (valvular pulmonary stenosis)

from the right-sided LV is deeply wedged between the two AV valves.

11.3.2 Associated Anomalies [34]

- Situs inversus with ccTGA (5–10%):
  - Morphologic RA on the left side connects to a left-sided LV, which gives off the pulmonary trunk.
  - Morphologic LA on the right side connects to a right-sided RV from which the aorta arises.
  - The aorta is positioned anterior and to the right of the pulmonary trunk.
- Dextrocardia or mesocardia (25%).
- **Ventricular Septal Defect** (80%) [33, 35]:
  - The most common type is perimembranous VSD with inlet or posterior extension—prone for partial occlusion by septal tricuspid leaflet.
  - Outlet and muscular VSDs are less common.
  - AV valve straddling needs to be looked for in cases of inlet VSD.
- ASD (12%).
- **Left AV (tricuspid) valve abnormalities** [36]:
  - Ebstein-like malformation: Dysplasia and apical displacement of the leaflet insertion.

  - Severe tricuspid regurgitation.
  - RV hypoplasia and subaortic obstruction may be associated with significant Ebsteinoid anomaly.
  - Posterior straddling of the left AV valve occurs in association with perimembranous inlet VSD.
  - Supravalvular stenosing ring.
- **Right AV (mitral) valve abnormalities** like isolated mitral valve cleft and parachute mitral valve are rare [37].
- **LVOT obstruction (valvular or subvalvular pulmonary stenosis):** Occurs in up to 50% of cases of ccTGA with situs solitus**.** Best demonstrated by anterior tilt of the transducer from the subcostal four-chamber view. The mechanism of obstruction may be valvular stenosis, prominent ventriculo-infundibular fold in the LVOT, and/or accessory AV valve tissue obstructing the LVOT [38].
- **RVOT (subaortic) obstruction** is unusual in ccTGA. The subaortic area is best imaged from the left parasternal or subcostal windows. The underlying mechanism of obstruction may be subaortic membrane, valvular stenosis, and/or accessory AV valve tissue in the RVOT [39]. Dynamic RVOTO has also been described. Obstruction or severe regurgitation of the left AV valve is a common association.
- Aortic arch and its branching pattern are better demonstrated from a high left parasternal view (ductal view). A right aortic arch (18%) or aortic coarctation may be seen.
- Right (systemic) ventricular dilation and dysfunction need to be documented.
- Coronary artery anomalies (24%) [40, 41]: Coronary arteries follow the ventricular morphology. A major CA branch coursing anterior to the pulmonary trunk (96%) and a large coronary branch crossing the RVOT (60%) are commonly encountered anomalies.

11.3.3 Post-Operative Evaluation [42]

Surgical management of ccTGA is variable. The various strategies adopted are:

(a) **Physiologic repair** of associated defects (VSD closure, TV repair/replacement, relief of LVOTO, etc.): In this surgical strategy, the discordant AV and VA connections are not corrected. Hence, the RV remains to be the systemic ventricle, with the attendant risk of progressive RV failure. PA banding may be performed in cases of significant tricuspid regurgitation, in order to minimize septal leaflet tethering.

(b) **Anatomic repair** of l-TGA (Double switch procedure): This involves correcting the AV discordance by atrial switch (Mustard/Senning) procedure, and rectifying the VA discordance by ASO or Rastelli procedure. Thus, the LV becomes the systemic ventricle.

(c) **Fontan palliation** (univentricular repair) is performed where a two-ventricle repair is not feasible.

The focus of post-operative evaluation of ccTGA varies depending on the type of surgery performed. [Post-operative evaluation of the atrial switch operation, ASO, and Rastelli operations have been discussed in the previous sections. Fontan repair is discussed in Chap. 12].

Points to Report

- Segmental analysis.
- Left AV (tricuspid) valve structure and function.
- LVOTO and/or RVOTO, if present.
- VSD, ASD.
- Ventricular size and function: RV (systemic ventricle) and LV.
- Aortic arch sidedness, coarctation.
- Coronary artery anomalies.
- Post-operative: Residual TR/VSD/outflow tract obstruction; systemic ventricular size/function; Baffle leak/stenosis, CA stenosis.

References

1. Wernovsky G. Transposition of the great arteries. In: Allen HD, Driscoll DJ, Shaddy RE, Feltes TF, editors. Moss and Adams' heart disease in infants, children, and adolescents. 8th ed. Philadelphia, PA: Lippincott Williams & Wilkins; 2013. p. 1097–146.
2. Perloff JK, Marelli AJ. Complete transposition of the great arteries. In: Clinical recognition of congenital heart disease. 6th ed. Philadephia, PA: Saunders; 2012. p. 472–91.
3. Wernovsky G, Sanders SP. Coronary-artery anatomy and transposition of the great-arteries. Coron Artery Dis. 1993;4:148–57.
4. Pasquini L, Sanders SP, Parness IA, et al. Conal anatomy in 119 patients with d-loop transposition of the great arteries and ventricular septal defect: an echocardiographic and pathologic study. J Am Coll Cardiol. 1993;21:1712–21.
5. Kirklin JW, Barratt-Boyes BG. Complete transposition of the great arteries. In: Kirklin JW, Barratt-Boyes BG, editors. Cardiac surgery. New York, NY: Churchill Livingston; 1993. p. 1383–467.
6. Cohen MS, Eidem BW, Cetta F, Fogel MA, Frommelt PC, Ganame J, Han BK, Kimball TR, Johnson RK, Mertens L, Paridon SM, Powell AJ, Lopez L. Multimodality imaging guidelines of patients with transposition of the great arteries: a report from the American Society of Echocardiography developed in collaboration with the Society for Cardiovascular Magnetic Resonance and the Society of Cardiovascular Computed Tomography. J Am Soc Echocardiogr. 2016 Jul;29(7):571–621.
7. Anderson RH, Weinberg PM. The clinical anatomy of transposition. Cardiol Young. 2005;15(Suppl 1):76–87.
8. Kim S-J, Kim W-H, Lim C, et al. Commissural malalignment of aortic-pulmonary sinus in complete transposition of great arteries. Ann Thorac Surg. 2003;76(6):1906–10.
9. Pasquini L, Parness IA, et al. Diagnosis of intramural coronary artery in transposition of the great arteries using two-dimensional echocardiography. Circulation. 1993;88:1136–41.
10. Freire G, Miller MS. Echocardiographic evaluation of coronary arteries in congenital heart disease. Cardiol Young. 2015;25:1504–11.
11. Pasquini L, Sanders SP, et al. Diagnosis of coronary artery anatomy by two-dimensional echocardiography in patients with transposition of great arteries. Circulation. 1987;75:557–64.
12. Brawn WJ, Barron DJ, Jones TJ, et al. The fate of the retrained left ventricle after double switch procedure for congenitally corrected transposition of great arteries. Semin Thorac Cardiovasc Surg Paediatr Card Surg Annu. 2008;11:69–73.
13. Lacour-Gayet F, Piot D, Zoghbi J, Serraf A, Gruber P, Macé L, et al. Surgical management and indication of left ventricular retraining in arterial switch for transposition of the great arteries with intact ventricular septum. Eur J Cardiothorac Surg. 2001;20(4):824–9.
14. Watanabe N, Mainwaring RD, Carrillo SA, Lui GK, Reddy VM, Hanley FL. Left ventricular retraining and late arterial switch for D-transposition of the great arteries. Ann Thorac Surg. 2015;99(5):1655–61.
15. Pasquali SK, Hasselblad V, Li JS, Kong DF, Sanders SP. Coronary artery pattern and outcome of arterial switch operation for transposition of the great arteries: a meta-analysis. Circulation. 2002;106:2575–80.
16. Legendre A, Losay J, Touchot-Kone A, Serraf A, Belli E, Piot JD, et al. Coronary events after arterial switch operation for transposition of the great arteries. Circulation. 2003;108(Suppl 1):II186–90.
17. Wong D, Golding F, Hess L, Caldarone CA, Van Arsdell G, Manlhiot C, et al. Intraoperative coronary artery pulse Doppler patterns in patients with complete transposition of the great arteries undergoing the arterial switch operation. Am Heart J. 2008;156:466–72.
18. Newfeld EA, Paul MM, Muster AJ, Idriss FS. Pulmonary vascular disease in complete transposition of the great arteries: a study of 200 patients. Am J Cardiol. 1974;34:75–82.
19. Wernovsky G, Bridges ND, Mandell VS, Castaneda AR, Perry SB. Enlarged bronchial arteries after early repair of transposition of the great arteries. J Am Coll Cardiol. 1993;21:465–70.
20. Brown JW, Park HJ, Turrentine MW. Arterial switch operation: factors impacting survival in the current era. Ann Thorac Surg. 2001;71:1978–84.
21. Losay J, Touchot A, Serraf A, Litvinova A, Lambert V, Piot JD, et al. Late outcome after arterial switch operation for transposition of the great arteries. Circulation. 2001;104:I121–6.
22. Marino BS, Wernovsky G, McElhinney DB, Jawad A, Kreb DL, Mantel SF, et al. Neo-aortic valvar function after the arterial switch. Cardiol Young. 2006;16:481–9.
23. Losay J, Touchot A, Capderou A, Piot JD, Belli E, Planche C, et al. Aortic valve regurgitation after arterial switch operation for transposition of the great arteries: incidence, risk factors, and outcome. J Am Coll Cardiol. 2006;47:2057–62.
24. Mustard WT, Chute AL, Keith JD, Sirek A, Rowe RD, Vlad P. A surgical approach to transposition of the great vessels with extracorporeal circuit. Surgery. 1954;36:31–51.
25. Senning A. Surgical correction of transposition of the great vessels. Surgery. 1959;45:966–80.
26. Armstrong WF, Ryan T. Feigenbaum's echocardiography. Philadelphia: Wolters Kluwer/Lippincott Williams & Wilkins Health; 2010.
27. Otto CM. Textbook of clinical echocardiography. Philadelphia: Saunders/Elsevier; 2009.
28. Silverman NH, Snider AR, Colo J, et al. Superior vena caval obstruction after Mustard's operation: detection by two-dimensional contrast echocardiography. Circulation. 1981;64:392–6.
29. Rastelli GC, McGoon DC, Wallace RB. Anatomic correction of transposition of the great arteries with

ventricular septal defect and subpulmonary stenosis. J Thorac Cardiovasc Surg. 1969;58:545–52.

30. Patel JK, Glatz AC, Ghosh RM, Jones SM, Natarajan S, Ravishankar C, et al. Intramural ventricular septal defect is a distinct clinical entity associated with postoperative morbidity in children after repair of Conotruncal anomalies. Circulation. 2015;132:1387–94.
31. Atallah J, Rutledge JM, Dyck JD. Congenitally Corrected Transposition of the Great Arteries (Atrioventricular and Ventriculoarterial Discordance). In: Allen HD, Driscoll DJ, Shaddy RE, Feltes TF, editors. Moss and Adams' heart disease in infants, children, and adolescents. 8th ed. Philadelphia, PA: Lippincott Williams & Wilkins; 2013. p. 1147–61.
32. Perloff JK, Marelli AJ. Congenitally corrected transposition of the great arteries. In: Clinical recognition of congenital heart disease. 6th ed. Philadephia, PA: Saunders; 2012. p. 56–73.
33. Becker A, Anderson R. Conditions with discordant atrioventricular connexions-anatomy and conducting tissues. In: Anderson RH, Macartney FJ, Shinebourne EA, editors. Pediatric cardiology. Churchill Livingstone: Edinburgh; 1978. p. 184–97.
34. Presbitero P, Somerville J, Rabajoli F, et al. Corrected transposition of the great arteries without associated defects in adult patients: clinical profile and followup. Br Heart J. 1995;74:57–9.
35. Allwork SP, Bentall HH, Becker AE, et al. Congenitally corrected transposition of the great arteries: morphologic study of 32 cases. Am J Cardiol. 1976;38:910–23.
36. Anderson KR, Danielson GK, McGoon DW, et al. Ebstein's anomaly of the left-sided tricuspid valve: pathological anatomy of the valvular malformation. Circulation. 1978;58:87–91.
37. Penny DJ, Somerville J, Redington AN. Echocardiographic demonstration of important abnormalities of the mitral valve in congenitally corrected transposition. Br Heart J. 1992;68:498–500.
38. Anderson RH, Becker AE, Gerlis LM. The pulmonary outflow tract in classically corrected transposition. J Thorac Cardiovasc Surg. 1975;69:747–57.
39. Miche E, Manneback H, Boyunovic N, et al. Right ventricular outflow obstruction due to accessory tricuspid valve tissue in corrected transposition of the great arteries with ventricular septal defect. Z Kardiol. 1991;80:468–70.
40. Dabizzi RP, Barletta GA, Caprioli G, et al. Coronary artery anatomy in corrected transposition of the great arteries. J Am Coll Cardiol. 1998;12:486–91.
41. McKay R, Anderson RH, Smith A. The coronary arteries in hearts with discordant atrioventricular connections. J Thorac Cardiovasc Surg. 1996;111:988–97.
42. Cui H, Hage A, Piekarski BL, Marx GR, Baird CW, Del Nido PJ, Emani SM. Management of Congenitally Corrected Transposition of the great arteries with intact ventricular septum: anatomic repair or palliative treatment? Circ Cardiovasc Interv. 2021;14(7):e010154.

12 Single Functional Ventricle

The term 'single functional ventricle' refers to all the congential cardiac defects in which one of the ventricles is small and nonfunctional [1, 2]. The major morphologic variants included under this terminology are:

- Univentricular AV connections (double inlet, common inlet, single inlet) and.
- Hearts with an atretic semilunar valve and intact ventricular septum (hypoplastic left and right heart syndromes).

Hearts with a 'single functional ventricle' are generally treated with a modified Fontan-type surgical repair.

12.1 Terminologies

An understanding of various terminologies used in relation to 'single ventricle physiology' is essential for appropriate diagnosis and nomenclature of this complex entity. Definitions of the commonly used terminologies, as suggested by Anderson et al., are presented below [3, 4].

Ventricle: A chamber receiving 50% or more of an inlet portion, regardless of the presence of an outlet portion. However, some authors prefer to consider any endocardial-lined chamber within the ventricular mass as a ventricle, regardless of the presence or absence of certain ventricular components.

Rudimentary chamber: A chamber receiving <50% of an inlet portion.

- Outlet chamber: Rudimentary chamber with an outlet portion.
- Trabecular pouch: Rudimentary chamber possessing only the trabecular portion.

Interventricular communication: Various terminologies used for the interventricular communication in univentricular AV connection are VSD, bulboventricular foramen, and outlet foramen.

Univentricular heart:

- LV type: The dominant chamber is morphologically LV, and the rudimentary chamber resembles the RV.
- RV type: The dominant chamber is morphologically RV, and the rudimentary chamber resembles the LV.

12.2 Univentricular AV Connection

Univentricular AV connection refers to the cardiac morphologies in which both the inlets, irrespective of whether they are patent or not, are committed primarily to one dominant ventricle

Supplementary Information The online version contains supplementary material available at https://doi.org/10.1007/978-981-96-0243-8_12.

A. Paul, *Manual of Echocardiography for Congenital Heart Diseases*,
https://doi.org/10.1007/978-981-96-0243-8_12

[1]. There are three basic types of univentricular AV connections, namely double-inlet, single-inlet, and common inlet AV connections.

12.2.1 Double Inlet AV Connection

In double inlet AV connection, both atria connect to the dominant ventricle by two distinct AV valves, which are labelled as the 'right' and 'left' AV valves.

12.2.1.1 Double Inlet Left Ventricle (DILV)

DILV is the most common form of double inlet AV connection (80–90%). In DILV, the LV is dominant, and the RV is hypoplastic. Ventriculo-arterial discordance (TGA) is present in 70% of the cases, most commonly l-TGA [1].

The three subtypes of DILV, in increasing order of frequency, are [2, 5]:

- DILV with VA concordance and normally related great arteries (**A-I** Single Ventricle, Holmes heart).
- DILV with right-sided subaortic hypoplastic right ventricle (**A-II** Single Ventricle, D-loop).
- DILV with left-sided subaortic hypoplastic right ventricle (**A-III** Single Ventricle, L-loop).

Interventricular communication: The VSD is usually in the muscular septum. Infundibular hypoplasia is associated with subaortic extension of the VSD [6]. A restrictive outlet foramen, in the setting of VA discordance, results in subaortic stenosis, leading to increased pulmonary blood flow and volume overload of the LV [7]. Coarctation of the aorta and/or interrupted aortic arch may be present.

Straddling:

In DILV, the chordae of the AV valve may straddle into the hypoplastic RV through the VSD. The valve that straddles is usually on the same side as the hypoplastic chamber. When the hypoplastic RV is on the right side, the right AV valve straddles into it. When the hypoplastic RV is located on the left side, it is the left AV valve that straddles [1]. Straddling AV valves are often redundant, with a predisposition to significant AV valve regurgitation.

In DILV and VA discordance, **subaortic obstruction** can result from [1, 6, 7]:

- Restrictive outlet foramen.
- Severe hypertrophy of the hypoplastic RV.
- Right AV valve atresia and previous PA banding are associated with a greater incidence of progressive subaortic obstruction.

Pulmonary outflow tract obstruction in DILV can result from [1, 5]:

- Obstruction by the right AV valve apparatus.
- Hypoplasia of the PV annulus or malformation of the PV leaflets.
- Posterior deviation of the infundibular septum.

Key point: When the hypoplastic RV is inverted [**i.e.** ***left-sided RV; l-loop], the left AV valve tends to be stenotic, and when the hypoplastic RV is non-inverted [*****i.e.** ***right-sided RV; d-loop], the right AV valve tends to be regurgitant*** [1].

12.2.1.2 Double Inlet Right Ventricle (DIRV)

DIRV is less common as compared to DILV (10–20%) [6]. DIRV is characterized by a dominant RV and a hypoplastic LV (usually a trabecular pouch without an outlet).

The ventriculo-arterial connection from the RV takes one of the following forms [1, 2]:

- Double outlet RV (most common).
- Common outlet (truncus arteriosus).
- Single outlet (with atresia of one of the semilunar valves).

12.2.1.3 Single Functional Ventricle of Undifferentiated or Indeterminate Type

This is seen in <10% of the cases of double inlet AV connection, where the morphology of neither of the ventricular chambers can be determined with certainty [1]. Most often, the hypoplastic chamber is a rudimentary pouch without an outlet. Van Praagh et al. have also described a double inlet ventricle of 'mixed morphology', where a tiny ridge of rudimentary ventricular septum is seen at the apex, separating distinct right and left ventricular zones of the heart [2, 8].

12.2.2 Common Inlet AV Connection

In common inlet AV connection, both atria empty via a common AV valve into the dominant ventricular chamber. This represents a type of unbalanced AV septal defect.

If 75% or more of the annulus of a common AV valve empties into one ventricular chamber, it is referred to as a common inlet ventricle [1]. Partial atresia or stenosis of a part of the common AV valve favours the predominant commitment of the valve to the contralateral ventricle. This may be associated with heterotaxy syndrome/asplenia.

The dominant ventricular chamber in a common inlet AV connection is usually a morphologic RV. The anomalies associated with the ventricular outflow are VA discordance, DORV, and pulmonary valve stenosis/atresia [2, 8].

12.2.3 Single Inlet AV Connection

Single inlet AV connection results when one of the AV valves is either absent (imperforate valve orifice) or atretic (e.g. tricuspid atresia and mitral atresia) (Fig. 12.1). In such cases, the segmental anatomy can vary as follows [1, 2, 8]:

AV connection: Concordant or discordant.

VA connection: Concordant or discordant.

Great arterial relationship: Normal/Transposed /Malposed.

Subaortic stenosis: Commoner in cases of VA discordance.

Subvalvular or valvular pulmonary stenosis: Can occur regardless of VA connection.

12.2.4 Echocardiography

12.2.4.1 Diagnosis of Univentricular AV Connection

The AV connection is best demonstrated in the A4C view (Fig. 12.2). This apical view allows for the assessment of the planes of the atrial and ventricular septa, which determines the degree of atrial commitment to a ventricular chamber.

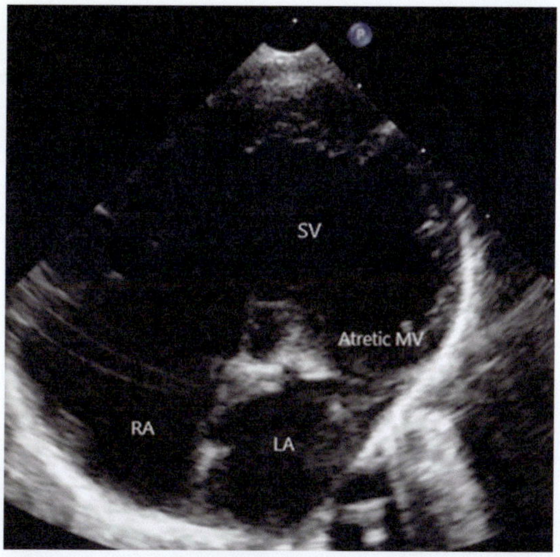

Fig. 12.1 Mitral atresia: A5C view showing an atretic mitral valve and a single ventricle (SV) of RV morphology. Not evident in the image is a large atrial septal defect with an obligatory left-to-right shunt

Dominant ventricle: A ventricular chamber receiving ≥50% of an inlet portion.

Hypoplastic ventricle: A ventricular chamber receiving <50% of an inlet portion.

Common-inlet ventricle: A ventricular chamber receiving ≥75% of the annulus of a common AV valve.

12.2.4.2 Morphology of the Dominant Ventricle and Function

Apical and subcostal four-chamber, PLAX, PSAX, and subcostal short-axis views demonstrate the ventricular morphology, chamber dimensions, wall thickness, and contractility.

- ***AV valve: Semilunar valve continuity*** aids in determining the morphology of the dominant ventricle (in addition to the usual morphologic features):
  - LV morphology: Direct continuity of the semilunar valve with one or both AV valves.
  - RV morphology: Interposition of myocardial tissue (conus) between the AV valves and the semilunar valve.
- Dominant ventricles with RV morphology are more likely to have [5]:
  - Systolic and diastolic dysfunction.
  - AV valve regurgitation.

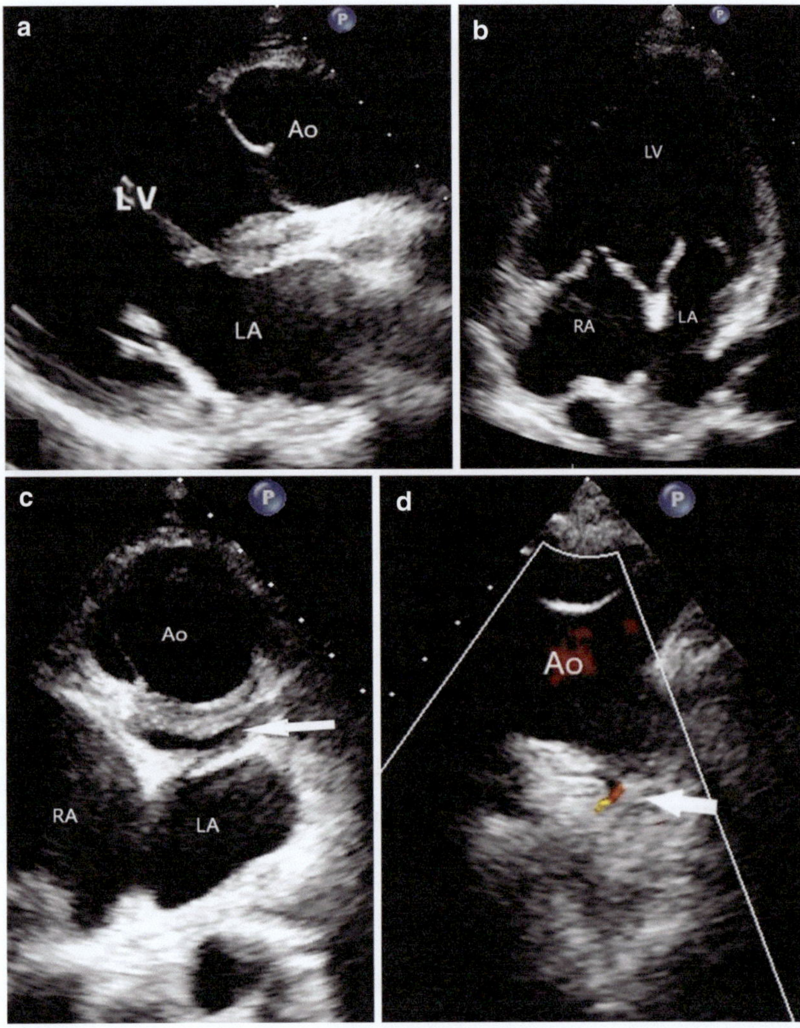

Fig. 12.2 Double inlet left ventricle: (**a**) PLAX view showing the anterior aorta arising from the single ventricle. (**b**) Apical view showing a single ventricle [SV] of LV morphology to which both the AV valves are committed. An interatrial communication is seen. (**c**) High PSAX view showing a large anterior aorta and a small posterior PA (denoted by arrow). (**d**) Suprasternal notch view depicting a small aortopulmonary collateral vessel (red flow denoted by arrow)

- Although the applicability of Simpson's biplane method in complex ventricular morphologies is not validated, it aids in a global assessment of ventricular function. Other indices, such as fractional area change, may also be used.
- Reduced preload may result in abnormal diastolic filling patterns, making them less specific for the diagnosis of diastolic dysfunction.

12.2.4.3 Morphology and Location of the Hypoplastic Ventricle

PSAX view and subcostal four-chamber view may aid in delineating the characteristics of the hypoplastic ventricle:

- Whether the hypoplastic ventricle is an outlet chamber or a rudimentary pouch (without an outlet) needs to be determined.
- The usual location of the hypoplastic ventricle is as follows [1, 2, 5]:
  - DILV: The hypoplastic RV is located basally, superior, and anterior to the LV.
  - DIRV: The hypoplastic LV is located inferiorly, on either side of the midline.

12.2.4.4 Adequacy of the Interventricular Communication

The size of the bulboventricular foramen is determined from short or long axis views.

- The ***foramen is considered stenotic if*** [9]:
  - Doppler flow velocity is more than 1.5 m/sec or.
  - Area of the foramen is <2 cm^2/m^2 BSA.
- A foramen with a size comparable to the aortic annulus is considered ideal.

12.2.4.5 AV Valve Morphology and Function

A4C, PSAX, subcostal coronal, and sagittal views are useful in the assessment of:

- AV valve leaflet configuration (trileaflet/bileaflet).
- Papillary muscle location and arrangement (multiple/double).
- Chordal attachments within the ventricular chambers (straddling).
- Valvular stenosis or regurgitation.

12.2.4.6 Ventriculo-Arterial Concordance and Ventricular Outflow Tracts

- PSAX view and subcostal coronal/sagittal sweeps are useful in determining the VA concordance and the great arterial relationship (as discussed in Chap. 3).
- Subaortic mean pressure gradients can be estimated using CWD interrogation of the ascending aorta from a high left parasternal view (near the second intercostal space) in the midclavicular line.
- When there is significant subaortic obstruction, concomitant obstruction of the LV outflow at valvular and/or supravalvular levels (owing to hypoplasia of the aortic annulus, stenosis of the ascending aorta, and/or aortic coarctation) needs to be ruled out [1].
- Subpulmonic and pulmonary valvular obstruction, as well as the adequacy of the pulmonary arteries, need to be determined.

12.2.4.7 Associated Abnormalities

- ASD, PDA, interrupted aortic arch, and aortic coarctation.

12.2.4.8 Post-Operative Evaluation

Subaortic obstruction secondary to progressive ventricular hypertrophy may occur in patients with DILV following pulmonary artery banding and the Fontan operation.

Key views	Key features assessed
Four chamber views	
Apical/subcostal	Delineate the crux of the heart Determine AV valve commitment to the ventricles Identify the ventricular morphology AV valve abnormalities: Atresia/stenosis/regurgitation Annular override; valvular straddling Location, commitment, and relationship of great arteries
Short axis views	
Parasternal/subcostal	Ventricular morphology and location of hypoplastic ventricle AV valve leaflet configuration (tricuspid/bicuspid) AV valve orifice stenosis Papillary muscle location and arrangement (multiple/double) Chordal attachment within the ventricles (straddling) Great arterial relationship Size of VSD (bulboventricular foramen) Ventricular wall thickness/chamber dimensions/contractility
Long axis views	
Parasternal/subcostal	Location, commitment, and relationship of great arteries Size of bulboventricular foramen and the pressure gradient across it Ventricular wall thickness/chamber dimensions/contractility
High left parasternal	CWD interrogation of ascending aorta to look for subaortic obstruction—Estimate mean gradient
Suprasternal window	Aorta and aortic arch hypoplasia/coarctation

12.3 Hypoplastic Left Heart Syndrome

Hypoplastic left heart syndrome [HLHS] encompasses a spectrum of congenital cardiac malformations characterized by underdevelopment of the left heart, with varying degrees of hypoplasia of the left ventricle, mitral valve, aortic valve, ascending aorta, and aortic arch. LV hypoplasia is accompanied by enlargement of the RA, PA, and pulmonary valve. RV is enlarged and hypertrophied [10].

12.3.1 HLHS with Intact Ventricular Septum

This is the most recognized form of HLHS, characterized by a dominant RV, hypoplastic LV outflow and aorta, and an intact ventricular septum, not amenable to two-ventricle repair. MV hypoplasia is usually less severe as compared to the degree of LVOTO. A localized coarctation of the aorta is present in 80% of patients [11].

12.3.1.1 Aortic Atresia with Mitral Atresia

This is the most extreme form of HLHS, where systemic output is duct-dependent. In aortic atresia, the role of the ascending aorta is to serve as a conduit for the retrograde flow of blood into the coronary arteries [10, 12].

12.3.1.2 Aortic Atresia with a Patent MV

LV inflow without outflow results in: (a) duct-dependent systemic output, (b) hypertrophy and endocardial fibroelastosis of the hypoplastic LV, and (c) distortion of RV inflow with TV dysplasia, RV subendocardial ischaemia, and fibroelastosis [10, 13].

12.3.1.3 Aortic Valve Stenosis with a Patent MV

This is a less severe form, characterized by a lesser degree of ascending aortic and arch hypoplasia and a variable degree of antegrade flow through the aortic valve.

12.3.2 HLHS with Ventricular Septal Defect

In addition to the variants described in HLHS with an intact septum, some specific anatomic variants of HLHS associated with VSD are [10, 11, 14]:

- Mitral atresia with a patent left ventricular outflow.
- DORV with or without obstruction to aortic outflow.
- Unbalanced AV septal defects with hypoplasia of the LV.

12.3.3 Echocardiography

12.3.3.1 Diagnosis

- The diagnosis of HLHS is often suspected from the PLAX view: the LV is seen as a small muscular chamber that does not extend to the cardiac apex.
- The features differentiating HLHS from critical aortic stenosis are [15, 16]:
  - LV cross-sectional area < 1.5 cm^2.
  - LV end-diastolic inflow dimension <25 mm (measured from the hinge-point of the PML to the apex).
  - Mitral annulus diameter of $\leq$6 mm.

12.3.3.2 Left Atrium and Atrial Septum

The LA is usually small in size. LA hypertension and hypertrophy may result from a restrictive ASD or an obstructed levoatrial cardinal vein [17]. An atrial septal aneurysm protruding into the RA may be present [10]. The septum primum may be anomalously attached to the LA wall [18].

12.3.3.3 LV Inflow

LV inflow is assessed from the PLAX, subcostal 4C, and A4C views. Mitral valve, when patent, is characterized by thickened leaflets, short chordae, or absent papillary muscle-chordal attachments.

The mitral annulus diameter should be documented, especially in cases of borderline LV size.

12.3.3.4 Left Ventricle and Ventricular Septum

LV cross-sectional area [PSAX view] and LV end-diastolic inflow dimension [PLAX view—measured as the distance from the PML attachment to the apex] should be measured. Endocardial fibroelastosis may be seen as echo-bright areas in the LV endocardium. A diminutive and/or dysfunctional LV may be unable to support the systemic circulation.

VSD is rare in the presence of aortic atresia. Colour Doppler interrogation of the ventricular septum may reveal ventriculo-coronary arterial connections.

12.3.3.5 LV Outflow

The patency of the aortic valve needs to be ascertained from the PLAX, A5C, and subcostal views. The ascending aorta is usually small (2 to 3 mm in diameter).

The transverse arch and DTA are best seen from the suprasternal view. Coarctation of the aorta and interruption of the aortic arch are common.

Key point: Retrograde systolic flow from the ductus into the aorta indicates duct-dependent systemic circulation, suggestive of the inadequacy of the LV for biventricular repair.

12.3.3.6 Right Heart

The RA, RV, and PA are usually dilated. A hypertrophic RV occupying the cardiac apex is suggestive of the inadequacy of the LV for biventricular repair. Tricuspid valve abnormalities such as bileaflet valve, TV dysplasia/prolapse, and abnormal papillary muscle arrangements are observed [10]. Neonates presenting with cardiac failure and acidosis following ductal closure are likely to have RV systolic dysfunction.

The pressure gradient from the PA to the aorta (across the PDA) should be quantified at the baseline and followed up when prostaglandin therapy is initiated.

12.3.3.7 Pulmonary Venous Anatomy and Drainage

Total or partial anomalous pulmonary venous drainage occurs in up to 10% of patients. This is best assessed from the subcostal and suprasternal views.

Levoatrial cardinal vein: In cases of an intact atrial septum, a persistent levoatrial cardinal vein (LACV) originating directly from the LA drains the pulmonary veins into the innominate vein. Stenosis of this anomalous vein leads to LA hypertension [17].

LACV can be identified from the suprasternal view as a vascular structure coursing posterior to the LPA, with cephalad flow into the innominate vein.

12.3.3.8 Coronary Arteries

The coronary arterial origin and direction of flow are best assessed from the PSAX view. ***Bidirectional flow in the coronary arteries suggests a left ventriculo-coronary arterial connection*** [10]. Colour flow Doppler may also reveal ventriculo-coronary arterial connections in the ventricular septum. The coronary artery anomalies described in HLHS are:

- Anomalous origin of either of the coronary arteries from the RPA.
- Coronary-cameral fistulae (in aortic atresia and patent MV).

Key views	Key features assessed
PLAX view	Morphology of left ventricular inflow, trabecular portion, outflow, aorta Endocardial fibroelastosis VSD, LV-coronary arterial connections
PSAX view	LV size and function MV papillary muscles; aortic valve size and anatomy LV-coronary arterial connections (Doppler interrogation of the coronary arteries) Pulmonary valve, MPA, branch PAs, PDA
A4C view	LV size and function MV anatomy and annulus size, subvalvular apparatus RV systolic function Tricuspid valve abnormalities
Subcostal views	Atrial septum: ASD, ASA Pulmonary venous drainage, levoatrial-cardinal vein
Suprasternal views	Aorta and aortic arch: Hypoplasia/ coarctation Proximal PAs, PDA Pulmonary venous drainage, levoatrial-cardinal vein

Points to Report

- Mitral apparatus: Valve anatomy; subvalvular apparatus; mitral annulus size.
- LV size and function.
- Aortic valve anatomy and function: Subvalvular/valvular/supravalvular stenosis.
- Aortic arch anatomy: Coarctation; interruption; dimensions.
- Atrial septum: Aneurysms, anomalous attachments, ASD.
- Ventricular septum: VSD; Ventriculo-coronary arterial connections.
- PDA: Direction of flow; Ao-PA gradient across the PDA.
- TV abnormalities and PA anatomy.
- RV size and function.
- Systemic venous anomalies: Persistent LSVC, levoatrial cardinal vein.
- Pulmonary venous anomalies: TAPVC/PAPVC.
- Coronary arteries: Anomalous origin from RPA; bidirectional coronary flow indicating ventriculo-coronary arterial connections; coronary-cameral fistulae.

12.4 Fontan Physiology

Univentricular Heart [Fontan] palliation is a surgical strategy adopted for the repair of congenital heart defects with univentricular heart (UH) physiology**.** Some of the cardiac defects that fall into this category are tricuspid/mitral/pulmonary atresia; hypoplastic left/right ventricle; single ventricle heterotaxy syndrome; double inlet left ventricle; double outlet right ventricle; unbalanced AV septal defect; and severe forms of Ebstein's anomaly.

12.4.1 Fontan Pathway

The 'Fontan circulation' is a univentricular palliation in which the systemic venous return is rerouted to the pulmonary arteries (PAs) without an intermediary ventricle, and all shunts (venous, atrial, ventricular, and arterial) are interrupted [19]. This is often done as a staged procedure to avoid overload of the systemic ventricle and to facilitate adaptation.

Stage I (neonatal period): Depending on the morphology and pathophysiology of the underlying heart, various procedures (most commonly a **systemic-pulmonary anastomosis**) may be done at this stage in anticipation of future Fontan repair. The aim of a procedure at this stage is to ensure:

- **Adequate blood flow to the aorta** [9].
  - **Damus-Kaye-Stansel + shunt**: The proximal PA trunk is connected to the ascending aorta to bypass the subaortic stenosis, and a conduit is placed between the RV and the distal PA, after closing the VSD (i.e. ***RV to PA + PA to Asc Ao***).
  - **Norwood procedure**: Consists of: (**a**) resection of the atrial septum to allow pulmonary venous return to the RV, (**b**) reconstruction of the aorta (neoaorta) using the MPA (side-to-side anastomosis between MPA and ascending aorta) to facilitate systemic blood supply from the RV, and (**c**) ensuring adequate pulmonary blood flow by means of either a modified BT shunt or an RV-to-PA conduit [Sano shunt].
  - **Coarctectomy.**
- **Balanced flow to the lungs.**
  - **Shunt (Modified Blalock-Taussig Shunt, central shunt), Stent in duct**—when PBF is inadequate.
  - **PA banding**—when pulmonary blood flow (PBF) is excessive.
- **Adequate return of blood to the ventricles** (Rashkind balloon septostomy).

If the above requirements are met naturally, a procedure is not warranted at this stage. Following the stage I procedure, there is volume overload of the heart, which if not excessive, aids in the development of the pulmonary vasculature.

Stage II (3–12 months)—Superior cavopulmonary connection: At this stage, a partial cavopulmonary anastomosis (SVC-to-PA) is established:

- **Bidirectional Glenn (BDG) shunt**: End-to-side anastomosis of the SVC to an undivided RPA.
- **Hemi-Fontan operation**: The RAA is connected to the RPA using a pulmonary allograft. An intraatrial patch is placed to direct SVC blood to the PAs [9].

Stage III (1–5 years)—Fontan completion: At this stage, total cavopulmonary anastomosis (i.e. IVC-to-PA and SVC-to-PA) is established. This is done either within 1–2 years after stage II operation or as a single-stage surgery, wherever appropriate. Fontan surgery has evolved over the decades as follows:

- **Classic Fontan operation**: **The** IVC drains to the RA; the RA is connected with a valved conduit to the proximal LPA. The SVC is connected to the distal RPA.
- **Atriopulmonary (AP) Fontan**: Both IVC and SVC drain to the RA, with the RA appendage anastomosed to the MPA or LPA.
- **Total cavopulmonary connection (TCPC)**: The RA is excluded from the system [19].
  - **Lateral tunnel (intraatrial baffle)**: This consists of: **(a)** end-to-side anastomosis of the SVC to an undivided RPA (bidirectional Glenn anastomosis), and **(b)** baffling of the IVC through the lateral wall of the RA into the RPA.
  - **Extracardiac Conduit**: This consists of: **(a)** end-to-side anastomosis of the SVC to an undivided RPA (bidirectional Glenn anastomosis), and **(b)** connection of the IVC to the RPA via a tube graft.
  - **Fenestrated Fontan**: **A** fenestration of up to 4 mm in diameter is made in the TCPC conduit as a safeguard mechanism. This allows a residual right-to-left shunt in patients with suboptimal PVR, especially in cases of increased pulmonary pressures that can occur early after surgery, and for Fontan failure with increased central venous pressures.

12.4.2 Echocardiography

The following details should be obtained before proceeding with the echocardiogram in a post-operative patient:

- Complete pre-operative diagnosis, including details about associated cardiac anomalies.
- Type of surgery performed, with the operation notes, wherever available.

12.4.2.1 Morphology of the Dominant Ventricle

Morphology of the dominant ventricle needs to be ascertained.

12.4.2.2 Ventricular Function

Standard modalities of assessment of ventricular function may be used based on the morphology of the dominant ventricle. Various indices for the assessment of RV function are discussed in Chap. 15.

AV valve systolic-to-diastolic duration ratio (S/D) is an index of ventricular dysfunction in patients with Fontan circulation [20].

- Calculated from CWD tracing of the systemic AV valve.
- Effective systolic duration (S) is measured from the onset to the end of AV valve regurgitation.
- Effective diastolic duration (D) is measured from the end of the AV valve regurgitation to the onset of the subsequent AV valve regurgitation signal.
- The normal value of the systolic-to-diastolic ratio (S/D) in patients with Fontan circulation is 0.4–1.6. ***A higher value may reflect ventricular dysfunction***.

Diastolic dysfunction of the systemic ventricle is observed in the majority of Fontan-palliated patients, with a restrictive pattern in more than a quarter [21]. The diastolic parameters that have been shown to correlate with elevated ventricular end-diastolic pressures by invasive measures are [22, 23]:

- Pulmonary vein A-wave/atrioventricular A-wave duration ratio.
- E velocity > 75 cm/s.
- E/A ratio > 1.7.
- DT < 135 ms.

Key point: Reduction in the late diastolic flow in the PA and diminished respiratory variation in PA flow velocity are indicators of systemic ventricular dysfunction.

12.4.2.3 Atrioventricular Valve

AV valve morphology depends on the underlying anatomic diagnosis. Significant AV valve regurgitation or stenosis may have a profound impact on the Fontan circulation.

12.4.2.4 Systemic Ventricular Outflow

- Aortic valve morphology.
- Great arterial origin and relationship.
- Presence of valvular, subvalvular, or supravalvular stenosis, or regurgitation.

12.4.2.5 Atria

Large ASD or a single atrium are common in most of these diseases. In the older AP Fontan system, the RA is incorporated into the Fontan system to act as a pumping chamber. This may result in marked enlargement of the RA, which can cause pulmonary vein obstruction. In TCPC, the intracardiac conduit can be visualized within the atria.

- Interatrial shunting [24]:
  - May be present in patients with RA–PA connection or fenestrated Fontan.
  - Seen as turbulent flow by colour Doppler.
  - Characterized by low-velocity flow (<4 m/sec).
- Atrial volume and size should be documented.
- Atrial thrombus should be searched for, especially in the setting of an AP Fontan.
- Pulmonary vein obstruction can be detected by assessment of the pulmonary venous flow pattern [25]:
  - Biphasic low-velocity flow is suggestive of unobstructed normal flow.
  - High velocity suggests obstruction.
  - Monophasic continuous low-velocity flow suggests downstream obstruction.

12.4.2.6 Fontan Circuit

Echocardiographic detection of these connections is challenging owing to their retrosternal location. High parasternal and subcostal views are the most effective.

SVC to RPA Connection (Bidirectional Glenn) (Fig. 12.3)

- Evaluated from the subcostal (bicaval) view, suprasternal view, and the right supraclavicular window (left supraclavicular for left-sided SVC connections).

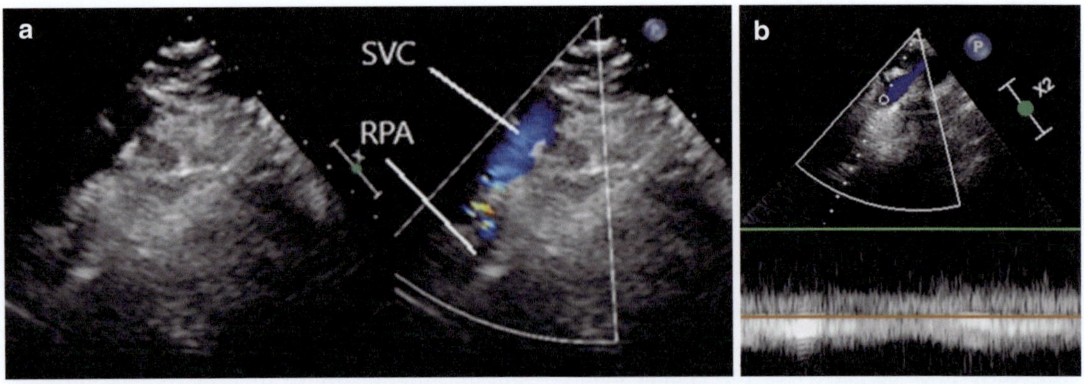

Fig. 12.3 (**a**) Suprasternal view of SVC-to-RPA anastomosis (Bidirectional Glenn). Mild turbulence at the SVC-RPA junction may be noted. (**b**) Continuous low-velocity laminar flow in an unobstructed BDG

- Nyquist limit needs to be reduced for optimal visualization [21, 26].
- A patent circuit is characterized by laminar, phasic, low-velocity flow in the SVC with respiratory variation [26].

Inferior Vena Cava Flow (Figs. 12.4 and 12.5)
The IVC can be evaluated from the subcostal view. Depending on the type of surgery, the IVC may be connected to the RA–PA Fontan, an extracardiac conduit, or an intracardiac conduit (lateral tunnel). In heterotaxy syndromes with interrupted IVC, the hepatic veins may be directed to the pulmonary arteries through an intraatrial conduit. The Fontan conduit within the RA in a lateral tunnel Fontan is best appreciated in the apical view, where the conduit appears as a circular structure within the RA.

- Laminar, phasic, low-velocity flow in the IVC with respiratory variation suggests a patent circuit.
- Higher velocities and absence of inspiratory collapse suggest [21, 26]:

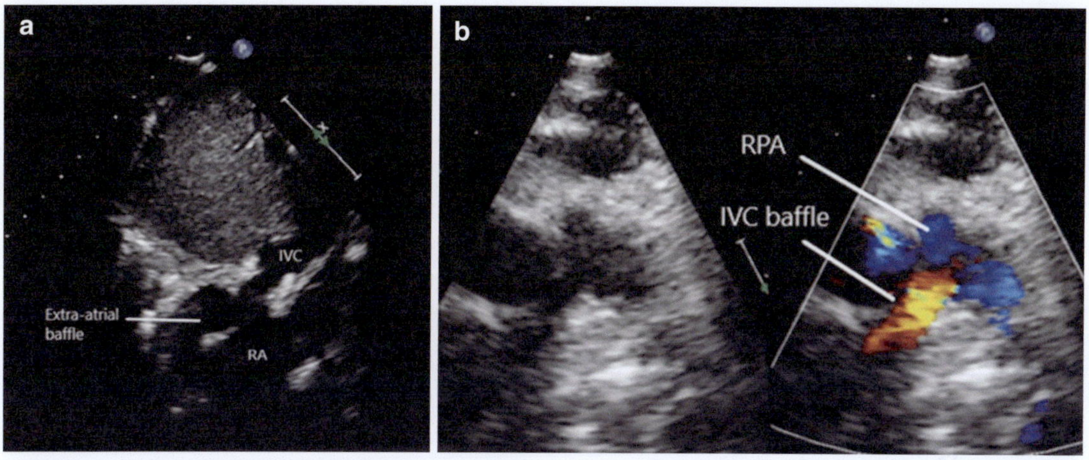

Fig. 12.4 Extracardiac Fontan circuit. The IVC is connected to the RPA via an extracardiac conduit, as seen from the subcostal (**a**) and suprasternal views (**b**)

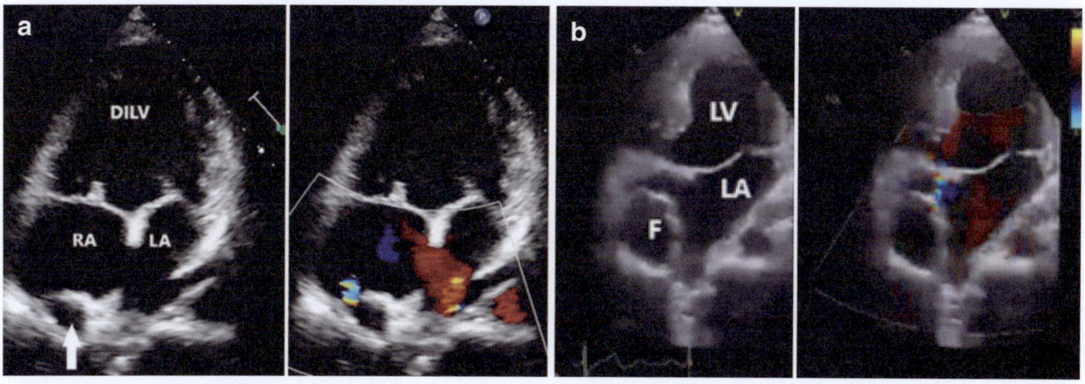

Fig. 12.5 Fenestrated lateral tunnel Fontan. (**a**) A case of DILV with an intraatrial baffle (lateral tunnel Fontan) connecting the IVC to the RPA. The Fontan circuit (marked by the vertical arrow) is seen within the right atrium. Note the interatrial septal defect with a left-to-right shunt. (**b**) A case of tricuspid atresia with a fenestrated lateral tunnel Fontan [F]. The atretic tricuspid valve, along with the interatrial and interventricular shunts, are seen. Turbulent flow across the fenestration is depicted by colour Doppler imaging in both cases

– High pulmonary pressures or.
– Obstruction (stenosis or thrombus).

Obstructed Fontan: As there is no pumping chamber, ***a mean gradient of > 2 mmHg*** between the Fontan circuit and the PA may be suggestive of significant obstruction [21, 26]. Significant pressure gradients in the Fontan circulation may be due to:

– Stenosis of the IVC, SVC, conduit, or PA.
– Pulmonary thromboembolism.

Fenestration in the lateral tunnel is better evaluated from the apical view (Fig. 12.5). In cases where device closure of the fenestration has been performed, the presence of residual flow needs to be looked for.

In the absence of a pressure gradient within the Fontan circuit, PA pressure can be estimated from the flow velocity across a fenestration.

12.4.2.7 Pulmonary Arteries

The RPA can be assessed from high PSAX, 'modified' A5C, and suprasternal views. The LPA may be visualized from the suprasternal sagittal view.

Normal PA flow in an unobstructed Fontan circulation is characterized by [26]:

- Low velocity, non-pulsatile flow.
- Exaggerated augmentation of flow velocity during inspiration.
- One peak in late systole and a larger peak in late diastole during atrial contraction.
- PA pressure and central venous pressure are nearly equal.
  – An elevated mean CWD gradient in the PA may suggest [21, 27]:
- Stenosis of the native PA or.
- Obstruction of the PA stent (if PA stenting has been performed earlier).
  – The presence of flow reversal in the MPA and/or branch PAs indicates [21, 27]:
- Pulmonary hypertension or.
- Aortopulmonary collaterals.
  – Systemic ventricular dysfunction is suggested by [21, 27]:
- Reduced late diastolic flow in the PA.
- Diminished respiratory variation in flow velocity.

Key point: Normal unobstructed flow in the Fontan circuit is indicated by biphasic, low-velocity, laminar flow with respiratory variation in the SVC and IVC; and a triphasic, non-pulsatile, low-velocity flow with exaggerated inspiratory augmentation in the pulmonary arteries.

12.4.2.8 Transoesophageal Echocardiography (TEE)

Advantages offered by TEE in Fontan patients include the identification of atrial morphology by assessing the appearance of the appendages, evaluation of patency of the Fontan anastomosis, and detection of thrombi in the atria or cavopulmonary anastomosis.

Points to Report

- Segmental description of the underlying cardiac anatomy.
- Ventricular morphology and function.
- AV valve function.
- Outflow of the systemic ventricle: Valvular/subvalvular stenosis, regurgitation.
- Status of the Fontan circuit.
  – Conduits/anastomosis: Patency, obstruction, gradients.
  – Fenestration: Gradient, closure device.

References

1. Earing MG, Hagler DJ, Edwards WD. Univentricular atrioventricular connection. In: Allen HD, Driscoll DJ, Shaddy RE, Feltes TF, editors. Moss and Adams' heart disease in infants, children, and adolescents. 8th ed. Philadelphia, PA: Lippincott Williams & Wilkins; 2013. p. 1175–94.
2. Van Praagh R, Van Praagh S, Vlad P, et al. Diagnosis of the anatomic types of single or common ventricle. Am J Cardiol. 1965;15:345–66.
3. Anderson RH, et al. The univentricular atrioventricular connection: getting to the root of a thorny problem. Am J Cardiol. 1984;54:822.
4. Anderson RH, et al. Problems in the nomenclature of the univentricular heart. Herz. 1979;4:184.

5. Perloff JK, Marelli AJ. Univentricular heart. In: Clinical recognition of congenital heart disease. 6th ed. Philadephia, PA: Saunders; 2012. p. 454–71.
6. Shiraishi H, Silverman NH. Echocardiographic spectrum of double inlet ventricle: evaluation of the interventricular communication. J Am Coll Cardiol. 1990;15:1401.
7. Bevilacqua M, et al. Double inlet single ventricle: echocardiographic anatomy with emphasis on morphology of the atrioventricular valves and ventricular septal defect. J Am Coll Cardiol. 1991;18:559.
8. Van Praagh R, Ongley PA, Swan HJC. Anatomic types of single or common ventricle in man: morphologic and geometric aspects of sixty necropsied cases. Am J Cardiol. 1964;13:367–86.
9. Park MK. Cyanotic congenital heart defects. In: Park's pediatric cardiology for practitioners. 5th ed. Philadephia, PA: Saunders; 2016. p. 178.
10. Tweddell SJ, Hoffman GM, Ghanayem NS, Frommelt MA, Mussatto KA, Berger S. Hypoplastic left heart syndrome. In: Allen HD, Driscoll DJ, Shaddy RE, Feltes TF, editors. Moss and Adams' heart disease in infants, children, and adolescents. 8th ed. Philadelphia, PA: Lippincott Williams & Wilkins; 2013. p. 1061–96.
11. Hawkins JA, Doty DB. Aortic atresia: morphologic characteristics affecting survival and operative palliation. J Thorac Cardiovasc Surg. 1984;88:620–6.
12. Lev M. Pathologic anatomy and interrelationship of hypoplasia of the aortic tract complexes. Lab Investig. 1952;1:61–70.
13. Noonan JA, Nadas AS. The hypoplastic left heart syndrome; an analysis of 101 cases. Pediatr Clin N Am. 1958;5:1029–56.
14. Bharati S, Lev M. The surgical anatomy of hypoplasia of aortic tract complex. J Thorac Cardiovasc Surg. 1984;88:97–101.
15. Leung MP, McKay R, Smith A, et al. Critical aortic stenosis in early infancy. Anatomic and echocardiographic substrates of successful open valvotomy. J Thorac Cardiovasc Surg. 1991;101:526–35.
16. Parsons MK, Moreau GA, Graham TP Jr, et al. Echocardiographic estimation of critical left ventricular size in infants with isolated aortic valve stenosis. J Am Coll Cardiol. 1991;18:1049–55.
17. Vlahos AP, Lock JE, McElhinney DB, et al. Hypoplastic left heart syndrome with intact or highly restrictive atrial septum: outcome after neonatal transcatheter atrial septostomy. Circulation. 2004;109:2326–30.
18. Chin AJ, Weinberg PM, Barber G. Subcostal two-dimensional echocardiographic identification of anomalous attachment of septum primum in patients with left atrioventricular valve underdevelopment. J Am Coll Cardiol. 1990;15:678–81.
19. de Leval MR, Kilner P, Gewillig M, et al. Total cavopulmonary connection: a logical alternative to atriopulmonary connection for complex Fontan operations. Experimental studies and early clinical experience. J Thorac Cardiovasc Surg. 1988;96:682–95.
20. Friedberg MK, Silverman NH. The systolic to diastolic duration ratio in children with hypoplastic left heart syndrome: a novel Doppler index of right ventricular function. J Am Soc Echocardiogr. 2007;20:749–55.
21. Buber J, Schwaegler RG, Mazor DE. Echocardiographic evaluation of univentricular physiology and cavopulmonary shunts. Echocardiography. 2019;36(7):1381–90.
22. Cordina R, Ministeri M, Babu-Narayan SV, Ladouceur M, Celermajer DS, Gatzoulis MA, Uebing A, Li W. Evaluation of the relationship between ventricular end-diastolic pressure and echocardiographic measures of diastolic function in adults with a Fontan circulation. Int J Cardiol. 2018;259:71–5.
23. Miranda WR, Warnes CA, Connolly HM, Taggart NW, O'Leary PW, Oh JK, Egbe AC. Echo-Doppler assessment of ventricular filling pressures in adult Fontan patients. Int J Cardiol. 2019;284:28–32.
24. Stumper O, Sutherland GR, Geuskens R, et al. Transesophageal echocardiography in evaluation and management after a Fontan procedure. J Am Coll Cardiol. 1991;17:1152–60.
25. White BR, Faerber JA, Katcoff H, Glatz AC, Mascio CE, Cohen MS. Venous flow variation predicts preoperative pulmonary venous obstruction in children with Total anomalous pulmonary venous connection. J Am Soc Echocardiogr. 2021;34(7):775–85.
26. Armstrong WF, Ryan T. Feigenbaum's echocardiography. Philadelphia: Wolters Kluwer/Lippincott Williams & Wilkins Health; 2010.
27. Zaragoza-Macias E, Schwaegler RG, Stout KK. Echocardiographic evaluation of univentricular physiology and cavopulmonary shunts. Echocardiography. 2015;32(Suppl 2):S166–76.

13 Venous Anomalies

The spectrum of congenital venous anomalies ranges from the anomalies involving the central systemic veins to the venous malformations at the peripheries. The clinical phenotypic diversity ranges from asymptomatic anatomical variants to complex abnormalities that can lead to central cyanosis. The prevalence of anomalous systemic venous connections in patients with congenital heart disease is nearly 10%. Clinically significant systemic venous anomalies are present in more than 90% of patients with heterotaxy syndrome [1, 2].

13.1 Anomalies of the Superior Venae Cavae

13.1.1 Bilateral Superior Venae Cavae (Persistent LSVC)

Failure of the embryonic left anterior cardinal vein and the adjacent part of the common cardinal vein to regress results in a persistent left SVC (PLSVC). The presence of PLSVC, in addition to a right SVC, constitutes bilateral SVC. The left SVC is formed at the confluence of the left jugular and left subclavian veins. It descends anterior to the aortic arch and the LPA, receives the left superior intercostal vein, descends obliquely along the posterior wall of the LA, and drains into the coronary sinus in the posterior AV groove. A left innominate vein (LIV) may be present in 60% of the cases, bridging the RSVC and the LSVC [1, 2].

The prevalence of PLSVC is 1–2% in the general population and up to 10% in individuals with congenital heart disease. The left SVC continues with the coronary sinus in most of the cases [3]. Persistent LSVC drains into:

- RA through the coronary sinus (80–90%)
- LA through a partially or completely unroofed coronary sinus (10–15%)
- IVC and hepatic veins [rarely]

When PLSVC is associated with LA drainage or an absent right SVC, associated congenital malformations are more common. Cardiac malformations that are commonly associated with persistent LSVC are [4–6]:

- Juxtaposition of the right atrial appendage (RAA) (35%)
- Atrioventricular septal defects (20%)
- Mitral atresia/Single ventricle
- Tetralogy of Fallot (10%)
- Left isomerism
- Cor triatriatum

An unrecognized LSVC may pose diagnostic and technical difficulties during cardiac catheterization, pacemaker/ICD implantation, and surgery.

Supplementary Information The online version contains supplementary material available at https://doi.org/10.1007/978-981-96-0243-8_13.

A. Paul, *Manual of Echocardiography for Congenital Heart Diseases*,
https://doi.org/10.1007/978-981-96-0243-8_13

Echocardiography

- ***A dilated coronary sinus is often the first clue to the diagnosis of LSVC*** (Fig. 13.1).
- Visualization of LSVC
  - LSVC can be seen from the suprasternal view as a vertical vascular structure to the left of the aortic arch (Fig. 13.2a).
  - The subcostal short-axis view demonstrates the LSVC and its drainage into the coronary sinus (Fig. 13.2b).
  - In the high PSAX view, the cross-section of the LSVC can be seen anterior to the LPA, while the DTA is seen posterior to the LPA.
  - The suprasternal view, high left parasternal view, and subclavicular view demonstrate the LSVC and the left innominate vein (LIV). In general, there is an inverse relationship between the calibre of the LSVC and the LIV when both are present together [1].
- Colour Doppler imaging demonstrates continuous venous flow of the LSVC in a cranio-caudal direction. The drainage of the LSVC needs to be ascertained by colour flow mapping from subcostal and/or suprasternal views.
- Demonstration of **retrograde flow in the LSVC** towards the LIV suggests obstructed flow in the coronary sinus, owing to stenosis or atresia of the coronary sinus orifice.

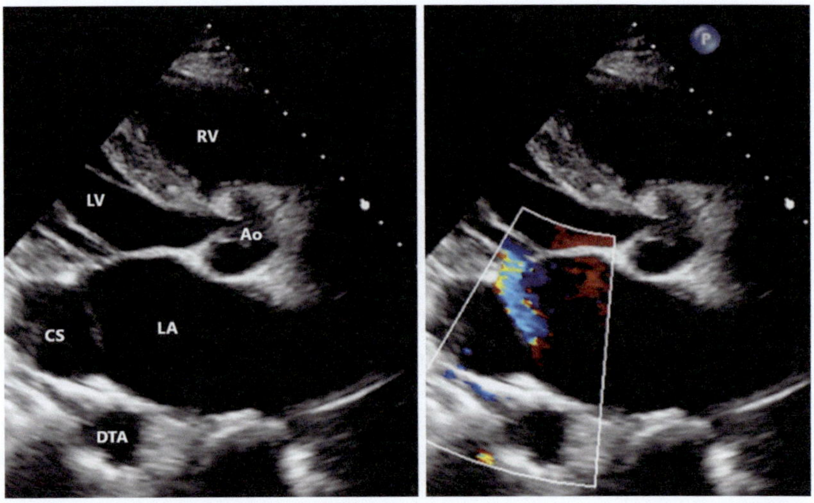

Fig 13.1 Dilated coronary sinus [CS] in a case of PLSVC as seen in the PLAX view. It may be noted that the CS is anterior to the pericardium in the posterior left AV groove, whereas the DTA is seen posterior to the pericardium. In the colour flow Doppler image, a posteriorly directed eccentric MR jet is seen bordering the wall of the CS

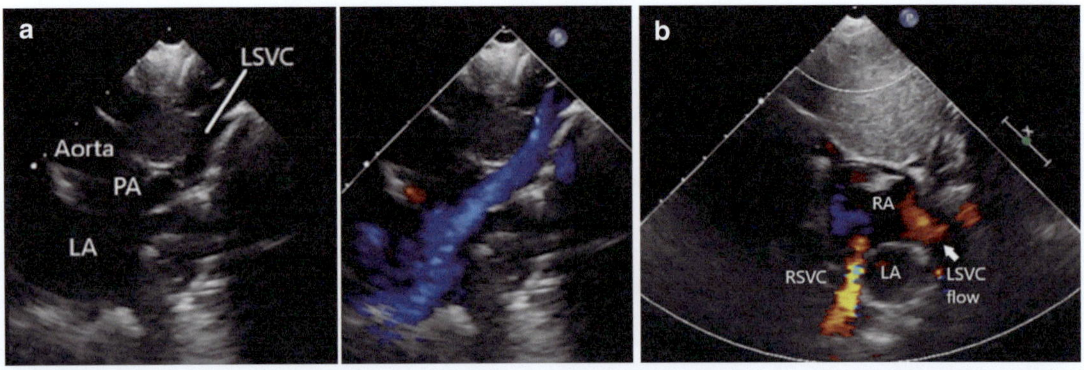

Fig 13.2 (**a**) Modified suprasternal short-axis view demonstrating LSVC flow passing obliquely posterior to the left atrium en route to the coronary sinus, (**b**) Subcostal short-axis view [with slight angulation to the left from a bicaval view] depicting the LSVC flow

- **Saline contrast-enhanced echocardiography**: If the injection of agitated saline into one of the left upper limb veins results in opacification of the coronary sinus before the right atrium, the diagnosis of a persistent LSVC draining into the coronary sinus is established.

Apart from the LSVC, ***the other veins that may connect with the LIV are*** [1]:

- Vertical vein
  - Present in the setting of anomalous pulmonary venous connection
  - Characterized by a pulmonary venous Doppler flow pattern and a caudo-cephalic direction of blood flow
- Left superior intercostal vein
  - A small systemic vein directed towards the anterior chest wall
- Levoatrial cardinal vein (LACV) [7]
  - An embryonic remnant that connects the LA or a pulmonary vein with the left innominate or other systemic veins, such as the right SVC
  - Functionally significant in cases of severe mitral stenosis/atresia with an intact atrial septum, as it provides an alternative drainage pathway for pulmonary venous blood
  - Unlike the PLSVC, which courses anterior to the LPA, an LACV typically ascends posterior to it

Key point: Colour Doppler imaging demonstrates that the direction of blood flow in the aforementioned veins is into the LIV (cephalad), in contrast to the LSVC, which drains away from the LIV into the coronary sinus (caudal). PWD demonstrates a typical systemic venous flow pattern in the LSVC.

13.1.2 Bilateral SVC (Persistent LSVC) with an Unroofed Coronary Sinus

Partial or complete absence of the common wall between the LA and the coronary sinus has been termed unroofing of the coronary sinus. This is otherwise known as ***Coronary sinus ASD***. An unroofed coronary sinus is associated with PLSVC draining into the LA (***Raghib syndrome***). PLSVC with an unroofed coronary sinus is commonly associated with visceral heterotaxy syndromes [2, 8].

Echocardiography

- PLSVC draining into the LA via an unroofed CS can be imaged from subcostal, suprasternal, and parasternal windows.
- In a completely unroofed coronary sinus, the termination of the LSVC is at the upper left posterior corner of the LA, between the LUPV posteriorly and the LAA anteriorly [1].
- Colour flow mapping demonstrates flow from the LSVC into the LA.
- **Saline contrast-enhanced echocardiography**: Injection of agitated saline into a left upper limb vein demonstrates the appearance of microbubbles in the LA before the RA. Early LA opacification can also result from abnormal drainage of the vena cava (either left or common) into the LA, in the absence of an unroofed coronary sinus.

13.1.3 Isolated Persistent LSVC

Persistent LSVC, with the concurrent absence of the right superior vena cava, is called isolated PLSVC. Isolated PLSVC occurs in 0.1% of individuals with CHD. Isolated PLSVC is often associated with other anomalies such as [9, 10]:

- Dextrocardia/Complete situs inversus/Left isomerism
- Atresia of the coronary sinus ostium
- Cor triatriatum
- Atrial septal defect
- Endocardial cushion defects
- Tetralogy of Fallot
- Coarctation of the aorta

Echocardiography demonstrates the absence of the right SVC in the subcostal bicaval view. An absent or atretic right SVC in the setting of viscero-atrial situs solitus is characterized by the presence of an LSVC draining into the RA and a left-sided azygos vein draining into the LSVC [10]. The right innominate vein drains into the LSVC.

13.1.4 Left Atrial or Biatrial Drainage of Right Superior Vena Cava

Abnormal drainage of the RSVC into the left atrium manifests as unexplained cyanosis and clubbing. This anomaly may represent a sinus venosus defect of the SVC type in which the RSVC orifice has become atretic. The sinus venosus defect results in unroofing of the RUPV and its branches into the SVC. The left atrial orifice of the RUPV functions as an interatrial communication. In this scenario, if the right atrial orifice of the RSVC becomes atretic, left atrial drainage results via the RUPV orifice. Biatrial drainage results when the RSVC ostium is stenotic without atresia. Left atrial or biatrial drainage of the right SVC can occur in cases of bilateral SVC as well [1, 11].

An echocardiogram from the subcostal short-axis view shows the RSVC and RUPV entering the roof of the LA via a common orifice. Contrast echocardiography from upper limb veins demonstrates opacification of the left atrium without opacification of the right atrium when the RSVC ostium is atretic.

13.1.5 Retroaortic (Postaortic) Innominate Vein

The normal course of the left innominate (brachiocephalic) vein is oblique from left to right, superoanterior to the aortic arch. A retroaortic innominate vein (RAIV) is characterized by an abnormal course of the LIV, which, after descending anterior to the LPA (similar to LSVC), takes a horizontal rightward turn, passes behind the ascending aorta, and joins the SVC below the insertion of the azygos vein, above the SVC–RA junction [1].

- Duplication of LIV refers to the coexistence of a retroaortic LIV and a normally positioned LIV.
- Right aortic arch and RVOTO are common associations.
- RAIV can be associated with complex congenital cardiac malformations such as TOF, pulmonary atresia, truncus arteriosus, heterotaxy syndrome, HLHS, AVSD, and coarctation of the aorta [12, 13].

Echocardiographic suspicion is often raised when an additional vascular structure is visualized in relation to the upper ascending aorta and aortic arch, apart from the right pulmonary artery. The differential diagnoses in such cases include PDA arising from the aortic arch, aortopulmonary collateral, vertical vein, and persistent fifth arch, in addition to RAIV [11, 12].

The diagnosis is established by delineating the course of the LIV from the left side of the neck through its retroaortic course to the SVC, by anterior and leftward tilting of the probe from the suprasternal long-axis view. The RAIV is visualized below the aortic arch, superior to the RPA. RAIV can be differentiated from PLSVC by its drainage into the SVC.

Key point: RAIV is characterized by low-velocity continuous flow towards the heart, whereas aortopulmonary collateral and PDA are associated with high-velocity continuous flow.

13.2 Anomalies of the Inferior Venae Cavae

13.2.1 Interrupted Inferior Vena Cava

Interrupted IVC can result from the failure of fusion of the embryologic components that constitute the IVC. This can occur at any level. Failure of the suprarenal segment (subcardinal vein) to fuse with the hepatic segment results in interruption of the hepatic/suprarenal segment of the IVC. Venous drainage beyond this level is facilitated by the infrarenal IVC, which continues cranially as a dilated azygos vein and drains into the RA via the right or left SVC. Rarely, the infrahepatic segment of the IVC may drain into both the right and left SVC via bilateral azygos veins [1].

Significance

- Usually associated with polysplenia syndrome; rarely in asplenia and normal hearts [2].
- Complicates cardiac catheterization and transfemoral interventional procedures.
- Awareness of the anomaly is extremely important when planning surgical therapy that involves redirection of the systemic venous return to the PAs (bidirectional Glenn and modified Fontan procedures).

Echocardiography

- The course of the IVC and the azygos vein can be imaged from the subcostal window.
- Normally, in the subcostal transverse abdominal view, the suprarenal segment of the IVC is seen as an oval blood vessel located anterior and to the right of the abdominal aorta. An intact IVC can be seen throughout most of its length from the subcostal sagittal plane. The diagnosis of interrupted IVC with azygos continuation is apparent in the transverse abdominal view:
  - There is conspicuous absence of the hepatic segment of the IVC anterior to the aorta (a hepatic vein should not be confused with the IVC).
  - There is a **large venous vessel immediately posterior to the aorta** (azygos or hemiazygos vein) (Fig. 13.3).
- The drainage of the enlarged azygos vein into the SVC can be imaged from the subcostal, parasternal, and suprasternal windows. Hepatic veins drain directly into the RA.

13.2.2 Bilateral Inferior Venae Cavae

Bilateral suprahepatic IVCs are constituted by a normal IVC on one side and a hepatic vein on the contralateral side. This is frequently associated with visceral heterotaxy with asplenia (relatively rare with polysplenia).

When bilateral IVCs occur in the setting of normal visceral situs, there is a normal IVC on the right side, while the left-sided hepatic vein drains into a normal coronary sinus [1].

13.2.3 Inferior Vena Cava Drainage to the Left Atrium

The Eustachian valve and the valve of the foramen ovale direct half of the right atrial blood into the LA during foetal life. Postnatally, a persistent and prominent Eustachian valve, in the presence

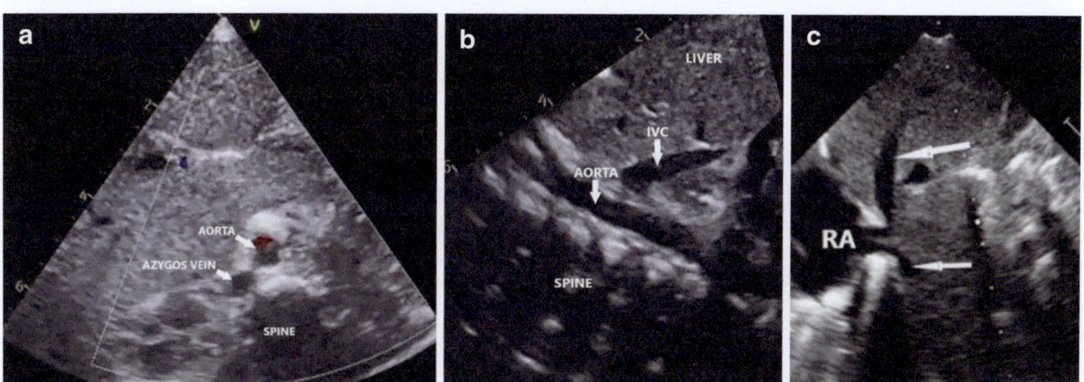

Fig 13.3 Interrupted IVC. (**a**) Subcostal transverse abdominal view showing Azygos vein, identified as a compressible vascular structure posterior to the aorta and anterior to the spine. The absence of the IVC anterior to the aorta is noteworthy. (**b**) Subcostal short-axis view showing the interrupted hepatic segment of the IVC. (**c**) Hepatic veins draining directly into the RA

of a PFO or an ostium secundum ASD, can direct the IVC blood into the LA. This effectively results in IVC drainage to the LA, although there is no direct connection between the IVC and the LA [1]. Contrast injected into the lower limb veins appears in the LA via the IVC and RA.

Key point: According to the current embryologic understanding, the right horn of the sinus venosus (i.e. the IVC and the SVC) cannot incorporate completely into the morphological LA, although some or all of the pulmonary veins can drain into the morphologic RA (as in visceral heterotaxy). When both the SVC and IVC are connected to the left-sided atrium, it will be a morphologic RA on the left side [14].

13.3 Anomalies of the Coronary Sinus

Coronary sinus dilation is usually the result of anomalous drainage into the sinus, either from a persistent LSVC or an anomalous pulmonary vein. A grossly dilated coronary sinus may be mistaken for pericardial effusion, pulmonary vein, or descending aorta.

Echocardiography

- The coronary sinus orifice can be visualized in the RV inflow view, as it opens into the posteroinferior aspect of the RA, adjacent to the orifice of the IVC. Chiari network may be seen in the posterior RA at the site of the coronary sinus orifice.
- A cross-section of the coronary sinus is visualized in the PLAX view as a circular structure in the posterior atrioventricular (AV) groove (Fig. 13.1). In contrast to the DTA, the CS is located anterior to the pericardium.
- The coronary sinus is visualized in the PSAX view as a tubular, crescent-shaped structure within the AV groove, communicating with the RA.
- A posterior angulation from the A4C view (coronary sinus view) demonstrates the coronary sinus in its long axis, coursing behind the LA and emptying into the RA (Fig. 13.3).

13.3.1 Coronary Sinus Defect and Unroofed Coronary Sinus

- Almost always associated with a persistent LSVC (see above).
- The orifice of the unroofed coronary sinus functions as the interatrial communication.
- The coronary sinus defect is physiologically obligatory in the case of:
  - Atretic coronary sinus ostium—for drainage of coronary venous blood.
  - Mitral atresia—as an alternative route for blood flow from the LA.
- Echocardiographic diagnosis can be made by imaging the defect in the coronary sinus septum in the posterior left AV groove and demonstrating blood flow through the defect by colour Doppler.

13.3.2 Coronary Sinus Aneurysm or Diverticulum

Coronary sinus diverticulum is a pouch originating from the coronary sinus and extending into the left ventricular wall. An association with Wolff-Parkinson-White syndrome has been identified [15].

Echocardiography

- The diverticulum is seen as an outpouching from the CS, which extends behind the left ventricle or into the ventricular myocardium.
- Colour Doppler demonstrates blood flow into the diverticulum from the CS.
- The other sites of connection, if any, can be demonstrated by colour flow mapping.

13.3.3 Atresia of Coronary Sinus Orifice

Atresia or severe stenosis of the right atrial orifice of the coronary sinus is rare [1]. When it occurs, an alternative exit for coronary venous return may exist in the form of:

- LSVC
- Large Thebesian vein
- Coronary sinus septal defect
- Communication with the IVC

Myocardial ischaemia, infarction, and death can occur when there is no alternative exit for the coronary sinus blood.

Key point: The demonstration of retrograde flow of blood in a persistent LSVC towards the left innominate vein suggests obstruction of the coronary sinus ostium.

13.4 Persistent Valves of the Sinus Venosus

The remnants of the embryologic valves of the sinus venosus are [16]:

- **Eustachian valve**: The superior portion of the right valve of the sinus venosus, along with part of the sinus venosus septum, persists as the Eustachian valve. This is normally represented by a ridge, located between the IVC orifice and the tricuspid valve orifice (Fig. 13.4).
- **Thebesian valve**: The inferior portion of the right valve of the sinus venosus, along with part of the sinus venosus septum, persists as the Thebesian valve, guarding the coronary sinus orifice.
- **Crista terminalis**: The remnant of the right valve of the sinus venosus, where it separated the RA into two portions—the anterior muscular portion and the posterior sinus (venosus) portion—persists as the crista terminalis.
- **Chiari networks**: Fine, filamentous structures that may represent persistence of either the right or left valves. Usually, Chiari networks are persistent right valves and extend from the crista terminalis to the Eustachian or Thebesian valves [16] (Fig. 13.5).

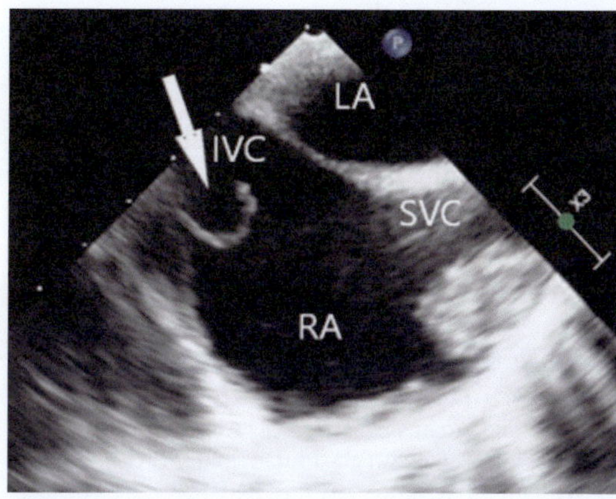

Fig 13.5 Transoesophageal echocardiographic bicaval view at the mid-oesophageal level, depicting a prominent Eustachian valve in the right atrium at the IVC orifice

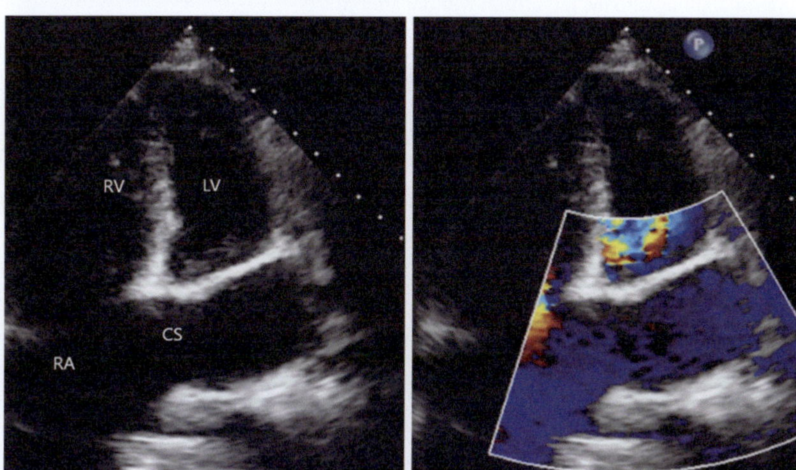

Fig 13.4 Apical coronary sinus view (obtained by posterior tilt of the imaging plane from the A4C view), depicting a dilated coronary sinus draining into the right atrium. It may be noted that the transverse flow (within the CS) from the left side of the heart to the RA is likely to create a false impression of an atrial septal defect

Abnormally persistent valves of the sinus venosus can rarely result in:

- RVOT obstruction:
  - Echocardiography reveals a thin, mobile, linear structure originating in the RA and traversing the tricuspid valve to obstruct pulmonary flow.
- Tricuspid Valve Obstruction:
  - A persistent right valve of the sinus venosus can occlude the TV orifice like a 'windsock' [1, 16].
  - Echocardiographic imaging from a high right parasternal longitudinal view demonstrates a linear structure attached to the RA near the IVC at one end and near the foramen ovale at the other end [1].
  - A four-chamber view demonstrates a linear, mobile, echogenic structure moving towards the tricuspid valve in diastole and towards the posterior RA wall in systole.

References

1. Geva T. Abnormal systemic venous connections. In: Allen HD, Driscoll DJ, Shaddy RE, Feltes TF, editors. Moss and Adams' heart disease in infants, children, and adolescents. 8th ed. Philadelphia, PA: Lippincott Williams & Wilkins; 2013. p. 840–63.
2. Van Praagh S, Santini F, Sanders SP. Cardiac malpositions with special emphasis on visceral heterotaxy (asplenia and polysplenia syndromes). In: Nadas AS, Fyler DC, editors. Nadas' pediatric cardiology. 4th ed. Philadelphia, PA: Hanley & Belfus; 1992. p. 589–608.
3. Meadows WR, Sharp JT. Persistent left superior vena cava draining into the left atrium without arterial oxygen unsaturation. Am J Cardiol. 1965;16:273–9.
4. DeLeval MR, Ritter DG, McGoon DC, et al. Anomalous systemic venous connection—surgical considerations. Mayo Clinic Proc. 1975;50:599–610.
5. Nash EN, Moore GW, Hutchins GM. Pathogenesis of persistent left superior vena cava with coronary sinus connection. Pediatr Pathol. 1991;11:261–9.
6. Van Praagh S, O'Sullivan J, Brili S, et al. Juxtaposition of the morphologically right atrial appendage in solitus and inversus atria: a study of 35 postmortem cases. Am Heart J. 1996;132:382–90.
7. Bernstein HS, Moore P, Stanger P, et al. The levoatrialcardinal vein: morphologic and echocardiographic identification of the pulmonary- systemic connection. J Am Coll Cardiol. 1995;26:995–1001.
8. Davis WH, Jordaan FR, Snyman HW. Persistent left superior vena cava draining into the left atrium as an isolated anomaly. Am Heart J. 1959;57:616–22.
9. Lenox CC, Zuberbuhler JR, Park SC, et al. Absent right superior vena cava with persistent left superior vena cava: implications and management. Am J Cardiol. 1980;45:117–22.
10. Bartram U, Van Praagh S, Levine JC, et al. Absent right superior vena cava in visceroatrial situs solitus. Am J Cardiol. 1997;80:175–83.
11. Van Praagh S, Geva T, Lock JE, et al. Biatrial or left atrial drainage of the right superior vena cava: anatomic, morphogenetic, and surgical considerations report of three new cases and literature review. Pediatr Cardiol. 2003;24:350–63.
12. Choi JY, Jung MJ, Kim YH, et al. Anomalous subaortic position of the brachiocephalic vein (innominate vein): an echocardiographic study. Br Heart J. 1990;64:385–7.
13. Gerlis LM, Ho SY. Anomalous subaortic position of the brachiocephalic (innominate) vein: a review of published reports and report of three new cases. Br Heart J. 1989;61:540–5.
14. O'Leary PW, Hagler DJ. Cardiac malpositions and abnormalities of atrial and visceral situs. In: Allen HD, Driscoll DJ, Shaddy RE, Feltes TF, editors. Moss and Adams' heart disease in infants, children, and adolescents. 8th ed. Philadelphia, PA: Lippincott Williams & Wilkins; 2013. p. 1195–216.
15. Guiraundon GM, Guiraundon CM, Klein GJ, et al. The coronary sinus diverticulum: a pathologic entity associated with the Wolff–Parkinson–white syndrome. Am J Cardiol. 1988;62:733–5.
16. Lucas RV Jr, Krabill KA. Abnormal systemic venous connections. In: Emmanouilides GC, Reimenschneider TA, Allen HD, editors. Heart disease in infants, children, and adolescents. 5th ed. Baltimore, MD: Williams & Wilkins; 1995. p. 874–902.

14 Coronary Artery Anomalies

Coronary artery anomalies (CAA) comprise a heterogeneous group of anatomical variants with diverse clinical implications. The abnormalities may involve the origin, course, or termination of the epicardial coronary arteries. The prevalence of coronary artery anomalies in the general population is around 1% [1]. Those anomalies involving the origins or proximal branches of the right or left coronary arteries may be diagnosed by a careful echocardiographic examination from the PSAX view at the coronary arterial level. Coronary arteries in the AV groove may also be demonstrated from parasternal and apical windows.

Echocardiographic Imaging of Coronary Arteries [Discussed in Detail in Chap. 3]

- High-frequency transducers (8, 10, or 12 MHz) provide better grey-scale resolution for the 2D imaging of coronary arteries.
- Lower-frequency transducers have better sensitivity for colour Doppler imaging of coronary arteries.
- The visualization of coronary arteries can be improved by optimizing the frame rate, especially in children. This can be done by reducing the depth and sector size.
- For colour Doppler imaging of coronary arteries, an ideal strategy would be to start at a Nyquist limit of 30 cm/s and gradually lower it until colour flow is detected in the coronary arteries [2].

14.1 Coronary Artery Anomalies of Origin

14.1.1 Anomalous Origin of Coronary Artery from Aorta

In this group of anomalies, one or more of the coronary arteries have an anomalous origin from the aorta. The various anomalies encountered in this category include [3–5]:

- Anomalous origin of LCA branches from right sinus of Valsalva
- Anomalous origin of RCA branches from left sinus of Valsalva
- LCA and/or RCA branches arising from the posterior sinus of Valsalva [rare]
- Single coronary artery
  - Can arise from either the right or the left sinus of Valsalva
  - Associated with TGA, TOF, truncus arteriosus, bicuspid aortic valve, and coronary-cameral fistulae [1]
- Absent LCA trunk with separate origin of LAD and LCx
- High origin of RCA and/or LCA from ascending aorta

A. Paul, *Manual of Echocardiography for Congenital Heart Diseases*,
https://doi.org/10.1007/978-981-96-0243-8_14

14.1.2 Anomalous Origin of Coronary Artery from Pulmonary Artery

14.1.2.1 Anomalous Left Coronary Artery from the Pulmonary Artery (ALCAPA)

The incidence of ALCAPA is about 1 in every 3,00,000 live births. This constitutes about 0.3% of all congenital heart disease. In this anomaly, the left coronary artery arises from the pulmonary artery, usually from the left posterior-facing sinus [6].

ALCAPA is pathophysiologically characterized by [7]:

- Perfusion of the LV with desaturated blood at low pressures
- Infarction of the anterolateral LV free wall, with resultant LV dysfunction
- Mitral regurgitation resulting from infarction and dysfunction of the anterolateral papillary muscle, or dilated mitral annulus

In a bid to improve the blood supply to the LCA territory, collateral vessels between the normal RCA and the abnormal LCA enlarge. The RCA also becomes dilated in this process. However, the ***collateral flow is directed into the low-pressure PA*** rather than into the high-resistance myocardial vasculature, resulting in pulmonary–coronary steal with a left-to-right shunt [7]. In patients with large post-tricuspid shunts, elevated pulmonary artery pressure improves myocardial perfusion by reversing the coronary steal. Closure of the shunt lesion in such cases would result in severe myocardial ischaemia and LV dysfunction [1].

Echocardiography

The echocardiographic clues to the diagnosis of ALCAPA are [8]:

1. ***Absence of aortic origin of the left coronary artery***
   (a) This needs to be confirmed from multiple (at least three) views.
   (b) ***A fluid-filled pericardial transverse sinus can mimic the origin of the LCA from the aorta***. However, the direction of flow towards the aortic side differentiates this from a coronary artery.
2. ***Abnormal origin of the left coronary artery from the PA***
   (a) The abnormal origin of the LCA from the PA may be visualized from the PSAX view (Fig. 14.1) or a high parasternal position in the parasagittal long-axis view of the MPA.
   (b) The LCA may falsely appear to connect to the aortic sinus due to poor lateral resolution.
3. ***Abnormal diastolic flow into the pulmonary artery***
   (a) Colour Doppler interrogation demonstrates ***diastolic flow from the left coronary artery into the pulmonary artery*** (the direction of flow may reverse in the presence of pulmonary hypertension).
4. ***Abnormal dilation of the right coronary artery***
   (a) RCA is dilated (often including the RPDA) owing to the excessive blood flow (Fig. 14.2). This serves as an indirect clue to the possibility of ALCAPA.

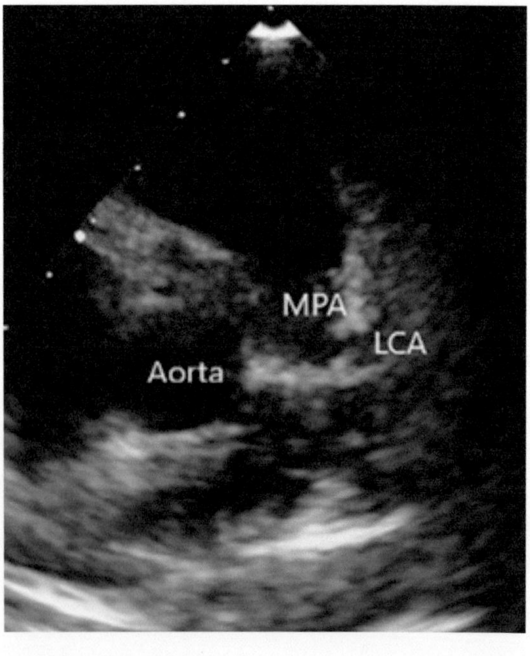

Fig. 14.1 Modified PSAX view showing the origin of the LCA from the MPA in a case of ALCAPA

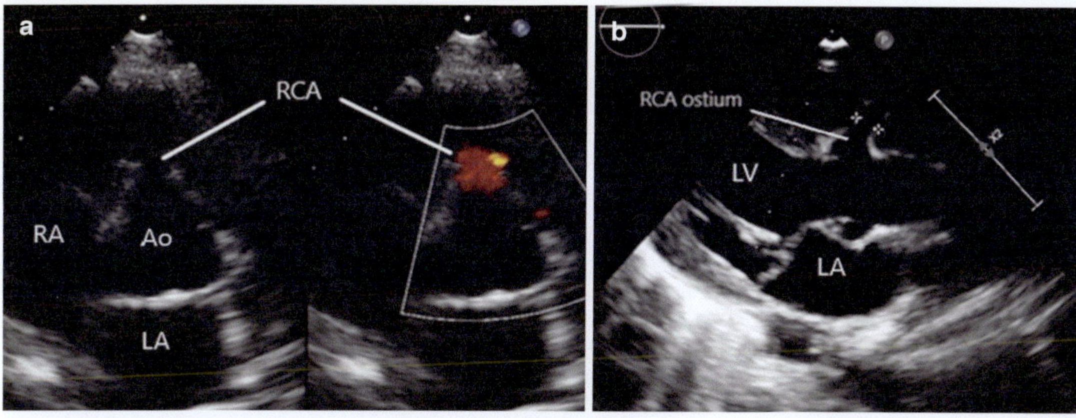

Fig. 14.2 Abnormally dilated RCA (owing to the excessive blood flow) in ALCAPA, visualized from PSAX (**a**) and PLAX (**b**) views

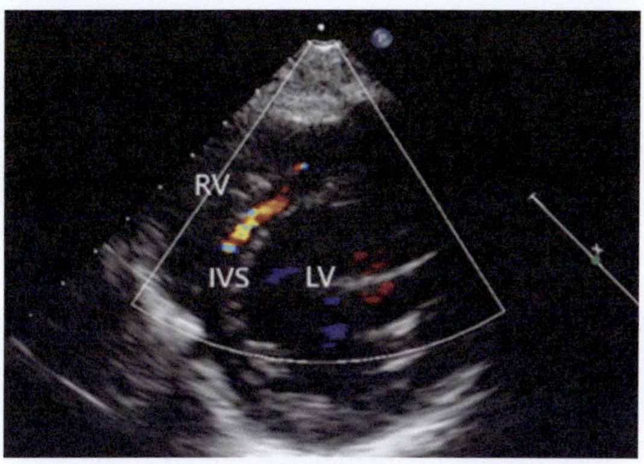

Fig. 14.3 PSAX view depicting turbulent collateral flow in the septal myocardial sinusoids in ALCAPA

5. ***Abnormal collateral blood flow***
   (a) Demonstration of abnormal collateral flow, especially in the septal myocardial sinusoids, may be the finding that raises the suspicion of ALCAPA in some cases (Fig. 14.3).
6. ***Complications of abnormal myocardial perfusion***
   (a) Atrophic and ***scarred anterolateral papillary muscle*** appears bright (increased echogenicity) on 2D examination.
   (b) Mitral regurgitation results from dysfunctional papillary muscle or dilated annulus.
   (c) Thinning and scarring of the anterolateral LV wall and apex, with or without mural thrombi.
   (d) LV systolic dysfunction resulting from regional wall motion abnormalities.
   (e) ***Endocardial fibroelastosis*** of a dilated and hypertrophied LV in infancy.

Key point: Atresia of the left coronary artery has echocardiographic features similar to ALCAPA, except that: (a) there is no shunt between the LCA and PA and (b) there is no reverse flow in the LCA.

14.1.2.2 Anomalous Right Coronary Artery from the Pulmonary Artery [ARCAPA]

- Rare as compared to ALCAPA
- Echocardiography with Doppler examination demonstrates:
  - Abnormal attachment of the RCA to the pulmonary trunk
  - Retrograde flow from the **RCA to the pulmonary artery**

14.1.2.3 Less Common Anomalies [1, 5]

- Origin of both the RCA and LCA, or a single coronary artery, from the pulmonary trunk
- Origin of the LAD or LCx from the main pulmonary artery or its branches

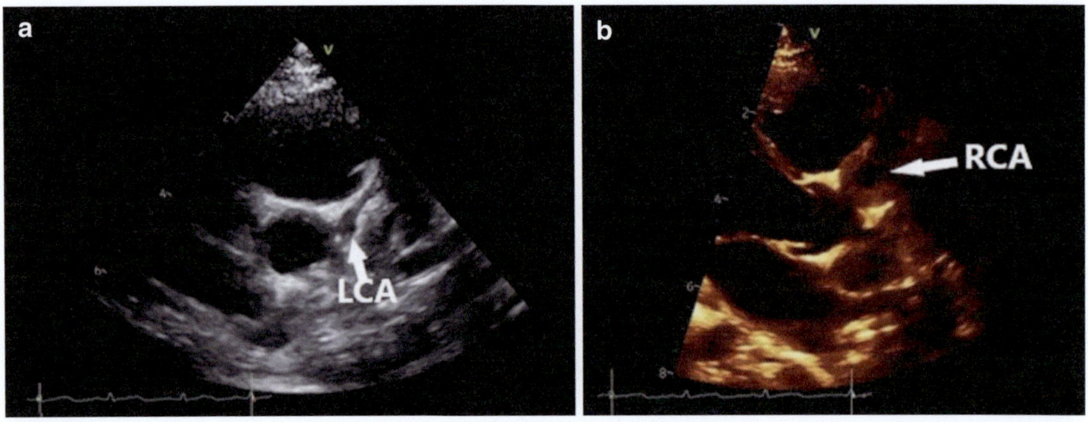

Fig. 14.4 (**a**) Proximal left coronary artery (LCA) aneurysm visualized from PSAX view at the level of coronary arteries. (**b**) Proximal right coronary artery (RCA) ectasia with thrombus seen within, visualized from PLAX view.

14.2 Anomalies of Intrinsic Coronary Arterial Anatomy

The various anomalies included in this category are:

- Coronary ostial stenosis or atresia [LCA, RCA, LAD, Circumflex]
- Absent coronary artery or coronary hypoplasia
- Coronary artery ectasia or aneurysm (Fig. 14.4)
- Abnormal origins of coronary artery branches [Split LAD, Split RCA]

14.3 Coronary Artery Anomalies of Termination

14.3.1 Coronary Artery Fistula

- A coronary artery fistula is an anomalous direct communication of a coronary artery with another blood vessel (called **coronary AV malformations**), or more commonly with a cardiac chamber (called **coronary-cameral fistula**).
- Based on the nature of the communication with the cardiac chamber, coronary-cameral fistulae are classified as arterioluminal (direct communication) and arteriosinusoidal (communication via sinusoidal network).
- The involvement of various coronary arteries in the order of frequency is as follows: LCA (40–60%), RCA (30–50%), both (10%) [9, 10].
- The most common sites of drainage are the RV (40%), RA (25%), and PA (15%). Drainage into the LA, LV, pulmonary vein, and SVC is rare [10].
- Significant shunt flow may result in volume overload of the recipient chamber.

Echocardiography

- The proximal part of the involved coronary artery appears **dilated or tortuous**.
- Colour Doppler examination demonstrates the shunt flow.
- **Continuous high-velocity flow** is seen when the coronary artery fistula drains into a low-pressure chamber, such as the RA, RV, or CS (Fig. 14.5).
- Shunt flow is limited to diastole when the fistula drains into a high-pressure chamber like the LV.
- The recipient cardiac chamber demonstrates features of volume overload in significant shunts.
- Differentiation of a coronary cameral fistula from RSOV is discussed in Chap. 4.

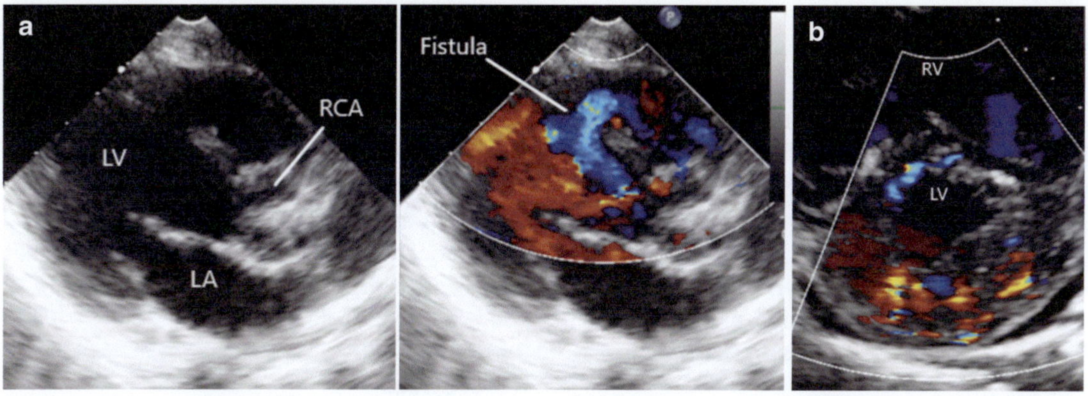

Fig. 14.5 (**a**) PLAX view demonstrating coronary AV fistula connecting RCA and LV; note the dilated LV. (**b**) Note the abnormal intramyocardial flow in the LV in a case of coronary AV fistulae.

14.4 Coronary Artery Anomalies in Complex Congenital Heart Defects

14.4.1 Tetralogy of Fallot

The prevalence of coronary artery anomalies in TOF is around 5%. About 70% of the anomalous coronary arteries in TOF cross the RVOT. The risk of encountering an anomalous vessel across the RVOT in TOF is 10%. The most common anomalous vessels crossing the RVOT in TOF are [1, 11]:

- LAD or accessory LAD arising from the RCA or the right aortic sinus
- Large conus artery [6% of TOF], particularly in those with significant RVH
- One of the major branches of a single coronary artery arising from either the right or the left sinus

The other coronary artery anomalies described in TOF are coronary artery fistulae, anomalous coronary artery origin from the pulmonary artery or its branches, and hypoplasia of the coronary tree [11]. Many of these anomalies can be detected by echocardiography. Aortic root or coronary angiography is necessary in uncertain cases.

14.4.2 Truncus Arteriosus [12]

- The coronary artery origins are usually normal.
- The left main coronary artery may arise from the posterior sinus.
- A single coronary artery is also described.

14.4.3 Complete Transposition of the Great Arteries

In d-TGA, coronary artery anomalies are more likely in the presence of a VSD or side-by-side alignment of the great vessels [1].

- Coronary arteries arise from the facing sinuses in most of the patients.
- The origin of the coronary arteries may be [13, 14]:
  - From the appropriate sinuses in 60%.
  - Completely inverted: RCA arising from the left anterior sinus, and LCA arising from the right posterior sinus.
  - Partially inverted: Left circumflex arising from the right posterior sinus, and LAD arising with the RCA from the left anterior sinus.

- The left circumflex may arise from the RCA and course posterior to the PA.
- A single coronary artery, an intramural course of one or more of the arteries, and branches passing between the great vessels are also seen.

14.4.4 Congenitally Corrected Transposition of the Great Arteries (L-Transposition)

- The coronary arteries arise from the facing sinuses (similar to d-TGA).
- The anterior sinus is usually the non-coronary sinus.
- The coronary artery is designated as left or right based on the territory of distribution, rather than the sinus of origin [15]:
  - LCA arises from the right-facing sinus and supplies the LV.
  - RCA arises from the left-facing sinus and supplies the systemic RV.
- A single coronary artery may arise from the right-facing sinus.

14.4.5 Double-Outlet Right Ventricle [16]

- The prevalence of CAAs in DORV is 50%.
- The origins of coronary arteries are usually normal.
- In cases of associated TGA, RCA arises from the right-facing sinus.
- An anomalous LAD may arise from the RCA and cross the RVOT.
- A single coronary artery is seen in 15%.

14.4.6 Double-Inlet Left Ventricle

When the outlet chamber is anterior and to the right [1, 17]:

- The aorta and pulmonary artery are related as in D-TGA.
- RCA arises from the right-facing aortic sinus.
- LMCA arises from the left-facing sinus.

When the outlet chamber is anterior and to the left [1, 17]:

- The aorta and pulmonary artery are related as in L-TGA.
- RCA arises from the right-facing aortic sinus.
- LMCA arises from the left-facing sinus.
- One or more large diagonal arterial branches may cross the RVOT.

References

1. Lim DS, Mathern GP. Congenital anomalies of the coronary vessels and the aortic root. In: Allen HD, Driscoll DJ, Shaddy RE, Feltes TF, editors. Moss and Adams' heart disease in infants, children, and adolescents. 8th ed. Philadelphia, PA: Lippincott Williams & Wilkins; 2013. p. 746–57.
2. Brown LM, Duffy CE, Mitchell C, Young L. A practical guide to pediatric coronary artery imaging with echocardiography. J Am Soc Echocardiogr. 2015;28:379–91.
3. Alexander RW, Griffi th GC. Anomalies of the coronary arteries and their clinical significance. Circulation. 1956;14:800–5.
4. Ogden JA. Congenital anomalies of the coronary arteries. Am J Cardiol. 1970;25:474–9.
5. Roberts WC. Major anomalies of coronary arterial origin seen in adulthood. Am Heart J. 1986;111:941–63.
6. Bland EF, White PD, Garland J. Congenital anomalies of the coronary arteries: report of an unusual case associated with cardiac hypertrophy. Am Heart J. 1933;8:787–801.
7. Neufeld HN, Schneeweiss A. Coronary artery disease in infants and children. Philadelphia, PA: Lea & Febiger; 1983.
8. King DH, Danford DA, Huhta JC, et al. Noninvasive detection of anomalous origin of the left main coronary artery from the pulmonary trunk by pulsed Doppler echocardiography. Am J Cardiol. 1985;55:608–9.
9. Fernandes ED, Kadivar H, Hallman GL, Reul GJ, Ott DA, Cooley DA. Congenital malformations of the coronary arteries: the Texas heart institute experience. Ann Thorac Surg. 1992;54(4):732–40.
10. Challoumas D, Pericleous A, Dimitrakaki IA, Danelatos C, Dimitrakakis G. Coronary arteriovenous fistulae: a review. Int J Angiol. 2014;23(1):1–10.
11. Fellows KE, Freed MD, Keane JF, et al. Results of routine preoperative coronary angiography in tetralogy of Fallot. Circulation. 1975;51:561–6.
12. Anderson KR, McGoon DC, Lie JT. Surgical signifi cance of the coronary arterial anatomy in truncus arteriosus communis. Am J Cardiol. 1978;41:76–81.
13. Wernovsky G, Sanders SP. Coronary artery anatomy and transposition of the great arteries. Coron Artery Dis. 1993;4:148–57.

14. Elliott LP, Amplatz K, Edwards JE. Coronary arterial patterns in transposition complexes. Anatomic and angiocardiographic studies. Am J Cardiol. 1966;17:362–78.
15. Kirklin JW, Barratt-Boyes BG. Congenitally corrected transposition of the great arteries. In: Barratt-Boyes B, Kirklin JW, editors. Cardiac surgery. New York, NY: John Wiley and Sons; 1993.
16. Gomes MM, Weidman WH, McGoon DC, et al. Double-outlet right ventricle without pulmonic stenosis. Surgical considerations and results of operation. Circulation. 1971;43:I31–136.
17. Lev M, Liberthson RR, Kirkpatrick JR, et al. Single (primitive) ventricle. Circulation. 1969;39:577–91.

Congenital Cardiomyopathies, Pulmonary Hypertension, and Other Anomalies

15

15.1 Congenital Cardiomyopathies

The genetic cardiomyopathies included in this section are:

- Hypertrophic cardiomyopathy (HCM).
- Arrhythmogenic right ventricular cardiomyopathy/dysplasia (ARVC/D).
- Left ventricular non-compaction (LVNC).

15.1.1 Hypertrophic Cardiomyopathy (HCM)

- Defined by the presence of LVH with a maximum LV wall thickness of [1]:
  - ≥15 mm in adults.
  - ≥13 mm in adults from families with a pathogenic sarcomere variant.
  - >2.5 standard deviations from the predicted mean in children [z-score > 2.5].
  - >2 standard deviations from the predicted mean in those children with a positive family history of HCM or a positive genetic test.
- Obstructive HCM (**HOCM**): Nearly one-third of HCM have detectable intracavitary obstruction at rest, and another one-third demonstrates obstruction with provocation.
- The aetiopathogenesis of HCM varies based on the age of the subject (Table 15.1).

15.1.1.1 Echocardiographic Diagnosis: Left Ventricular Hypertrophy (Fig. 15.1)

- LVH in HCM may involve one or more regions of the LV myocardium. The hypertrophy may be symmetric or asymmetric. The RV wall may also be involved.
- Asymmetrical septal hypertrophy is the most frequently encountered variant, characterized by [1]:
  - Septal thickness of ≥15 mm in any myocardial segment (see above) or.
  - Septal to posterior free wall thickness ratio (IVS/PW ratio) >1.3 (in normotensives) or > 1.5 (in hypertensives).
- Various patterns of LVH observed in HCM are [2]:
  - Sigmoidal [40–50%]: Basal septal protuberance with concave septum.
  - Reverse curve [30–40%]: Convex septum and concentric LV cavity.
  - Apical HCM [10%]: Apical hypertrophy [Ace of Spades].
  - Neutral HCM [10%]: Straight septum.

Supplementary InformationThe online version contains supplementary material available at https://doi.org/10.1007/978-981-96-0243-8_15.

A. Paul, *Manual of Echocardiography for Congenital Heart Diseases*,
https://doi.org/10.1007/978-981-96-0243-8_15

Table 15.1 Common causes of HCM in children

Neonates and infants [Age < 1 year]	Children [1–18 years]
Hyperinsulinism – Maternal diabetes mellitus Primary hyperinsulinism Insulin-resistance syndromes Metabolic disorders – Glycogen storage disorders [Pompe disease, Danon disease, Cori-Forbes disease, PRKAG2 syndrome] – Lysosomal storage diseases [Mucopolysaccharidoses] – Fatty acid oxidation disorders RAS-MAPK disease [malformation syndromes] – Noonan syndrome – Costello syndrome – Cardio-facio-cutaneous syndrome – LEOPARD Mitochondrial disorders Sarcomeric HCM and idiopathic HCM [<25%]	Sarcomeric HCM [60%] Idiopathic HCM [25%] RAS-MAPK disease [malformation syndromes] – Cornelia de Lange Neuromuscular syndromes – Friedrich ataxia Glycogen storage disorders [Pompe disease, Danon disease, Cori-Forbes disease, PRKAG2 syndrome] Mitochondrial disorders

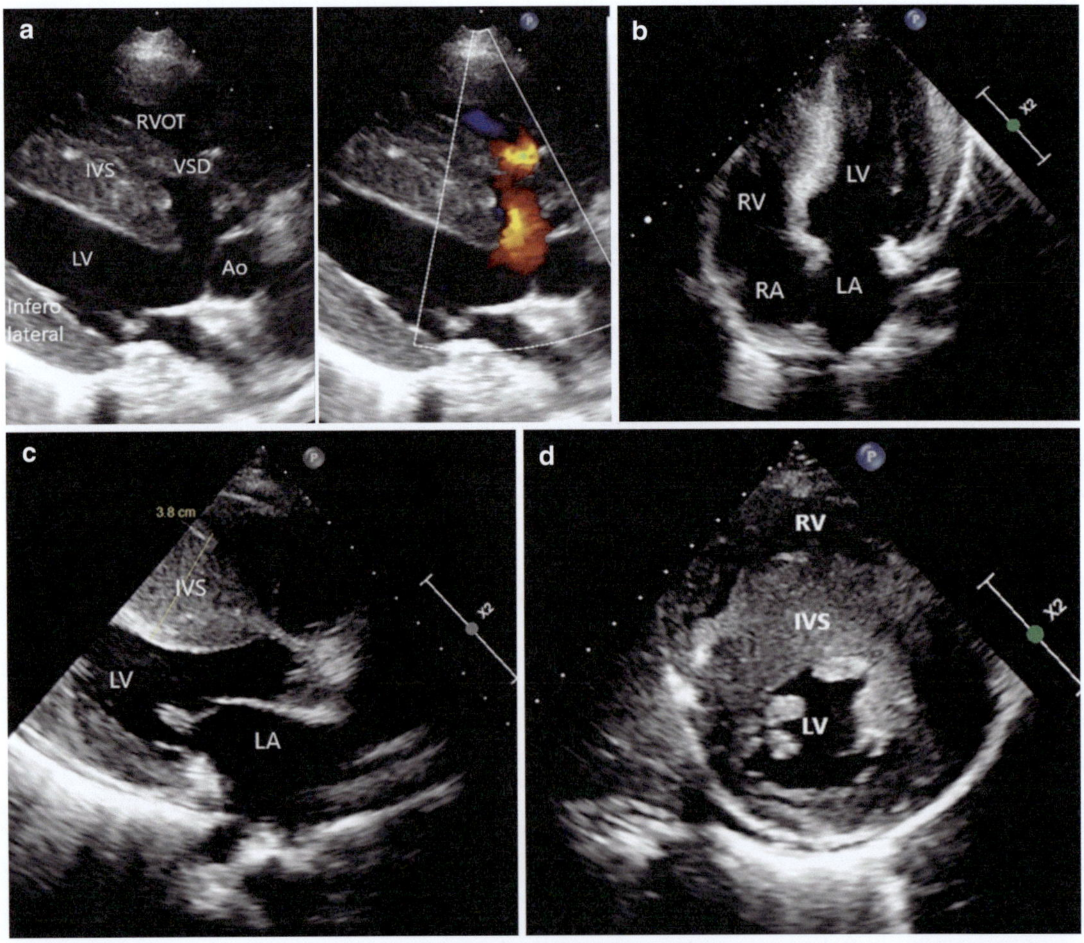

Fig. 15.1 (**a**) PLAX view depicting asymmetric septal hypertrophy (ASH) and VSD in a case of neonatal HCM; (**b**) A4C view of adult HCM; (**c**) PLAX view of adult HCM showing ASH; (**d**) PSAX view of ASH in adult HCM

15.1.1.2 Systolic Anterior Motion (SAM) of Anterior Mitral Leaflet (Fig. 15.2)

- Anterior motion of the AML during mid-late systole may result in LVOT obstruction.
- The severity of SAM, graded on the basis of M-mode imaging, has been found to correlate with the peak LVOT pressure gradient [3]:
  - **Grade I**: No AML-septal contact; minimal systolic distance between AML and septum >10 mm.
  - **Grade II**: No AML-septal contact; minimal systolic distance between AML and septum <10 mm.
  - **Grade III**: Brief mitral leaflet-septal contact [<30% of systolic time].
  - **Grade IV**: Prolonged mitral leaflet-septal contact [>30% of systolic time].

15.1.1.3 LVOT Obstruction

- Severe LVH and significant SAM can contribute to dynamic obstruction at the level of LVOT, mid-cavity, and/or apex.
- Peak LVOT pressure gradient is estimated using CWD [discussed in Chap. 9].
- CWD-derived peak gradient of >50 mmHg and/or mean gradient >30 mmHg are considered significant.
- The exact location of the obstruction should be confirmed using colour Doppler and PWD.

15.1.1.4 Mitral Regurgitation

- MR in HOCM is often multifactorial.
- SAM-related MR is characterized by a predominantly posteriorly directed jet.
- Intrinsic abnormalities of the mitral valve apparatus, such as redundant mitral leaflets, flail PML, and abnormal papillary muscles, are common [1, 2].
- Rupture of chordae tendineae results in severe MR and reduction of SAM/LVOTO.

15.1.1.5 LV Function

- Diastolic dysfunction occurs early in HCM, while systolic dysfunction is often a feature of end-stage disease [1].

Key point: Normal anterior excursion of the AML towards the LVOT during diastole should not be mistaken for SAM. M-mode echocardiography facilitates confirmation of temporality.

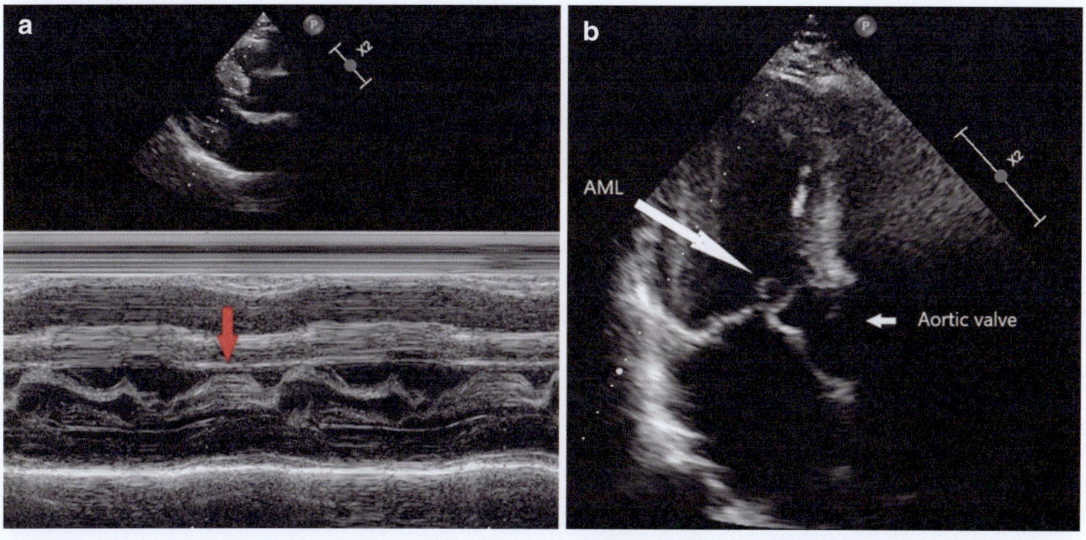

Fig. 15.2 (**a**) M-mode echocardiogram in HOCM demonstrating systolic anterior motion of AML [red arrow], resulting in transient contact between the AML and the interventricular septum; (**b**) A3C view showing AML obstructing the LVOT during systole

15.1.2 Arrhythmogenic Right Ventricular Cardiomyopathy/Dysplasia (ARVC/D)

ARVC/D is pathologically characterized by progressive fibrofatty infiltration of the RV myocardium, although it may also involve the LV. The phenotypic variants are:

- Classic Arrhythmogenic Right Ventricular Cardiomyopathy (dominant right) phenotype.
- Arrhythmogenic biventricular cardiomyopathy phenotype.
- Arrhythmogenic Left Ventricular Cardiomyopathy (dominant left) phenotype.

Anatomic abnormalities of classic ARVC include varying degrees of global RV dilation, aneurysms, and segmental hypokinesia. The classic sites of involvement (triangle of dysplasia) are the RV infundibulum, RV apex, and the inferoposterior wall of the RV inflow (under the tricuspid valve).

15.1.2.1 Echocardiographic Diagnosis

Major criteria [4]:

- Regional RV akinesia, dyskinesia, or aneurysm.
  AND
- One of the following:
  - PLAX RVOT dimension (end-diastole): ≥ 32 mm (19 mm/mm^2 BSA).
  - PSAX RVOT dimension (end-diastole): ≥ 36 mm (21 mm/mm^2 BSA).
  - RV fractional area change of ≤33%.

  Minor criteria [4]:

- Regional RV akinesia/dyskinesia.
  AND
- One of the following parameters:
  - PLAX RVOT dimension (end-diastole): ≥ 29 to <32 mm (≥ 16 to <19 mm/mm^2 BSA).
  - PSAX RVOT dimension (end-diastole): ≥ 32 to <36 mm (≥ 18 to <21 mm/mm^2 BSA).
  - RV fractional area change of ≥33% to <40%.

15.1.2.2 Other Morphologic Abnormalities (Fig. 15.3) [4]

- Derangement/complete loss of trabeculations.
- Hyper-reflectiveness/hypertrophy of moderator band.
- Sacculations of RV free wall.
- RV aneurysm with or without associated thrombus.
- Regional wall motion abnormality (RWMA) involving the apex and/or anterior wall.

15.1.2.3 RV Function Assessment

ARVC is characterized by progressive RV dysfunction and failure, followed by LV dysfunction and biventricular failure. The various indices used in 2D echocardiography for the assessment of RV function are:

(a) **Fractional area change (FAC)**:
   - FAC is a two-dimensional measure of RV global systolic function.
   - RV area is estimated in end-systole and end-diastole by tracing the endocardial border from an RV-focused A4C view.

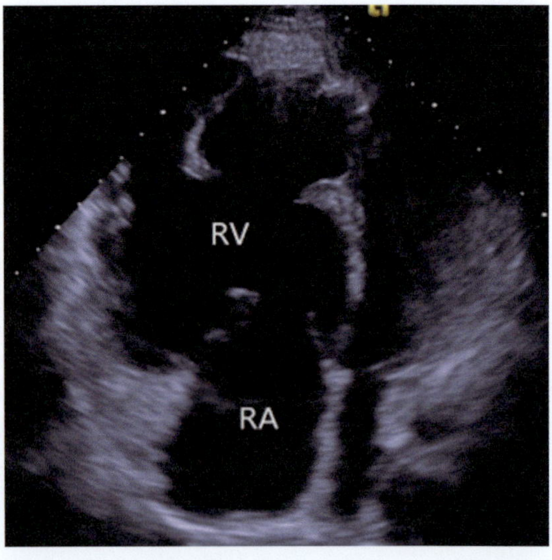

Fig. 15.3 Apical view of ARVD showing dilated RV with trabecular derangement, hyperreflective moderator band, and free wall sacculations

- FAC is calculated as the difference between end-diastolic area and end-systolic area, divided by the end-diastolic area, and expressed as a percentage.
- Fractional area change **>35%** is considered normal in adults.

(b) **Tricuspid annular plane systolic excursion (TAPSE)** (Fig. 15.4a).

- TAPSE is a measure of longitudinal RV systolic function [5].
- M-mode image is obtained through the lateral tricuspid annulus from the A4C view.
- The excursion of the tricuspid valve annulus from the base of the heart towards the apex during systole is measured.
- In adults, TAPSE **≥17 mm** is considered normal.
- Normal values range from 0.9 to 2.5 mm in neonates and children.

(c) **Tissue Doppler-derived systolic velocity of the tricuspid annulus (TDI S')** (Fig. 15.4b).

- TDI S′ is another index of RV systolic function.
- Tissue Doppler velocity of the lateral tricuspid annulus is obtained from the A4C view.
- The S′ wave represents the systolic velocity, which reflects the RV longitudinal function.
- An S′ value **≥9.5 cm/s** is considered normal [6].

(d) **RV Myocardial Performance Index (RV MPI) or Tei Index** (Fig. 15.4c).

- RV MPI is a global index of RV systolic and diastolic function.

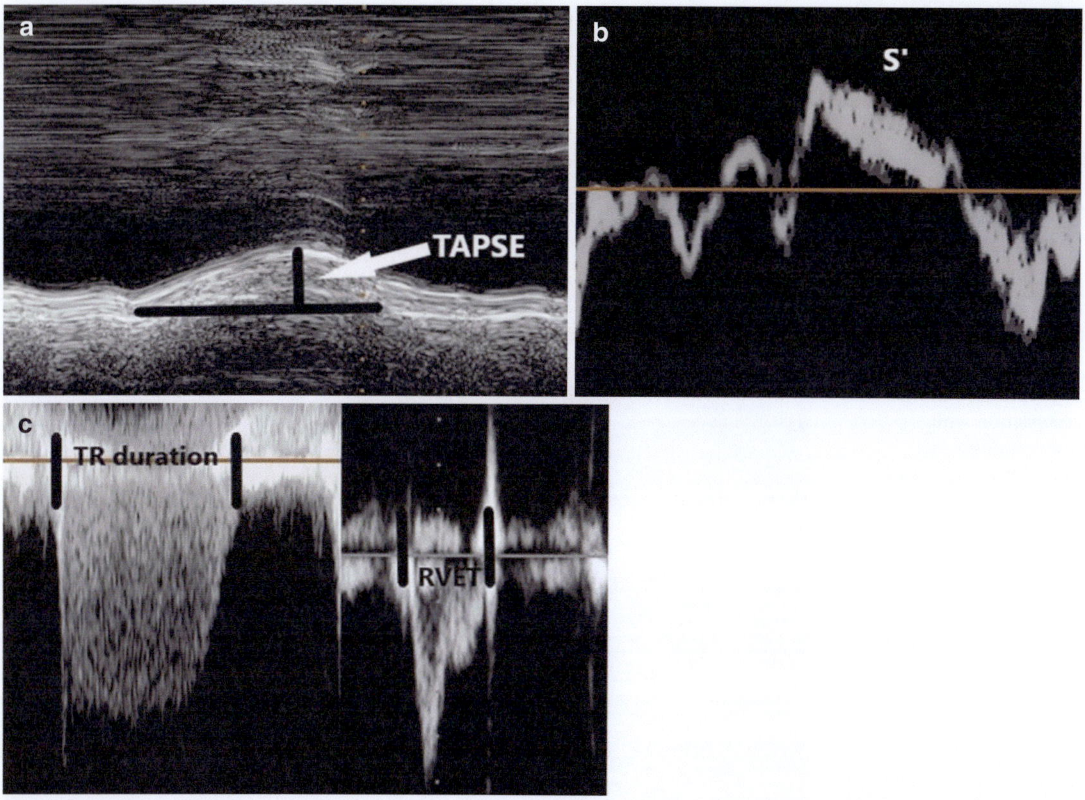

Fig. 15.4 Indices of RV function. (**a**) M-mode imaging of lateral tricuspid annulus from A4C view. The vertical displacement of the annulus [denoted by arrow] corresponds to the longitudinal excursion of tricuspid annulus during RV systole [TAPSE]. (**b**) Tissue Doppler imaging [TDI] of lateral tricuspid annulus from A4C view. The peak systolic velocity of lateral tricuspid annulus [S′] is measured. (**c**) MPI can be calculated from TR duration and RV ejection time [RVET] as explained in the text

- MPI is the ratio of isovolumic (relaxation + contraction) time to ejection time.
  - MPI = [Isovolumic Relaxation Time + Isovolumic Contraction Time]/ Ejection time.
  - TR duration = Isovolumic Relaxation Time + Isovolumic Contraction Time + RV Ejection time.
  - Hence, Isovolumic time can be calculated as: TR duration—RV ejection time.
  - Therefore, **MPI = [TR duration—RV ejection time]/RV ejection time.**
- TR duration is obtained from CWD tracing at the tricuspid valve, and RV ejection time is obtained from PWD tracing of RVOT.
- Normal value is **<0.43** [when measured using PWD] and **< 0.55** [when estimated using TDI] [7].

15.1.3 Left Ventricular Non-Compaction (LVNC)

LV non-compaction results from the arrest of foetal development of the heart. This leads to a bi-layered myocardial architecture, which is composed of [8]:

- Thin, compacted epicardial layer.
- Thick, non-compacted endocardial layer, which is comprised of:
  - Prominent trabeculations.
  - Deep intertrabecular recesses that directly communicate with the LV cavity.

15.1.3.1 Echocardiographic Diagnosis

LVNC is characterized by segmental thickening of the LV myocardium, with prominent trabeculations and deep intertrabecular recesses, in the absence of any other major cardiac abnormality (Fig. 15.5) [8]. The most frequently involved segments are the apical and mid-ventricular (inferior and lateral) segments.

Benign LV trabeculations can be distinguished from pathological LVNC based on the echocardiographic criteria for the diagnosis of LVNC:

- **Jenni criteria** [8]:
  - Thicknesses of compacted epicardial myocardium (C) and non-compacted endocardial myocardium (N) are measured from the PSAX view in end-systole.
  - LVNC is defined by a ratio of N/C > 2.
  - Absence of coexisting cardiac abnormalities, presence of multiple prominent trabeculations in end-systole, and communication of the trabecular recesses

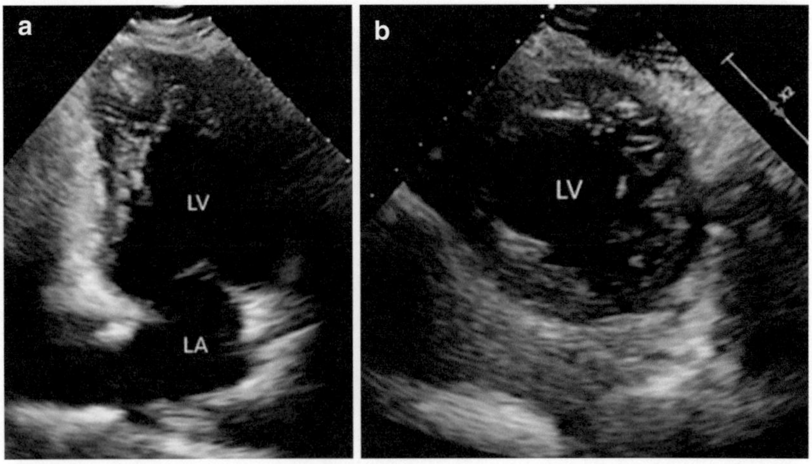

Fig. 15.5 LV non-compaction with trabeculations and intertrabecular recesses in the myocardium—A2C view (**a**), PSAX view (**b**)

with the ventricular cavity are the other components of the diagnostic criteria.

- **Chin index** [9]:
  - LVNC is defined by a ratio of $X/Y \leq 0.5$. [X = Distance from the epicardial surface to the trough of the trabecular recess. Y = Distance from the epicardial surface to the peak of trabeculation].
  - Measurements are taken from the LV free wall in end-diastole.
  - LV apical trabeculations seen in PSAX and apical views are considered for measurement.
- **Sollberger criteria** [10]:
  - More than three trabeculations protruding from the LV wall (with the same echogenicity as the myocardium), apically to the papillary muscles, visible in a single imaging plane in end-diastole.
  - Perfusion of intertrabecular spaces from the LV cavity, visualized by colour Doppler.

The following points merit consideration while performing echocardiography for LVNC:

- Optimal image resolution and positioning of the focus in the apical region are important.
- When the acoustic windows are suboptimal, intravenous ultrasound contrast agents can be used for better depiction of trabecular spaces.
- Differential diagnostic possibilities to be considered include intramyocardial haematoma/abscess, thrombi, fibroma, false tendons, and apical HCM.

Key point: Demonstration of direct flow from the ventricular cavity into intertrabecular recesses using colour Doppler or echo contrast is helpful in differentiating LVNC from other apical echocardiographic abnormalities such as apical HCM and LV apical thrombus.

15.1.3.2 LV Systolic Function

LV systolic dysfunction usually ensues with time.

15.1.3.3 LV Thrombus

The intertrabecular recesses may act as nidi for thrombus formation and subsequent embolization.

15.1.3.4 Other Cardiac Anomalies

- Non-compacted myocardium may also be seen in Ebstein's anomaly, other cardiomyopathies, and cyanotic congenital heart diseases, although the diagnosis of LVNC cardiomyopathy can be made only in the absence of other significant cardiac anomalies.

15.2 Pulmonary Hypertension

The aetiopathogenesis of pulmonary hypertension (PH) may be localized to two principal anatomic domains:

- Pre-capillary PH: Caused by pulmonary vascular remodelling and increased PVR.
- Post-capillary PH: Caused by left-sided heart diseases and increased pulmonary venous pressure.

15.2.1 Echocardiographic Diagnosis

15.2.1.1 Elevated Pulmonary Artery Systolic Pressure (PASP)

'Pulmonary hypertension' implies that the PA pressures are elevated. The gold standard for confirming the diagnosis of pulmonary hypertension is documentation of pulmonary artery pressures by right heart catheterization. The probability of pulmonary hypertension can be estimated by various echocardiographic parameters. In the absence of RVOT obstruction at any level, PASP can be estimated from peak TR velocity using the modified Bernoulli equation as follows:

$$\text{PASP} = \text{RVSP} = 4\text{V}^2 + \text{RA pressure} \left(\text{where V} = \text{Peak systolic velocity of TR jet}\right)$$

RA pressure can be estimated from the diameter of the IVC and its respiratory collapse as follows [11]:

- **Low (3 mmHg)**:
  - IVC diameter < 21 mm + >50% collapse with sniff.
- **Intermediate (8 mmHg)**:
  - IVC diameter < 21 mm + <50% collapse with sniff.
  - IVC diameter > 21 mm + >50% collapse with sniff.
- **High (15 mmHg)**:
  - IVC diameter > 21 mm + <50% collapse with sniff.

Based on peak TR velocity and other echocardiographic signs (see below), the echocardiographic probability of pulmonary hypertension is determined as follows [11]:

- **Low**:
  - Peak TR velocity ≤ 2.8 m/s with no other echo signs of pulmonary hypertension.
- **Intermediate**:
  - Peak TR velocity ≤ 2.8 m/s with other signs of pulmonary hypertension.
  - Peak TR velocity 2.9–3.4 m/s with no other signs of pulmonary hypertension.
- **High**:
  - Peak TR velocity 2.9–3.4 m/s with other signs of pulmonary hypertension.
  - Peak TR velocity > 3.4 m/s with or without other signs of pulmonary hypertension.

Key point: Before making the diagnosis of pulmonary hypertension based on the TR gradient, RVOT obstruction needs to be ruled out. RVSP = PASP, if and only if there is no significant RVOT obstruction.

15.2.1.2 Other Signs of Pulmonary Hypertension

The other echocardiographic signs of pulmonary hypertension are [11]:

- **Ventricles** (Fig. 15.6)
  - Dilated RV, as evidenced by RV/LV basal diameter ratio > 1.0.
  - D-shaped septum: Flattening of the interventricular septum.
  - LV eccentricity index >1.1 (in end-systole and/or end-diastole) (Fig. 15.6).

    Measured as D2/D1 (where D2 is the length of the minor axis of the LV parallel to the septum and D1 is the length of the minor axis of the LV perpendicular to the septum].

    Measured from the PSAX view at the level of LV papillary muscles.
- **Pulmonary artery**
  - PWD tracing obtained from the RVOT has the following characteristics (Fig. 15.6):

    Acceleration time < 105 ms and/or.

    Mid-systolic notching.
  - Early diastolic PR velocity > 2.2 m/s.
  - PA diameter > 25 mm.
- **Inferior Vena Cava and Right atrium**
  - IVC diameter > 21 mm with decreased inspiratory collapse [<50% with sniff or < 20% with quiet inspiration].
  - Right atrial area [end-systole] > 18 cm^2.

15.2.1.3 Right Ventricular Size and Function

Evaluation of pulmonary hypertension is incomplete without assessing RV size and function, as pulmonary hypertension can eventually result in diastolic/systolic RV dysfunction ['Assessment of RV dysfunction' is discussed in the section on ARVC/D].

15.2.2 Pre-Capillary Pulmonary Hypertension (Pulmonary Arterial Hypertension)

The aetiological possibilities for pulmonary arterial hypertension (PAH) can be approached as follows [11, 12]:

1. **In the absence of congenital heart disease**
   - Persistent pulmonary hypertension of the newborn.
   - Developmental lung diseases (e.g. bronchopulmonary dysplasia).
   - Primary [idiopathic] pulmonary hypertension.

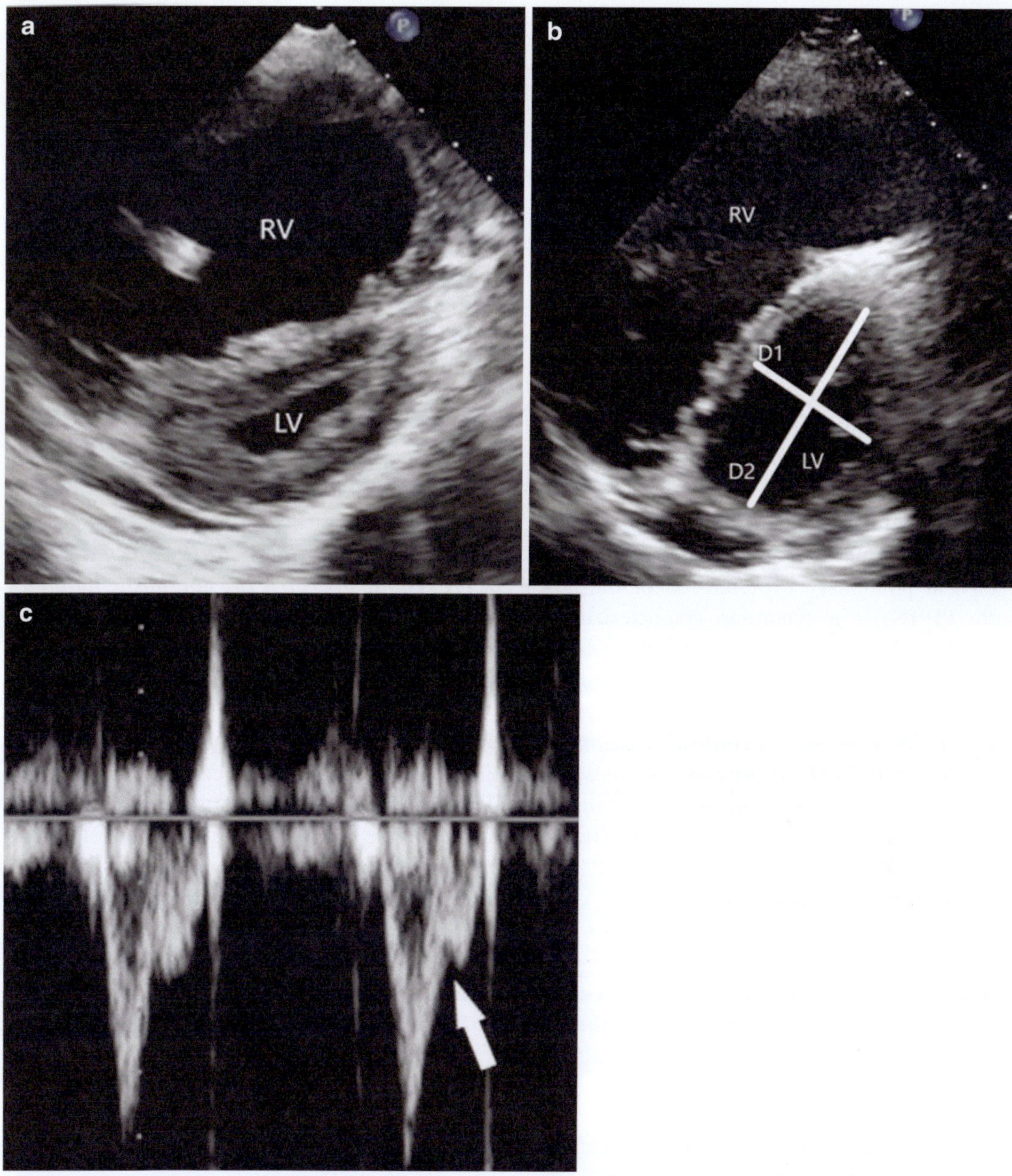

Fig. 15.6 Pulmonary hypertension. (**a**) PSAX view showing dilated RV and flattened interventricular septum. (**b**) Calculation of LV eccentricity index from PSAX view at the papillary muscle level. (**c**) PWD tracing from the RVOT showing reduced acceleration time (steeper down-stroke) and mid-systolic notch (denoted by arrow)

2. **In the presence of congenital heart disease**
   - **Left-to-right shunts**: Moderate to large defects with a mild to moderate increase in PVR and no reversal of the shunt.
   - **Eisenmenger syndrome**: Shunt/Admixture lesions with a severe increase in PVR, resulting in bidirectional shunting or shunt reversal.
   - **Complex congenital heart disease.**
     - Segmental pulmonary hypertension.
       Isolated pulmonary artery of ductal origin.

Absent pulmonary artery.
Pulmonary atresia with VSD and MAPCAs.
Hemitruncus.
  - Single ventricle.
  - Scimitar syndrome.
- **PAH with coincidental congenital heart disease**: The cardiac defects that are unlikely to account for the elevated PVR (e.g. ASD in the first decade of life is unlikely to cause PAH).
- **Post-operative PAH**: Persistent, recurrent, or incident PAH after surgical repair of congenital heart disease, in the absence of significant residual lesions.

15.2.2.1 Persistent Pulmonary Hypertension of the Newborn

Persistent pulmonary hypertension of the newborn (PPHN) is a syndrome characterized by marked pulmonary hypertension that causes hypoxaemia secondary to right-to-left shunting of blood at the foramen ovale and ductus arteriosus [13]. ***The diagnosis is confirmed regardless of the value of PASP, as long as it is accompanied by a right-to-left shunt, in the absence of significant congenital structural heart disease***.

Echocardiographic Diagnosis of Pulmonary Hypertension in Neonates

- It needs to be remembered that PASP, which is normally high at birth, decreases over the next 24–72 h. The PASP in a normal neonate usually remains in the range of [14]:
  - < 80 mmHg (within 2 h after birth)
  - < 70 mmHg (within 12 h after birth)
  - < 60 mmHg (within 24 h after birth)
  - < 50 mmHg (within 48 h after birth)
  - < 40 mmHg (within 72 h after birth).
- Unusual RV dilation and bowing of the interventricular septum towards the LV are indicative of significant pulmonary hypertension.

Key point: It is not unusual to find PASP in the range of 30–40 mmHg even at 14 days after birth.

Right-to-Left or Bidirectional Shunting

Right-to-left or bidirectional shunting across the ductus and/or PFO is a mandatory feature of PPHN.

Rule Out Alternative Diagnostic Possibilities

- Obstructed total anomalous pulmonary venous return can have a similar clinical presentation.
- Other cyanotic defects should be ruled out.

RV Systolic Function

Diminished tricuspid annular plane systolic excursion (TAPSE) is a predictor of poor outcome.

Key point: Regardless of the absolute value of PA systolic pressure, the diagnostic hallmark of PPHN is the presence of right-to-left shunting of blood across the PFO/PDA, in the absence of associated anomalies.

15.2.2.2 Eisenmenger Syndrome

Eisenmenger syndrome refers to the progressive development of pre-capillary pulmonary hypertension in the setting of a large congenital cardiac defect, which exposes the pulmonary circulation to increased blood flow and elevated pressure.

- Eisenmenger syndrome is characterized by a fixed and markedly elevated PVR [15].
- With the onset of Eisenmenger physiology, the direction of shunt flow at the atrial, ventricular, or great arterial levels becomes either balanced or predominantly right-to-left.

Echocardiography

1. ***Diagnosis of pulmonary hypertension*** (see above).
2. ***Identification of the underlying cardiac defect***:
   - Site: The common cardiac defects that predispose to Eisenmenger physiology are:
     - Ventricular septal defect.
     - Complete AV septal defect.
     - Patent ductus arteriosus.

    - Aorto-pulmonary window.
    - d-TGA with VSD,
    - Truncus arteriosus.
    - Common atrium.
    - Single ventricle.
    - Iatrogenic: systemic-to-PA shunts like Waterson or Pott shunts.
    - Large ASD [less often; not until the second or third decade of life].
  - Size: Eisenmengerization is less likely in cases of small defects such as [12, 15]:
    - Atrial level defects: < 2–3 cm.
    - Ventricular level defects: < 1–1.5 cm.
    - Great arterial level defects: < 0.5–0.7 cm.
3. ***Shunt quantification and direction of shunt flow (bidirectional or right-to-left).***
4. **Right ventricular size and function.**

Key points: If no underlying cardiac defect is identified, the diagnosis of Eisenmenger syndrome is not tenable; alternative aetiologic possibilities for pulmonary hypertension need to be considered. It may be difficult to identify the right-to-left shunt flow across a PDA or an AP window, as the shunt flow is in the same direction as pulmonary arterial blood flow.

15.2.3 Post-Capillary Pulmonary Hypertension

Post-capillary pulmonary hypertension results from pathologies distal to the pulmonary capillaries. The possibilities include [11, 12]:

- Congenital post-capillary obstructive lesions, such as:
  - Pulmonary vein stenosis.
    - Isolated
    - Associated with bronchopulmonary dysplasia or prematurity
  - Obstructed total anomalous pulmonary venous return.
  - Cor triatriatum.
  - Mitral/aortic stenosis (including supra/subvalvular).
  - Coarctation of the aorta.
- Left ventricular dysfunction: systolic/diastolic.

Echocardiography reveals features of pulmonary hypertension. The underlying cause needs to be identified.

Key point: It is inaccurate to report the echocardiographic diagnosis of pulmonary hypertension as 'PAH', as the term 'PAH' implies pre-capillary pulmonary hypertension alone.

15.3 Criss-Cross Heart (CCH) and Superior–Inferior Ventricle (SIV)

Criss-cross heart (CCH) is a rare congenital anomaly of cardiac rotation characterized by [16]:

- Criss-crossing of ventricular inlets and inflow streams, and drainage of the atria into contralaterally located ventricles.
- Atrio-ventricular (AV) and ventriculo-arterial (VA) connections may be concordant or discordant.
- Ventricular alignment can be either side-by-side or superior–inferior (superior–inferior ventricles, SIV).

SIV occurs due to displacement of the ventricular mass along the horizontal plane, resulting in [16]:

- Horizontal alignment of the interventricular septum and.
- Superior–inferior alignment of the ventricles.

Although CCH and SIV frequently co-exist, ***the two are not synonymous***, and either can be found in isolation.

15.3.1 Echocardiographic Diagnosis

The relationship of the ventricular inflows is assessed in apical and subcostal four-chamber views (Fig. 15.7).

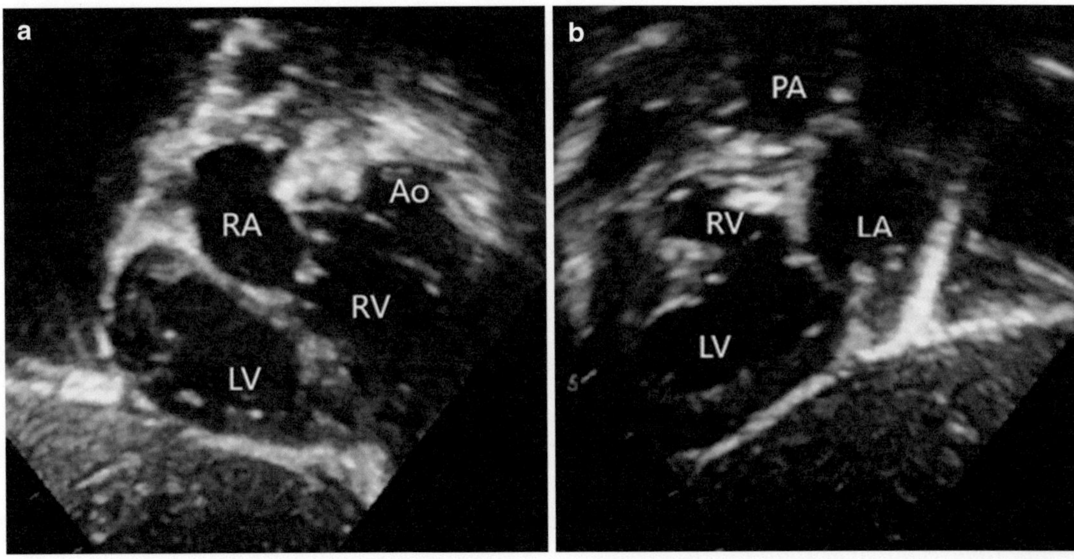

Fig. 15.7 Subcostal view of criss-cross heart with horizontal orientation of the interventricular septum and superior–inferior ventricular relationship. (**a**) RA drains into morphologic RV, which is superior and to the left of morphologic LV. (**b**) LA drains into morphologic LV, which is inferior to RV

The characteristic findings in CCH are [17, 18]:

- ***Inability to visualize both the AV valves in a single imaging plane in a four-chamber view.***
- Demonstration of cross-over of AV inflow blood streams by colour Doppler, with each atrium draining into the ventricle located contralateral to it.
- Anterior sweep of the imaging plane from apical or subcostal four-chamber views demonstrates the connection of the left-sided LA and the right-sided LV through the mitral valve.
- More anterior angulation reveals the drainage of the right-sided RA into the left-sided RV through the tricuspid valve.
- Basal-to-apical sweeping of the transducer in the PSAX view also facilitates visualization of drainage from each atrium to the contralateral ventricle.

The defining feature of SIV is horizontal alignment of the interventricular septum in mid-ventricular parasternal and/or subcostal short-axis views.

15.3.2 Associated Anomalies [16, 18]

- VSD.
- TGA (the aortic valve may be positioned leftward and anterior to the pulmonary valve).
- RV hypoplasia.
- Subpulmonary stenosis.
- Straddling AV valves.

Key point: Inability to visualize both the AV valves in a single imaging plane in a four-chamber view should alert the echocardiographer towards the possibility of criss-cross AV connection.

15.4 Juxtaposition of Atrial Appendages

Juxtaposition of atrial appendages is defined as the condition in which two atrial appendages lie adjacent to each other on the same side of the great arteries instead of encircling their roots.

Dixon's classification [based on the anatomical position of the two atrial appendages in relation to the great arteries] [19]:

- Left juxtaposition: Both appendages are positioned on the left side of the great arteries.
- Right juxtaposition: Both appendages lie on the right side of the great arteries.

Van Praagh's classification [based on the morphology of the appendages] [20]:

- Left juxtaposition of the right atrial appendage:
  - Morphologic RAA [broad, triangular, more trabeculated] is abnormally positioned above the morphologic LAA [thin, slender, finger-like] on the left side.
  - Commoner than right juxtaposition of the LAA.
  - Leftward juxtaposition can mimic an interatrial communication.
  - Commonly associated with transposition of the great arteries, tricuspid atresia.
- Right juxtaposition of the left atrial appendage:
  - Morphologic LAA is located on the right side, superior to the RAA.
  - May be seen in a structurally normal heart or with cardiac anomalies such as LV hypoplasia.
- Partial juxtaposition:
  - The atrial appendage on one of the sides may be bifid, with one component in the normal position and the other component juxtaposed.

Echocardiography

Visualization of normal atrial appendages is discussed in Chap. 3.

- The juxtaposed atrial appendage always lies above the native atrial appendage, i.e. the juxtaposed RAA lies above the native LAA in left juxtaposition of RAA, and the juxtaposed LAA lies above the native RAA in right juxtaposition of LAA [18].
- ***Horizontal orientation of the anterior atrial septum is the most diagnostic echocardiographic finding***. The plane of the atrial septum is seen to be parallel to the anterior chest wall in the anterior part and normally oriented in the posterior part. This variation in the orientation of the atrial septum in its anterior and posterior parts is a clue to the diagnosis [21].
- Subcostal long-axis view shows the atrial septum in a perpendicular orientation to the diaphragm (in contrast to the normal, where the atrial septum is parallel to the diaphragm). An anterior sweep of the probe reveals the ostium of the right atrial appendage [21].
- RAA, which is located posterior to the pulmonary artery, can be visualized in subcostal long axis, A4C, and PLAX views.

Key point: Differential orientation of the anterior and posterior parts of the atrial septum should alert the echocardiographer towards the possibility of juxtaposed atrial appendages.

References

1. Maron BJ, Desai MY, Nishimura RA, Spirito P, Rakowski H, Towbin JA, Rowin EJ, Maron MS, Sherrid MV. Diagnosis and evaluation of hypertrophic cardiomyopathy: JACC state-of-the-art review. J Am Coll Cardiol. 2022;79(4):372–89.
2. Weissler-Snir A, Crean A, Rakowski H. The role of imaging in the diagnosis and management of hypertrophic cardiomyopathy. Expert Rev Cardiovasc Ther. 2016;14(1):51–74.
3. Pollick C, Rakowski H, Wigle ED. Muscular subaortic stenosis: the quantitative relationship between systolic anterior motion and the pressure gradient. Circulation. 1984;69:43–9.
4. Corrado D, van Tintelen PJ, McKenna WJ, Hauer RNW, Anastastakis A, Asimaki A. Et al; international experts. Arrhythmogenic right ventricular cardiomyopathy: evaluation of the current diagnostic criteria and differential diagnosis. Eur Heart J. 2020;41(14):1414–29.
5. Kaul S, Tei C, Hopkins JM, Shah PM. Assessment of right ventricular function using two-dimensional echocardiography. Am Heart J. 1984;107(3):526–31.
6. Meluzin J, Spinarova L, Bakala J, et al. Pulsed Doppler tissue imaging of the velocity of tricuspid annular systolic motion. Eur Heart J. 2001;22:340–8.
7. Wang J, Prakasa K, Bomma C, Tandri H, Dalal D, James C, et al. Comparisonof novelechocardiographic parameters of right ventricular functionwith ejection fraction by cardiac magnetic resonance. J Am Soc Echocardiogr. 2007;20:1058–64.

8. Jenni R, Oechslin E, Schneider J, Attenhofer Jost C, Kaufmann PA. Echocardiographic and pathoanatomical characteristics of isolated left ventricular noncompaction: a step towards classification as a distinct cardiomyopathy. Heart. 2001;86(6):666–71.
9. Chin TK, Perloff JK, Williams RG, Jue K, Mohrmann R. Isolated noncompaction of left ventricular myocardium. A study of eight cases. Circulation. 1990;82(2):507–13.
10. Stöllberger C, Gerecke B, Finsterer J, Engberding R. Refinement of echocardiographic criteria for left ventricular noncompaction. Int J Cardiol. 2013;165(3):463–7.
11. Galiè N, Humbert M, Vachiery JL, Gibbs S, Lang I, Torbicki A, Simonneau G, Peacock A, Vonk Noordegraaf A, Beghetti M, Ghofrani A, Gomez Sanchez MA, Hansmann G, Klepetko W, Lancellotti P, Matucci M, McDonagh T, Pierard LA, Trindade PT, Zompatori M, Hoeper M, ESC Scientific Document Group. 2015 ESC/ERS guidelines for the diagnosis and treatment of pulmonary hypertension: the joint task force for the diagnosis and treatment of pulmonary hypertension of the European Society of Cardiology (ESC) and the European Respiratory Society (ERS): endorsed by: Association for European Paediatric and Congenital Cardiology (AEPC), International Society for Heart and Lung Transplantation (ISHLT). Eur Heart J. 2016;37(1):67–119.
12. Del Cerro MJ, Abman SH, Diaz G, et al. A consensus approach to the classification of pediatric pulmonary hypertensive vascular disease: report from the PVRI pediatric taskforce, Panama 2011. Pulm Circ. 2011;1:286–98.
13. Sukys MC, Tyson JE, Wright LL, Bauer CR, Korones SB, Stevenson DK, Verter J, Stoll BJ, Lemons JA, Papile LA, Shankaran S, Donovan EF, Oh W, Ehrenkranz RA, Fanaroff AA. Persistent pulmonary hypertension of the newborn in the era before nitric oxide: practice variation and outcomes. Pediatrics. 2000;105(1 Pt 1):14–20.
14. EmmanouilidesGC MAJ, Duffie ERJr, Adams FH. Pulmonary arterial pressure changes in human newborn infants from birth to 3 days of age. J Pediatr. 1964;65:327–33.
15. Wood P. The Eisenmenger syndrome or pulmonary hypertension withreversed central shunt. Br Med J. 1958;2:701–9.
16. Anderson K, Lie J, Sieg K, et al. A criss-cross heart: detailed anatomicdescription and discussion of morphogenesis. Mayo Clin Proc. 1977;52:569–75.
17. Manuel D, Ghosh G, Joseph G, Lahiri A, George PV. Criss-cross heart: transthoracic echocardiographic features. Indian Heart J. 2018;70(1):71–4.
18. Kimball TR, Michelfelder EC. Echocardiography: basic principles and imaging. In: Allen HD, Driscoll DJ, Shaddy RE, Feltes TF, editors. Moss and Adams' heart disease in infants, children, and adolescents. 8th ed. Philadelphia, PA: Lippincott Williams & Wilkins; 2013. p. 134–71.
19. Dixon AS. Juxtaposition of the atrial appendages: two cases of an unusual congenital cardiac deformity. Br Heart J. 1954;16:153–64.
20. Van Praagh S, et al. Juxtaposition of the morphologically right atrial appendage in solitus and inversus atria: a study of 35 postmortem cases. Am Heart J. 1996;132:382–90.
21. Rice MJ, Seward JB, Hagler DJ, Edwards WD, Julsrud PR, Tajk AJ. Left juxtaposed atrial appendages: diagnostic two-dimensional echocardiographic features. J Am Coll Cardiol. 1983;1:1330–6.

16 Diagnostic Probabilities at the Bedside

Two-dimensional echocardiography has evolved as one of the primary modalities for the evaluation of heart disease. It is routinely being performed for the evaluation of neonatal or paediatric clinical presentations, such as cyanosis, cardiomegaly, respiratory distress, and murmurs. A fair idea of the differential diagnostic possibilities in each scenario empowers the echocardiographer to answer the clinical questions more specifically.

16.1 Cyanosis

The cardiac causes of cyanosis at various timelines after birth are as follows:

- **At birth**:
  - Transposition physiology.
    TGA.
    Taussig Bing anomaly.
  - Admixture physiology.
    TAPVC.
    Tricuspid atresia with VSD.
    Truncus arteriosus.
    Single ventricle physiology (univentricular AV connections, pulmonary atresia with intact IVS).
  - Tetralogy of Fallot (one-third).
  - Ebstein's anomaly.
- **At the first week of life**:
  - Duct-dependent lesions (tricuspid atresia/pulmonary atresia with intact IVS, HLHS).
- **After 1 month of life**:
  - Fallot's physiology.
    Tetralogy of Fallot (one-third).
    DORV with VSD-PS.
    TGA with VSD-PS.
    DILV-PS.
- **After 2 years of age**:
  - Tetralogy of Fallot (one-third).
  - VSD with Eisenmenger syndrome.
- **After the first decade**:
  - VSD with Eisenmenger syndrome.
  - PDA with Eisenmenger syndrome.
- **After the second decade**:
  - PDA with Eisenmenger syndrome.
  - ASD with Eisenmenger syndrome.

16.2 Congestive Heart Failure (Cardiac Causes)

- **At birth** (within 72 h):
  - Premature infant with large PDA.
  - HLHS.
  - Critical stenosis/atresia of pulmonary, mitral, or aortic valves.
  - Severe tricuspid or pulmonary regurgitation.

A. Paul, *Manual of Echocardiography for Congenital Heart Diseases*,
https://doi.org/10.1007/978-981-96-0243-8_16

  - Endocardial fibroelastosis.
  - Myocardial dysfunction secondary to sepsis, asphyxia, hypoglycaemia.
  - Acute myocarditis or Cardiomyopathy.
- **Within 1 week**:
  - Any of the above.
  - TGA with inadequate mixing of blood.
  - Obstructed TAPVC.
- **1 week to 1 month**:
  - Any of the above.
  - Coarctation of the aorta, critical aortic stenosis.
  - Large shunt lesions (VSD, PDA).
  - AV septal defect.
- **1 to 2 months of life**:
  - Any of the above.
  - ALCAPA.
  - TAPVC, truncus arteriosus, DORV, tricuspid atresia.
- **Children or adults with predominant**:
  - **Right ventricular failure**: ASD, TAPVC, PS, Ebstein's anomaly, aortic atresia.
  - **Left ventricular failure**: PDA, COA, AS, tricuspid atresia, LV inflow obstruction, MR.
  - **Biventricular failure**: VSD, AP window, TGA-VSD, AV septal defect, myocarditis.

16.3 Massive Cardiomegaly in the Newborn

- Ebstein's anomaly with anatomic or functional pulmonary atresia.
- Anatomic or functional aortic atresia with AV and VA discordance and severe left AV valve regurgitation.
- Neonatal dilated cardiomyopathy.
- Congenital tricuspid insufficiency.
- Intrapericardial teratoma/lipoma.

16.4 Cardiogenic Shock

- **Left ventricular outflow tract obstruction:**
  - Critical aortic stenosis.
  - Interrupted aortic arch.
  - Severe coarctation of the aorta.
  - Hypoplastic Left Heart Syndrome.
- **Myocardial dysfunction:**
  - Acute myocarditis.
  - Asphyxia, Hypoglycaemia, Sepsis, Acidosis.
  - Cardiomyopathy due to various causes.
  - Myocardial ischaemia/infarction as in ALCAPA.
- **Rhythm disturbances:**
  - Tachyarrhythmias.
  - Complete AV block.

16.5 Asymptomatic Neonate with a Murmur

- Shunt lesions.
- Obstructive lesions.
- Valvular regurgitation.
- Physiological murmurs.

16.6 Syncope

- Obstructive lesions.
  - Outflow tract obstruction: LVOTO, RVOTO.
  - Inflow obstruction.
- Pulmonary hypertension.
- Myocardial dysfunction/arrhythmia.
  - Cardiomyopathies: HOCM, ARVD.
  - MVP.
  - Coronary artery anomalies causing myocardial ischaemia.
  - Post-operative (arrhythmia in repaired TOF).

16.7 Chest Pain

- Severe AS, severe PS, HOCM.
- MVP.
- Pulmonary hypertension.
- Coronary artery anomalies: ALCAPA, anomalous course of the left coronary artery between the aorta and PA, Kawasaki disease sequelae.

16.8 Duct-Dependent Lesions

Emergency echocardiographic examinations are sought in neonatal ICUs to look for duct-dependent lesions when clinicians contemplate the initiation of prostaglandin E1 therapy to maintain ductal patency. The common duct-dependent lesions are listed in the section on patent ductus arteriosus (Chap. 4).

16.9 Congenital Syndromes

An echocardiogram is often ordered to rule out congenital defects of the heart in children with primary non-cardiac abnormalities, such as dysmorphic features. Although any cardiac defect may be possible in any given patient, an idea about the common cardiac associations of such congenital syndromes aids in conducting a focused echocardiographic study.

Apert syndrome (Craniofacial dysostosis): VSD, ASD, PDA, TOF.

Alagille syndrome (Arteriohepatic Dysplasia): Peripheral PA stenosis, ASD, VSD, COA, TOF with or without pulmonary atresia.

Cardiofaciocutaneous syndrome: Valvular PS, OS ASD, dysplastic valves, HCM.

Carpenter Syndrome: PDA, VSD, PS, TGA.

Cat eye syndrome: TAPVC, PAPVC.

Char syndrome: PDA, muscular VSD.

CHARGE Association (Coloboma, heart defects, choanal atresia, growth or mental retardation, genitourinary anomalies, ear anomalies): Conotruncal anomalies (TOF, truncus arteriosus, DORV), aortic arch anomalies (e.g. vascular ring, interrupted aortic arch), AVSD.

Cornelia de Lange (de Lange) Syndrome: VSD, ASD, PS, HCM.

Costello syndrome: Valvular PS, dysplastic valves, HCM, aortic aneurysm.

Cri Du Chat (Cat's cry) syndrome: VSD, PDA, ASD, Ebstein's anomaly, COA.

Crouzon syndrome (Craniofacial dysostosis): PDA, COA, VSD, AVSD, absent pulmonary valve.

DiGeorge Syndrome/Velocardiofacial Syndrome/CATCH 22: Interrupted aortic arch type B, truncus arteriosus, VSD, PDA, TOF.

Down's syndrome: AV septal defects (complete/partial), VSD, OS ASD, PDA, TOF.

Edward Syndrome (Trisomy 18): Conoventricular VSD, PDA, PS, AVSD, polyvalvar dysplasia, TOF, DORV.

Ehlers-Danlos syndrome: ASD, aortic aneurysm, MVP.

Ellis-van Creveld Syndrome (Chondroectodermal dysplasia): AVSD (complete/partial) including single atrium, OS ASD.

Fanconi syndrome: PDA, VSD.

Friedreich Ataxia: Hypertrophic cardiomyopathy.

Goldenhar syndrome (Occuloauriculovertebral dysplasia): VSD, TOF.

Glycogen Storage Disease II (Pompe Disease): Cardiomyopathy, LVH.

Holt-Oram Syndrome: ASD, VSD, PAPVC.

Jacobsen syndrome: VSD, HLHS.

Kabuki syndrome: ASD, VSD, LVOTO (bicuspid aortic valve, COA, HLHS).

Keutel syndrome: Peripheral PA stenosis, VSD, HCM.

Kartagener syndrome: Dextrocardia with situs inversus.

LEOPARD (Multiple Lentigenes) syndrome: Valvular PS, HCM.

Laurence Moon Biedl syndrome: VSD, TOF.

Marfan's syndrome: Aortic aneurysm/dissection, aortic and/or mitral regurgitation, MVP.

McKusick-Kaufman syndrome: AVSD (complete/partial/single atrium).

MELAS: HCM (asymmetric LV hypertrophy of inferolateral wall).

Mucopolysaccharidosis [Hurler syndrome (Type I), Hunter syndrome (Type II), Morquio syndrome (Type IV)]: Aortic and/or mitral regurgitation.

Neurofibromatosis: Valvular PS, AS, COA, HCM.

Noonan Syndrome: PS (dysplastic pulmonary valve), HCM, OS ASD, VSD, partial AVSD, AS, COA.

Osteogenesis Imperfecta: Aortic aneurysm, aortic regurgitation, MVP.

Patau Syndrome (Trisomy 13): VSD, ASD, PDA, AVSD, dextrocardia, conotruncal anomalies (TOF, DORV), polyvalvar dysplasia.

Pierre Robin syndrome: VSD, PDA, ASD, COA, TOF.

Radius (TAR) syndrome: TOF, ASD, dextrocardia.

Ritscher-Schinzel syndrome: Conotruncal anomalies (TOF, DORV), AVSD, ASD, VSD.

Rubinstein-Taybi syndrome: PDA, VSD, ASD, COA, HLHS.

Simpson-Golabi-Behmel syndrome: OS ASD, VSD, cardiomyopathy.

Smith-Lemli-Opitz syndrome: OS ASD, VSD, PDA, AVSD, TAPVC.

Townes-Brocks syndrome: Truncus arteriosus, TOF, ASD, VSD.

Treacher Collins syndrome: VSD, PDA, ASD.

Turner syndrome: LVOTO (COA, bicuspid aortic valve with or without valvular aortic stenosis), mitral valve anomalies, MVP, PAPVC, LSVC, aortic aneurysm/dissection.

VATER Association (VATER/ VACTERL Syndrome—Vertebral anomalies, Anal atresia, Congenital heart defects, Tracheoesophageal fistula, Renal dysplasia, Limb anomalies): VSD.

Williams-Beuren (Williams) Syndrome: Supravalvular AS, valvular AS, PS, PA stenosis, MVP, COA.

Zellweger Syndrome (Cerebrohepatorenal syndrome): PDA, VSD, or ASD.

16.10 Common Teratogens

The following is a list of reportedly teratogenic drugs and the congenital cardiac anomalies associated with their teratogenicity. (It should be remembered that the list is based on observations and case reports, and a causal association between the drug and the anomaly cannot be established beyond doubt in many of these cases.)

Alcohol: ASD, VSD, PDA, TOF.

Amphetamine: VSD, ASD, PDA, TGA.

Bupropion: LVOTO.

Ibuprofen: TGA, VSD, bicuspid AV.

Indomethacin: Premature closure of ductus arteriosus.

Lithium: Ebstein's anomaly, mitral atresia, COA.

Nitrofurantoin: ASD, HLHS.

Organic solvents: Conotruncal anomalies (TOF), PS, VSD, AP window, HLHS, COA, TAPVC.

Paroxetine: ASD, VSD.

Retinoic acid: Conotruncal anomalies, PS.

Sulfonamides: HLHS, COA.

Thalidomide: Conotruncal anomalies (TOF, truncus arteriosus).

Trimethadione: TGA, VSD, TOF, HLHS.

Valproic acid: ASD, VSD, AS, COA, TOF, pulmonary atresia-intact IVS, hypoplastic right heart.

Warfarin: VSD, COA, PDA, AP window, Situs inversus totalis.

16.11 Maternal Illnesses

Diabetes mellitus: TGA, VSD, COA, AVSD, HLHS, PDA, cardiomyopathy; persistent pulmonary hypertension.

Phenylketonuria: VSD, PDA, TOF, univentricular heart.

Rubella syndrome: PDA, VSD, PA stenosis.

Index

A. Paul, *Manual of Echocardiography for Congenital Heart Diseases*,
https://doi.org/10.1007/978-981-96-0243-8

M

N

O

P

MIX
Papier aus verantwortungsvollen Quellen
Paper from responsible sources
FSC® C105338

If you have any concerns about our products,
you can contact us on
ProductSafety@springernature.com

In case Publisher is established outside the EU,
the EU authorized representative is:
Springer Nature Customer Service Center GmbH
Europaplatz 3, 69115 Heidelberg, Germany

Printed by Libri Plureos GmbH
in Hamburg, Germany